MONTE CARLO DEVICE SIMULATION:
Full Band and Beyond

THE KLUWER INTERNATIONAL SERIES IN ENGINEERING AND COMPUTER SCIENCE

VLSI, COMPUTER ARCHITECTURE AND DIGITAL SIGNAL PROCESSING

Consulting Editor

Jonathan Allen

Latest Titles

Introduction to Analog VLSI Design Automation, M. Ismail, J. Franca, ISBN: 0-7923-9102-0
Gallium Arsenide Digital Circuits, O. Wing, ISBN: 0-7923-9081-4
Principles of VLSI System Planning, A.M. Dewey , S.W. Director ISBN: 0-7923-9102-0
Mixed-Mode Simulation, R. Saleh, A.R. Newton, ISBN: 0-7923-9107-1
Automatic Programming Applied to VLSI CAD Software: A Case Study, D. Setliff, R.A. Rutenbar, ISBN: 0-7923-9112-8
Models for Large Integrated Circuits, P. Dewilde, Z.Q. Ning ISBN: 0-7923-9115-2
Hardware Design and Simulation in VAL/VHDL, L.M. Augustin, D.C..Luckham, B.A.Gennart, Y.Huh, A.G.Stanculescu ISBN: 0-7923-9087-3
Subband Image Coding, J. Woods, editor, ISBN: 0-7923-9093-8
*Low-Noise Wide-Band Amplifiers in Bipolar and CMOS Technologies,*Z.Y.Chang, W.M.C.Sansen, ISBN: 0-7923-9096-2
Iterative Identification and Restoration of Images, R. L.Lagendijk, J. Biemond ISBN: 0-7923-9097-0
VLSI Design of Neural Networks, U. Ramacher, U. Ruckert ISBN: 0-7923-9127-6
Synchronization Design for Digital Systems, T. H. Meng ISBN: 0-7923-9128-4
Hardware Annealing in Analog VLSI Neurocomputing, B. W. Lee, B. J. Sheu ISBN: 0-7923-9132-2
Neural Networks and Speech Processing, D. P. Morgan, C.L. Scofield ISBN: 0-7923-9144-6
Silicon-on-Insulator Technology: Materials to VLSI, J.P. Colinge ISBN: 0-7923-9150-0
Microwave Semiconductor Devices, S. Yngvesson ISBN: 0-7923-9156-X
A Survey of High-Level Synthesis Systems, R.A. Walker, R. Camposano ISBN: 0-7923-9158-6
Symbolic Analysis for Automated Design of Analog Integrated Circuits,, G. Gielen, W. Sansen, ISBN: 0-7923-9161-6
High-Level VLSI Synthesis, R. Camposano, W. Wolf, ISBN: 0-7923-9159-4
Integrating Functional and Temporal Domains In Logic Design: The False Path Problem and Its Implications, P.C. McGeer, R.K. Brayton , ISBN: 0-7923-9163-2
Neural Models and Algorithms for Digital Testing, S.T. Chakradhar, V.D. Agrawal,M.L. Bushnell, ISBN: 0-7923-9165-9
Monte Carlo Device Simulation: Full Band and Beyond, edited by Karl Hess ISBN: 0-7923-9172-1

MONTE CARLO DEVICE SIMULATION:
Full Band and Beyond

edited by

Karl Hess
University of Illinois

Kluwer Academic Publishers
Boston/Dordrecht/London

Distributors for North America:
Kluwer Academic Publishers
101 Philip Drive
Assinippi Park
Norwell, Massachusetts 02061 USA

Distributors for all other countries:
Kluwer Academic Publishers Group
Distribution Centre
Post Office Box 322
3300 AH Dordrecht, THE NETHERLANDS

Library of Congress Cataloging-in-Publication Data

Monte Carlo device simulation : full band and beyond / edited by Karl Hess.
p. cm. -- (Kluwer international series in engineering and computer science ; SECS 144)
Includes index.
ISBN 0-7923-9172-1
1. Semiconductors--Mathematical models. 2. Monte Carlo method. 3. Semiconductors-Computer simulation. I. Hess, Karl, 1945- . II. Series.
TK7871.85.M5736 1991
621.381'52--dc20

91-4199
CIP

Printed on acid-free paper.

Printed in the United States of America

TABLE OF CONTENTS

CONTRIBUTORS

D. W. Bailey
Department of Physics
University of Florida
Gainesville, FL 32611

K. F. Brennan
School of Electrical Engineering
Georgia Tech
Atlanta, GA 30332

J. Bude
The Beckman Institute for
Advanced Science and Technology
and the Department of Electrical
and Computer Engineering
University of Illinois
Urbana, IL 61801

D. K. Ferry
Center for Solid State
Electronics Research
Arizona State University
Tempe, AZ 85287-6206

M. V. Fischetti
IBM Research Division
Thomas J. Watson Research Center
P.O. Box 218
Yorktown Heights, NY 10598

J. M. Higman
The Beckman Institute for
Advanced Science and Technology
University of Illinois
405 N. Mathews Avenue
Urbana, IL 61801

R. P. Joshi
Department of Electrical and
Computer Engineering
Old Dominion University
Norfolk, VA 23529-0246

D. Jovanovic
The Beckman Institute for
Advanced Science and Technology
and the Department of Electrical
and Computer Engineering
University of Illinois
Urbana, IL 61801

M. -J. Kann
Department of Electrical and
Computer Engineering
Old Dominion University
Norfolk, VA 23529-0246

K. W. Kim
Electrical and Computer
Engineering Department
North Carolina State University
Raleigh, NC 27695-7911

A. M. Kriman
Center for Solid State
Electronics Research
Arizona State University
Tempe, AZ 85287-6206

S. E. Laux
IBM Research Division
T. J. Watson Research Center
P.O. Box 218
Yorktown Heights, NY 10598

J. P. Leburton
The Beckman Institute for
Advanced Science and Technology
and the Department of Electrical
and Computer Engineering
University of Illinois
Urbana, IL 61801

M. A. Littlejohn
Electrical and Computer
Engineering Department
North Carolina State University
Raleigh, NC 27695-7911

U. Ravaioli
The Beckman Institute for
Advanced Science and Technology
and Coordinated Science Laboratory
University of Illinois
Urbana, IL 61801

H. Shichijo
Texas Instruments
Central Research
Dallas, TX 75265

C. J. Stanton
Department of Physics
University of Florida
Gainesville, FL 32611

J. Y. Tang
1315 Miraveille Avenue
Los Altos, CA 94024

H. Tian
Electrical and Computer
Engineering Department
North Carolina State University
Raleigh, NC 27695-7911

R. J. Trew
Electrical and Computer
Engineering Department
North Carolina State University
Raleigh, NC 27695-7911

D. L. Woolard
Electrical and Computer
Engineering Department
North Carolina State University
Raleigh, NC 27695-7911

D. Yoder
The Beckman Institute for
Advanced Science and Technology
and Department of Electrical
and Computer Engineering
University of Illinois
Urbana, IL 61801

Preface

Monte Carlo simulation is now a well established method for studying semiconductor devices and is particularly well suited to highlighting physical mechanisms and exploring material properties. Not surprisingly, the more completely the material properties are built into the simulation, up to and including the use of a full band structure, the more powerful is the method. Indeed, it is now becoming increasingly clear that phenomena such as reliability related hot-electron effects in MOSFETs cannot be understood satisfactorily without using full band Monte Carlo. The IBM simulator DAMOCLES, therefore, represents a landmark of great significance. DAMOCLES sums up the total of Monte Carlo device modeling experience of the past, and reaches with its capabilities and opportunities into the distant future. This book, therefore, begins with a description of the IBM simulator.

The second chapter gives an advanced introduction to the physical basis for Monte Carlo simulations and an outlook on why complex effects such as collisional broadening and intracollisional field effects can be important and how they can be included in the simulations. References to more basic introductory material can be found throughout the book. The third chapter describes a typical relationship of Monte Carlo simulations to experimental data and indicates a major difficulty, the vast number of deformation potentials required to simulate transport throughout the entire Brillouin zone. The fourth chapter addresses possible further extensions of the Monte Carlo approach and subtleties of the electron-electron interaction. Chapter five offers a resolution to the deformation potential problem: It appears possible to perform a complete transport simulation with little more input than a few Fourier components of the crystal potential that are fairly well known. This approach represents an important step toward standardization of the input parameters, and a complete device simulation based on this approach appears to be the ultimate solution for semiclassical electronic transport in semiconductor devices.

Chapters six and seven provide examples of applications to heterolayer structures and systems of lower dimensionality. Chapter 8 describes the connection between Monte Carlo and more conventional device simulations, particularly Hydrodynamic models. Chapter 9 discusses numerical considerations and, finally, chapter 10 provides the complete commented code for a full band Monte Carlo program.

This book is written for the researcher who wants to know where we stand and "whither we are tending". It also contains much introductory material that cannot be found elsewhere. It is hoped that, as the power of computers increase, this material will serve an ever increasing group of scientists and engineers who are interested in the detailed mechanisms that underlie semiconductor device function.

Much of the original impetus to create this collection arose from work around the NSF National Center for Computational Electronics.

Urbana, Illinois

Karl Hess

1

NUMERICAL ASPECTS AND IMPLEMENTATION OF THE *DAMOCLES* MONTE CARLO DEVICE SIMULATION PROGRAM

Steven E. Laux and Massimo V. Fischetti

IBM Research Division
T. J. Watson Research Center
P.O. Box 218
Yorktown Heights, NY 10598 USA

ABSTRACT

We discuss our numerical model for electronic transport in semiconductor devices from three perspectives: (i) physical basis of the model, (ii) issues related to numerics and program implementation, and (iii) device simulation results. The emphasis in this chapter will be on numerics and program implementation, while the other two topics are only reviewed for completeness. Our Monte Carlo device simulation program is called DAMOCLES (Device Analysis using Monte Carlo et Poisson solver). To date, DAMOCLES has been used to simulate Si, GaAs, InP, $Ga_{0.47}In_{0.53}As$ and InAs MOSFETs and bipolar homojunction transistors and GaAs MESFETs.

INTRODUCTION

The combination of a statistical solution of the Boltzmann transport equation (BTE) and a self-consistent solution of the Poisson equation continues to gain popularity as a viable approach to device simulation. This technique is often simply referred to as "Monte Carlo device modeling" [1,2], as compared to "drift-diffusion" [3] or "hydrodynamic" [4] formulations of the device simulation problem. While the Monte Carlo method is stochastic, the latter two formulations are deterministic and involve the solution of coupled systems of partial differential equations. A larger difference between the two approaches, however, is the way semiconductor physics is introduced into the formulation. In the Monte Carlo technique, the physics enters through the description of the semiconductor band structure (*i.e.*, kinematics) and the scattering behaviors (*i.e.*, dynamics). In the deterministic approaches, the physics is lumped into parameterized mobilities, diffusion constants, and lifetimes. For this reason, access to the physics is more straightforward in a Monte Carlo approach. However, in exchange for this advantage, significantly longer computer execution times are required. This computational appetite of Monte Carlo methods is unwelcome in an engineering context, and precludes direct application of the model to engineering device design. Today, then, Monte Carlo device modeling codes are used to attack key questions relating to a device technology where understanding and insight into the device physics are crucial. The code also serves as a computational reference against which other, less cpu-intensive, device analyses can be compared. In the foreseeable future, massively parallel computers may render codes like DAMOCLES standard engineering design tools.

Our approach to Monte Carlo device simulation is twofold: improve upon the physics conventionally employed in such models, and preserve program flexibility and extensibility. With respect to physics, our Monte Carlo transport model improves upon other similar device modeling approaches by: (i) using the full semiconductor band-structure obtained from empirical pseudopotentials in order to permit the study of hot carrier transport beyond the parabolic band approximation; (ii) employing carrier-lattice, carrier-impurity and short-range Coulomb intercarrier collisions obtained consistently with the full band structure of the semiconductor in order to account for density-of-states effects in the scattering rates; (iii) treating the long-range (plasma) Coulomb interactions and degenerate statistics in heavily doped regions, thus being able to treat carrier-plasmon interactions and band-gap narrowing effects. This transport model is united with a self-consistent solution of the Poisson equation and a flexible representation for device geometry and doping in two space dimensions.

With respect to program flexibility and extensibility, the evolution of DAMOCLES over the last four years is chronicled in the literature [5-17]. DAMOCLES has been used to simulate Si, GaAs, InP, $Ga_{0.47}In_{0.53}As$ and InAs MOSFETs and bipolar homojunction transistors, and GaAs MESFETs. Today, the code consists of approximately 80,000 lines of FORTRAN.

By emphasizing the physics of semiconductor devices simultaneously with the requirement of simulating realistic device structures, we hope to show that *physics matters in device design.* In this chapter, we will review the physics contained in the DAMOCLES program and summarize device results obtained to date. However, the central focus will be on numerics and program implementation.

PHYSICS AND NUMERICS

The transport physics implemented within DAMOCLES continues to be well described by reference [5], with only small additions and enhancements detailed in [12]. The first subsection on homogeneous transport is is based on these prior publications. The final three subsections are more related to program numerics, namely, particle-mesh coupling, boundaries and statistical enhancements. However, given the close coupling between physics and numerics in a program such as DAMOCLES, any attempt to discuss physics without numerics (or vice versa) becomes difficult. As a result, both topics are freely mixed throughout this section.

Homogeneous Transport

The basic Monte Carlo procedure for evolving a solution to the BTE is given by Fawcett, *et al.* [18], and roughly consists of computing intervals of ballistic transport for carriers punctuated by scattering events. The duration of the ballistic flight and final state after scattering is calculated probabilistically, sampling the appropriate distribution in each case. For the ballistic free-flight intervals, the following equations of motion (for the general case of a graded alloy $A_{\mathrm{x}}B_{1-\mathrm{x}}C$) are integrated:

$$\frac{d\mathbf{r}}{dt} = \frac{1}{\hbar}\nabla_{\mathbf{k}}\{E_\nu[\mathbf{k}, \mathrm{x}(\mathbf{r})]\} \tag{1a}$$

$$\frac{d\mathbf{k}}{dt} = \frac{e\mathbf{F}(\mathbf{r})}{\hbar} - \frac{1}{\hbar}\frac{\partial E_\nu[\mathbf{k}, \mathrm{x}(\mathbf{r})]}{\partial \mathrm{x}}\nabla_{\mathbf{r}}\mathrm{x}(\mathbf{r}) \ , \tag{1b}$$

where $E_\nu[\mathbf{k}, \mathrm{x}(\mathbf{r})]$ gives the kinetic energy of band ν as a function of wavevector $\mathbf{k}$ and alloy mole fraction $\mathrm{x}(\mathbf{r})$ which depends on real space coordinate $\mathbf{r}$ (*i.e.*, the band structure) and $\mathbf{F}$ is the local electric field value. Also note e carries with it the sign of the charged particle. In DAMOCLES, this integration is performed numerically. It is easy to see these two equations are consistent with total energy conservation. Let $W(\mathbf{r}, \mathbf{k}) = U(\mathbf{r}) + E_\nu[\mathbf{k}, \mathrm{x}(\mathbf{r})]$ be the total energy, which is the sum of potential and kinetic energy. For both electrons and holes, $\nabla_{\mathbf{r}}U(\mathbf{r}) = -e\mathbf{F}(\mathbf{r})$, provided the usual care is taken to define increasing energy in opposite directions for electrons and holes. By constancy of total energy, we can write

$$0 \equiv \frac{dW(\mathbf{r}, \mathbf{k})}{dt} = \frac{d}{dt}\{U(\mathbf{r}) + E_\nu\{\mathbf{k}, \mathrm{x}(\mathbf{r})]\} = \nabla_{\mathbf{r}}U(\mathbf{r}) \bullet \frac{d\mathbf{r}}{dt} + \nabla_{\mathbf{k}}E_\nu(\mathbf{k}, \mathbf{r}) \bullet \frac{d\mathbf{k}}{dt} +$$

$$\nabla_{\mathbf{r}}E_\nu(\mathbf{k}, \mathbf{r}) \bullet \frac{d\mathbf{r}}{dt} = -e\mathbf{F}(\mathbf{r}) \bullet \frac{d\mathbf{r}}{dt} + \nabla_{\mathbf{k}}E_\nu(\mathbf{k}, \mathbf{r}) \bullet \frac{d\mathbf{k}}{dt} + \frac{\partial E_\nu(\mathbf{k}, \mathbf{r})}{\partial \mathrm{x}}\nabla_{\mathbf{r}}\mathrm{x}(\mathbf{r})\frac{d\mathbf{r}}{dt} \tag{2}$$

The final right-most expression in Eqn. (2) can be obtained by simply multiplying Eqns. (1a) and (1b) together, demonstrating that total energy conservation is obeyed. In cases where only one semiconductor is present in the model, x is independent of position and considerable simplification occurs.

Conventionally, the form taken for the kinetic energy is

$$\frac{\hbar^2\mathbf{k}^2}{2m^*} = E(\mathbf{k})[1 - \alpha E(\mathbf{k})] \ , \tag{3}$$

where m^* is the effective mass at the bottom of the valley and α is the nonparabolicity factor. With $\alpha = 0$, the kinetic energy is purely parabolic; for ellipsoidal valleys, the left-hand side of Eqn. (3) is modified is the usual way. For examples of the use of parabolic or first-order non-parabolic bands, see [18-29]. We rather adopt a band structure obtained from the local empirical pseudopotential approach of Cohen and Bergstresser [30], to permit a better description of energetic carriers. This approach was pioneered by Hess and co-workers for both Si and GaAs [31-33], and subsequently extended to other III-V compounds by Brennan and co-workers [34-35]. This band structure can be tabulated on a mesh of 916 points in the irreducible wedge (1/48) of the

Brillouin Zone (BZ), with a mesh size of $0.05(\frac{2\pi}{a})$, where a is the lattice constant. In DAMOCLES, however, the tabulation is extended to cover an octant of the Brillouin Zone. This gives a better trade-off between additional storage required for the table and the decreased compute time associated with fewer symmetry operations associated with mapping to the octant. Our interpolation scheme is quadratic for the kinetic energy, and linear for the velocity. There are two exceptions where this tabulation of the band structure is not used, however.

First, errors in the interpolation become unacceptably large near valley minima, especially when small effective mass materials are considered. For this reason, below a certain kinetic energy threshold, electrons revert to an analytic band structure (precisely Eqn. (3) with constant nonparabolicity and ellipsoidal valley shape as required). This energy threshold is set as high as possible for a given material (for "analytic" electrons consume much less computer time than do "numeric" electrons), limited by the extent of the deviation between analytic and tabulated band structure. Secondly, while this procedure of analytically representing the valley extrema can conceptually be carried over to holes in the valence band, the true valence band shape near $\mathbf{k} = \mathbf{0}$ is very complex and defies accurate analytic description. For this reason, for holes (to date in silicon only), a finer interpolation mesh (4 times finer mesh in each $\mathbf{k}$ direction) is used in the central portion of the valence band BZ.

For simulating heterostructure devices, the band structure is linearly interpolated in mole fraction $\mathrm{x}(\mathbf{r})$; for the alloy $A_{\mathrm{x}}B_{1-\mathrm{x}}C$, the band structure is given by:

$$E_{alloy}(\mathbf{k}, \mathrm{x}) = \mathrm{x}E_{AC}(\mathbf{k}) + (1 - \mathrm{x})E_{BC}(\mathbf{k}) - E_0(\mathrm{x}) \quad , \tag{4}$$

where E_0 shifts the conduction band minimum to zero and E_{AC} and E_{BC} are both measured from their respective conduction band minima.

Fig. 1 gives as an example the conduction band structure for silicon, and compares it to the first-order non-parabolic approximation to the bands employed in more conventional Monte Carlo device models. The pseudopotential band structures for all of the materials we have considered to date, and more importantly, the local pseudopotential form factors necessary to derive the band structures, are spelled out in great detail in [16].

As mentioned in the Introduction, scattering rates in our transport model are calculated consistently with the density-of-states obtained from the band structure. More precisely, for phonons we calculate scattering rates from Fermi's golden rule together with the true density of final states, but with phenomenological matrix elements. This amounts to adopting values for deformation potentials and phonon coupling constants that appear in the matrix elements. This topic is discussed at length elsewhere [5,16]. An alternative to adopting phenomenological matrix elements in scattering rate calculations is described in another chapter of this book [36]. Suffice it to say, the "usual" carrier-phonon processes are included in our model: transverse and longitudinal acoustic phonons (TA and LA), and either polar or nonpolar transverse and

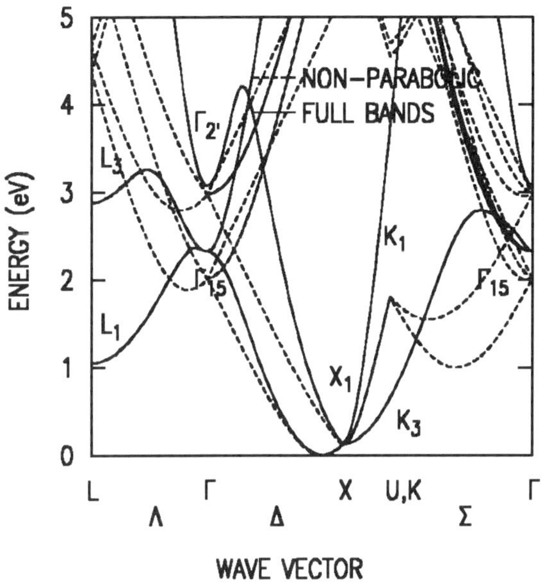

Fig. 1 The conduction band structure for silicon obtained from empirical pseudopotential calculations compared to a first-order non-parabolic approximation of the band structure.

longitudinal optical phonons (TO and LO), depending on the semiconductor. To assist in finding final states quickly which satisfy energy conservation in the tabulated band structure $E_n(\mathbf{k})$, the inverse function $\mathbf{k}(E_n)$ is also tabulated and used by the program as explained in [5]. This greatly speeds up the determination of the locus of **k** locations in the BZ which will fufill energy conservation, given an initial energy and known phonon energy.

Interaction of a carrier with an ionized dopant in a screened Coulomb field is modeled by either a Brooks-Herring or Ridley's statistical screening formulation. The screening parameter β is calculated self-consistently during program execution according to

$$\beta = \sqrt{\frac{e^2}{\varepsilon}\left(\frac{\partial n}{\partial E_{Fn}} + \frac{\partial p}{\partial E_{Fp}}\right)} \quad , \tag{5}$$

where $E_{Fn(Fp)}$ is the quasi-Fermi level for electrons (holes) at the location of interest. Eqn. (5) reduces to the Debye-Hückel and Thomas-Fermi expressions in the nondegenerate and degenerate limits, respectively. Our implementation of impurity scattering is discussed at more length in [5,12].

Short-range carrier-carrier scattering (electron-electron, hole-hole or electron-hole) are conditionally included in our simulations. This scattering mechanism adds a significant computational expense, and is often neglected from calculations whenever deemed appropriate. Carrier-carrier scattering is evaluated in the Born approximation; the golden rule expression for the total scattering rate is given in [5]. This scattering is so computationally expensive because it involves a double sum over final states which are consistent with energy and momentum conservation in the full band structure (with final states weighted by the density-of-final states). The inverse tabulated band structure is employed in this double search, to make the search as efficient as possible. The screening parameter in the matrix element is precisely Eqn. (5), which serves to define the (somewhat arbitrary) threshold between "short-range" and "long-range" Coulomb interactions. Clearly then, for particles separated by more then β^{-1}, we depend on the Poisson solver to resolve the "long-range" Coulomb forces. This immediately implies that the spatial mesh resolution for the Poisson solution should be matched to the local screening length $L_S = \beta^{-1}$, in order to avoid double counting of the Coulomb interaction. Finally, the scattering rate is rescaled by the probability that two partner carriers in 2D real space that are within a screening circle are actually within a screening sphere in 3D real space. This rescaling will work provided there are always at least as many particles in a screening circle as is nominally expected in a screening sphere, for the rescaling can only preclude scattering in 2-space, but not create partner particles should none be present in the first place. The number of actual super-particles present in a 2D screening circle is intimately related to the amplitude of fluctuations in the potential solution, *i.e.*, the amplitude of the plasma oscillations will vary according to the number of super-particles present. Too many particles in a screening circle will result in potential solutions that are too smooth; conversely, too few particles in a screening circle will yield potential solutions that are too noisy. Further discussion of mesh design and the selection of a "proper" number of particles to represent fluctuations correctly is deferred to the subsection on Statistical Enhancements.

Impact ionization in DAMOCLES is included via a simple Keldysh formula. The threshold energy and prefactor in the Keldysh formula are fitted for each material available in DAMOCLES, in order to reproduce experimental data where available [5,16]. We are aware of a few groups that have been actively pursuing improved impact ionization formulations[37,38], and we hope to incorporate their advances into DAMOCLES in the near future.

Alloy scattering is included in DAMOCLES following a treatment for completely random alloys. Inclusion of alloy scattering has a relatively small influence on velocity-field characteristics in the cases we have examined; therefore, no conclusions can be drawn concerning the appropriateness of our formulation [16].

In realistic device structures, doping levels very often dictate at least a portion of the device cross-section will be greatly influenced by degeneracy effects. By assuming that carrier-carrier scattering drives the carriers towards a Fermi-Dirac distribution, the local distribution function in degenerate regions is approximated by:

$$f_{\text{app}}(E, \mathbf{r}, t) = \frac{1}{1 + \exp\left(\dfrac{E - E_{Fn,Fp}(\mathbf{r}, t)}{k_B T_{n,p}}\right)} . \tag{6}$$

Any of the aforementioned scattering are then rejected if the final state energy E' satisfies the the inequality $1 - f_{app}(E', \mathbf{r}, t) \leq \xi$, where ξ is uniform on (0,1). This approach has the advantage of avoiding expensive tabulations of the distribution function in **r** and **k** as it evolves in the simulation. This approach breaks down where deviations from a Fermi-Dirac distribution are large; however, this is expected to be precisely where degeneracy is no longer important.

The preceding discussion of scattering rates and outline of the basic Monte Carlo procedure suffice to construct a model capable of solving space-homogeneous transport problems. By matching model predictions of velocity-field curves with experimental data at 300 and 77 K, a consistent set of phonon coupling constants can be found. Note since most bulk experimental data deal with undoped material, this calibration has been performed without any Coulomb scattering included (either ionized impurity or carrier-carrier). This set of coupling constants obtained is not unique. However, in lieu of belaboring this point, we set out our parameters in great detail for the benefit of others who wish to pursue this issue [16]. We can say, however, that the calibration of the transport model involves only experimental data and simulation performed in a space-homogeneous context; subsequent device simulations are dependent only on bulk measurements and modeling calibrations. This point is discussed at length in [5,16].

Particle-Mesh Coupling

To construct a full fledged device simulation program, the preceding transport model must be united with a self-consistent solution of the electric field in the given device cross-section. The actual coupling between particle transport and calculation of forces on the mesh has been dealt with at length [1]. The four steps in a particle-mesh calculation are:

(i) assign charge to the Poisson mesh;
(ii) solve the Poisson equation for $\psi(\mathbf{r})$;
(iii) calculate the force $e\mathbf{F}(\mathbf{r}) = -e\nabla_{\mathbf{r}}\psi$ and interpolate it to the particle locations; and
(iv) solve the equations of motion Eqn. (1).

Each step will be discussed in turn, beginning with step (ii).

Solution of Poisson Equation. We discretize the Poisson equation $-\nabla_{\mathbf{r}} \bullet [\varepsilon(\mathbf{r})\nabla_{\mathbf{r}}\psi] = \rho(\mathbf{r})$ in standard fashion on a two-dimensional, tensor-product (no internal mesh-line terminations) finite-difference mesh with a five-point stencil. Specific material properties are associated with the rectangle bounded by four nodes in the mesh; doping, density and potential are associated with the nodes of the mesh. The resultant algebraic equations can be written in matrix form, where the divergence operator becomes a five-diagonal matrix. In some cases (such as a n-MOSFET where electrons are represented as particles and holes are included in a continuum, zero-current approximation), the charge density is of the form $\rho(\mathbf{r}, \psi)$. This renders the Poisson equation non-linear. After discretization, the set of nonlinear algebraic equations are solved iteratively by Newton's method, with a damping procedure proposed by Bank and Rose [39].

To select the best solution strategy for the (now) linear system of equations, we reason as follows. First, the dielectric constant $\varepsilon(\mathbf{r})$ is conditionally a function of position in the device domain. This precludes expressing the divergence operation in the following separable form:

$$-\nabla_{\mathbf{r}} \bullet [\varepsilon(\mathbf{r})\nabla_{\mathbf{r}}\psi] \neq a(x)\frac{\partial^2\psi}{\partial x^2} + b(x)\frac{\partial\psi}{\partial x} + c(y)\frac{\partial^2\psi}{\partial y^2} + d(y)\frac{\partial\psi}{\partial y} \quad . \tag{7}$$

This lack of a separable form disqualifies rapid elliptic solvers such as FACR [1] from consideration. Secondly, we maintain a desire to permit nonuniform mesh spacings (all Δx_i not necessarily equal, similarly for all Δy_i). This also precludes the use of a rapid elliptic solver. Finally, we choose to solve the Poisson equation very frequently, in order to time resolve the field changes accurately. A typical run could consist of >10000 Poisson solves, separated in time by as little as 0.2 fsec. This implies virtually *any* indirect solution method (an iterative solution method which benefits from an initial solution guess, *e.g.*, successive overrelaxation or conjugate gradients) should perform better than all but the very fastest direct methods (a solution method which is guaranteed to complete in a set number of arithmetic operations, *e.g.*, Gaussian elimination, FACR) because of the availability of an excellent initial guess for the potential solution. The indirect solver we use is based on a polynomial preconditioned conjugate gradient technique [40].

Charge Assignment. We use the cloud-in-cell (CIC) method [1] to assign particle charge to the mesh. This represents a good trade-off between computational burden and smoothness in charge assignment. In the presence of nonuniform meshes (in the sense of the previous paragraph), the charge cloud appears to change spatial extent across the mesh, as the charge cloud inherits its extent from the underlying mesh size. This is somewhat unfortunate, and could be addressed by fixing the charge cloud size independent of mesh spacing. One approach would be to fix the charge cloud size equal to the biggest mesh element present. This introduces additional computational burden and looses spatial resolution in areas where the mesh spacing is reduced compared to the cloud size. To maintain simplicity when using nonuniform meshes, we permit the cloud size to change according to the mesh spacing.

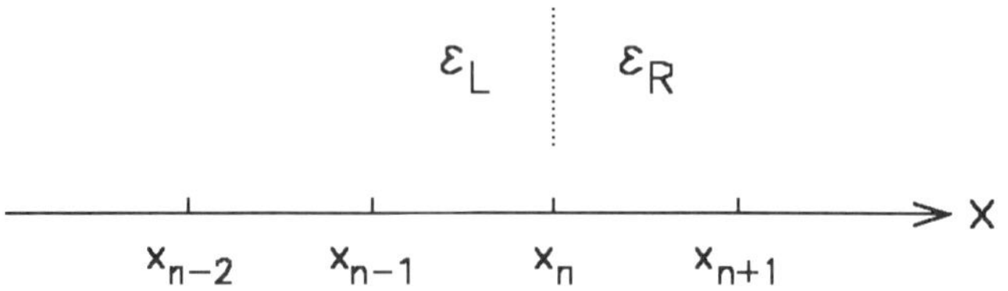

Fig. 2 Dielectric discontinuity on a simple one-dimensional mesh.

Field Calculation. We modify slightly the CIC approach for field calculation and interpolation of the field to particle positions. The original CIC method did not consider a spatially dependent dielectric constant $\varepsilon(\mathbf{r})$, nor did it permit a nonuniform mesh structure. To explain our extension of CIC to allow dielectric interfaces, consider the simple one-dimensional mesh in Fig. 2. The extension to two dimensions is self evident. The one-dimensional mesh has a dielectric discontinuity at $x = x_n$ and ε is piecewise constant. Suppose a particle is located to the left of the interface at $x = x_p$, $x_{n-1} < x_p < x_n$. Standard CIC requires the x-directed electric field F_x be evaluated at x_{n-1} and x_n and linearly combined to yield the force at location x_p as follows:

$$F_x(x_p) = \left\{ F_x(x_n)\left[\frac{x_p - x_{n-1}}{x_n - x_{n-1}}\right] + F_x(x_{n-1})\left[\frac{x_n - x_p}{x_n - x_{n-1}}\right]\right\} \tag{8}$$

The field at a mesh location x_i is calculated according to

$$F_x(x_i) = -\frac{\psi_{i+1} - \psi_{i-1}}{x_{i+1} - x_{i-1}} \quad , \tag{9}$$

that is, a standard centered-difference approximation at x_i. While our modified CIC method leaves Eqn. (8) intact, Eqn. (9) is improperly applied at $x_i = x_n$ due to the dielectric discontinuity and hence discontinuity in F_x. This is repaired by differencing the x-directed electric flux density rather than the electric field, and then calculating the electric field. This implies two values of x-directed electric field should be used at $x = x_n$ for interpolation purposes, one for interpolation to the left (F_x^L) and one for the right (F_x^R) of x_n. Two different pairs of interpolation fields are:

$$F_x^L(x_n) = \frac{(\psi_n - \psi_{n-1}) + \frac{\varepsilon_R}{\varepsilon_L}(\psi_{n+1} - \psi_n)}{x_{n+1} - x_{n-1}} \quad \text{and}$$

$$F_x^R(x_n) = \frac{\frac{\varepsilon_L}{\varepsilon_R}(\psi_n - \psi_{n-1}) + (\psi_{n+1} - \psi_n)}{x_{n+1} - x_{n-1}} \quad ; \quad \text{or} \tag{10a}$$

$$F_x^L(x_n) = \frac{1}{2\varepsilon_L}\left\{\varepsilon_L\left[\frac{\psi_n - \psi_{n-1}}{x_n - x_{n-1}}\right] + \varepsilon_R\left[\frac{\psi_{n+1} - \psi_n}{x_{n+1} - x_n}\right]\right\} \quad \text{and}$$

$$F_x^R(x_n) = \frac{1}{2\varepsilon_R}\left\{\varepsilon_L\left[\frac{\psi_n - \psi_{n-1}}{x_n - x_{n-1}}\right] + \varepsilon_R\left[\frac{\psi_{n+1} - \psi_n}{x_{n+1} - x_n}\right]\right\} . \tag{10b}$$

Both pairs in Eqn. (10a) and (10b) are identical on a uniform mesh and solve the dielectric discontinuity problem in that case. However, the pairs are quite different on a nonuniform mesh. In general, the use of a nonuniform mesh with CIC has an *important* failing: the particle self-force is no longer zero. Self-force is discussed at length in [1], and is one criterion useful in evaluating the particle-mesh coupling employed. When using a nonuniform mesh together with CIC for both force interpolation and charge assignment, the particle can literally feels itself and is unphysically accelerated. For a *one-dimensional* mesh, however, Eqn. (10b) maintains zero self force exactly. Both Eqn. (10a) and (10b) result in nonzero self force in two dimensions; however, the self force using Eqn.

(10b) will always be less. For this reason, we prefer Eqn. (10b). A general solution to the problem of nonzero self force when nonuniform meshes are used is currently lacking. Until this problem is solved, we use nonuniform meshes as a last resort, or in regions of minimal particle population, together with the use of Eqn. (10b) at dielectric interfaces.

Integration of the Equations of Motion. We generally use a second order Runge-Kutta numerical integrator to solve the equations of motion Eqn. (1). However, there are two fine points to be discussed: first, why use a Runge-Kutta technique, and secondly, the need to alter multistep integration schemes to maintain zero self force. While there are many possible integration schemes from which to choose in order to solve Eqn. (1), we contrast five schemes which are representative to address the relevant issues involved in selecting a specific integrator. We consider Eqn. (1) written for one dimension in position ($\mathbf{r} \rightarrow x$) and wavevector ($\mathbf{k} \rightarrow k$) and a parabolic $E(k)$ relation for electrons ($e < 0$), to simplify the discussion. Our starting point is then

$$\frac{dx}{dt} = \frac{\hbar k}{m^*} \tag{11a}$$

$$\frac{dk}{dt} = \frac{eF(x)}{\hbar} \quad . \tag{11b}$$

The five integrators we consider are given in Eqns. (12-16). A first order (FO) integrator from time level n to $n+1$ is (δ is the time step):

$$x^{n+1} = x^n + \frac{\delta \hbar k^n}{m^*} \tag{12a}$$

$$k^{n+1} = k^n + \frac{\delta e F(x^n)}{\hbar} \quad . \tag{12b}$$

A second order (SO) centered differences integration gives:

$$x^{n+1} = x^{n-1} + \frac{2\delta \hbar k^n}{m^*} \tag{13a}$$

$$k^{n+1} = k^{n-1} + \frac{2\delta e F(x^n)}{\hbar} \quad . \tag{13b}$$

A second order "leap-frog" (LF) centered differences integration reads:

$$x^{n+1} = x^n + \frac{\delta \hbar k^{n+½}}{m^*} \tag{14a}$$

$$k^{n+½} = k^{n-½} + \frac{\delta e F(x^n)}{\hbar} \quad . \tag{14b}$$

A second order Runge-Kutta (RK) integration yields:

$$x^{n+1} = x^n + \frac{\delta}{2}\frac{\hbar}{m^*}k^n + \frac{\delta}{2}\frac{\hbar}{m^*}\left[k^n + \frac{\delta e F(x^n)}{\hbar}\right] \tag{15a}$$

$$k^{n+1} = k^n + \frac{\delta}{2}\frac{e}{\hbar}F(x^n) + \frac{\delta}{2}\frac{e}{\hbar}\left[F\left(x^n + \frac{\delta\hbar k^n}{m^*}\right)\right] . \tag{15b}$$

Finally, to show one of the combinations possible, a second order Runge-Kutta integration of Eqn. (11a) and a first order integration of Eqn. (11b), identified as SF (second-first) is:

$$x^{n+1} = x^n + \frac{\delta}{2}\frac{\hbar}{m^*}k^n + \frac{\delta}{2}\frac{\hbar}{m^*}\left[k^n + \frac{\delta eF(x^n)}{\hbar}\right] \tag{16a}$$

$$k^{n+1} = k^n + \frac{\delta eF(x^n)}{\hbar} . \tag{16b}$$

We consider three evaluation criteria: consistency, accuracy and stability [1]. Each of the integrators is consistent, *i.e.*, each scheme tends to the original differential equation in the limit $\delta \to 0$. Accuracy refers to the size of truncation errors introduced by approximating differentials as differences. We have already identified the accuracy of each scheme above: the first order scheme (FO) has truncation errors proportional to δ, while the three second order schemes (SO, LF, RK) have truncation errors proportional to δ^2. The mixed scheme (SF) is both first order accurate in wavevector and second order accurate in position. The five schemes also differ in their stability, *i.e.*, how roundoff errors are propagated for different values of δ. Stability analysis is reviewed in Chap. 4 of [1]; we only summarize the results here for these five methods. We seek the eigenvalues of the amplification matrix λ_{amp}, as we must ascertain the conditions where these eigenvalues have magnitude less than or equal to unity. The amplification matrix is found by substituting $x^n = X^n + \zeta_x^n$ (position at time level n is the sum of position without roundoff error plus the roundoff error) and $k^n = K^n + \zeta_k^n$ (similarly interpreted) and canceling all terms without roundoff error to obtain evolution equations for ζ_x^n and ζ_k^n. In essence, this is a linearization procedure to first order. The amplification matrix G is defined as the 2×2 matrix $(\zeta_x^{n+1}, \zeta_k^{n+1})^T = G\,(\zeta_x^n, \zeta_k^n)^T$, where the superscript T denotes the matrix transpose. The eigenvalues obtained from the stability analysis are given in Table 1. The loci of λ_{amp} in the complex plane are shown in Fig. 3. Two of the schemes are numerically stable ($|\lambda_{amp}| = 1$) over a limited range of timesteps, namely, SO (for $\delta\omega_p \leq 1$) and LF (for $\delta\omega_p \leq 2$). FO, RK and SF are formally unstable unless $\delta = 0$.

Unfortunately, the preceding stability analysis can be misleading for two reasons. First, SO is in fact a very poor scheme, for although the scheme is indeed stable for $\delta\omega_p \leq 1$, this stability analysis really only measures whether the method can integrate a dissipationless harmonic oscillator without numerically growing errors. Unfortunately, adding even the slightest damping to the test problem demonstrates the SO method is wholly unsuitable, immediately becoming numerically unstable with large error growth rates. This is readily seen by adding a term $-\gamma\dot{x}$ to the right-hand side of Eqn. (11b), for such a term will damp the oscillatory solutions for a proper range of γ. We disqualify SO from further consideration as a result. Secondly, although FO, RK and SF are un-

Table 1. Stability results for five integration schemes

scheme[a]	λ_{amp}^{b}	$\mid \lambda_{amp} \mid^{b}$
FO	$1 \pm j\delta\omega_p$	$\sqrt{1 + \delta^2\omega_p^2}$ for all δ
SO	$1 - 2\delta^2\omega_p^2 \pm j2\delta\omega_p\sqrt{1 - \delta^2\omega_p^2}$	1 for $\delta \leq \omega_p^{-1}$
LF	$1 - \dfrac{\delta^2\omega_p^2}{2} \pm j\delta\omega_p\sqrt{1 - \dfrac{\delta^2\omega_p^2}{4}}$	1 for $\delta \leq 2\omega_p^{-1}$
RK	$1 - \dfrac{\delta^2\omega_p^2}{2} \pm j\delta\omega_p$	$\sqrt{1 + \dfrac{\delta^4\omega_p^4}{4}}$ for all δ
SF	$1 - \dfrac{\delta^2\omega_p^2}{4} \pm j\delta\omega_p\sqrt{1 - \dfrac{\delta^2\omega_p^2}{16}}$	$\sqrt{1 + \dfrac{\delta^2\omega_p^2}{2}}$ for $\delta \leq 4\omega_p^{-1}$

[a] See Eqns. (12)-(16) for definitions.

[b] The square of the plasma radian frequency $\omega_p^2 = \dfrac{\mid e \mid}{m^*}\dfrac{dF}{dx} = \dfrac{\mid e\rho \mid}{m^*\varepsilon}$, where ρ is the charge density.

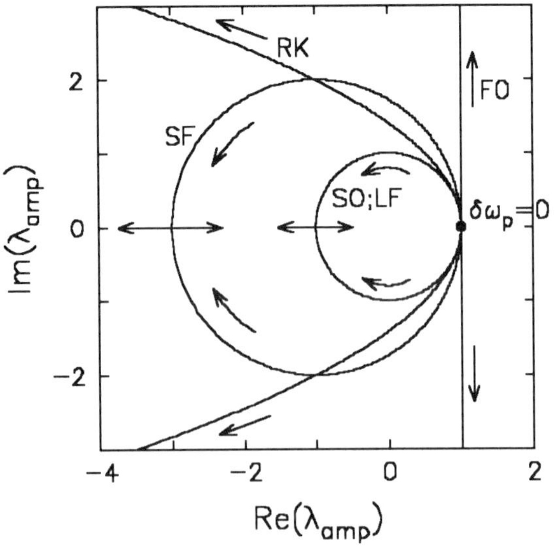

Fig. 3 The loci of the eigenvalues of the amplification matrix λ_{amp} in the complex plane for the five different integration schemes listed in Table 1. The arrows indicate how the eigenvalues move as $\delta\omega_p$ increases from zero.

stable, can we quantify how fast the errors grow in time? This is readily answered, for the growth of errors over s timesteps is simply $| \lambda_{amp} |^s$. For example, the number of timesteps required for an anomalous doubling of amplitude of the error is found by solving $| \lambda_{amp} |^s = 2$, *i.e.*, $s = \ln 2 / \ln | \lambda_{amp} |$. The actual simulation time T_D required to double the error amplitude is the time spanned by these s timesteps, *i.e.*, $T_D = \delta s = \delta \ln 2 / \ln | \lambda_{amp} |$. Consider T_D for the FO, RK and SF methods. Using $| \lambda_{amp} |$ from Table 1, we find for the FO scheme:

$$T_{D(FO)} = \frac{\delta \ln 2}{\ln(1 + \delta^2 \omega_p^2)^{1/2}} \sim \frac{2\delta \ln 2}{\delta^2 \omega_p^2} \sim \frac{1.39}{\omega_p} (\delta\omega_p)^{-1} \quad \text{for } \delta\omega_p \ll 1 \; ; \qquad (17)$$

for the RK scheme:

$$T_{D(RK)} = \frac{\delta \ln 2}{\ln\left(1 + \dfrac{\delta^4 \omega_p^4}{4}\right)^{1/2}} \sim \frac{8\delta \ln 2}{\delta^4 \omega_p^4} \sim \frac{5.55}{\omega_p} (\delta\omega_p)^{-3} \quad \text{for } \delta\omega_p \ll 1 \; ; \qquad (18)$$

and for the SF scheme:

$$T_{D(SF)} = \frac{\delta \ln 2}{\ln\left(1 + \dfrac{\delta^2 \omega_p^2}{2}\right)^{1/2}} \sim \frac{4\delta \ln 2}{\delta^2 \omega_p^2} \sim \frac{2.77}{\omega_p} (\delta\omega_p)^{-1} \quad \text{for } \delta\omega_p \ll 1 \, . \qquad (19)$$

Consider a worst case situation for silicon with an electron density of $n = 10^{20}\ \text{cm}^{-3}$. In this case, $\omega_p \sim 2.8 \times 10^{14}\ \text{sec}^{-1}$. With a timestep $\delta = 0.2$ fsec (typical for DAMOCLES), we obtain $\delta\omega_p \sim 0.056$ which gives $T_{D(FO)} \sim 0.089$ psec, $T_{D(RK)} \sim 110$ psec and $T_{D(SF)} \sim 0.177$ psec. Respectively, these times represent approximately 4, 4900 and 8 oscillator periods at a frequency of 44.6 THz. For device simulation with DAMOCLES, FO and SF are acceptable to simulate plasma oscillations (recall collisions in the system will thwart the growth of errors). Method RK has an extremely slow error growth.

While four of the methods are synchronous in time, scheme LF is not, for positions and wavevectors are never evaluated at the same time instant. This can be addressed by integrating Eqn. (11b) forward for a time 0.5δ by any method whenever true time synchronization is required. Furthermore, both SO and LF are not self-starting, *i.e.*, given x^0 and k^0 at $t = 0$ we cannot immediately evaluate Eqn. (13) or (14). With LF, for example, we must "start" the method by integrating Eqn. (11b) forward by 0.5δ to $k^{1/2}$ by a self-starting method. This issue of self-starting also comes into play just after a collision of a particle with a phonon, impurity, another particle, or even a device boundary, since the final state of the particle after the collision is not related to the initial state by a continuous time evolution. In these cases, we must abandon LF integration temporarily, as it presumes a continuous motion from $n \rightarrow (n+1)$ in position and $(n-1/2) \rightarrow (n+1/2)$ in wavevector.

There is one problem with scheme RK relative to all others considered here: it is a multistep scheme, and calls for two field evaluations while advancing from time level $n \rightarrow (n+1)$. As seen in Eqn. (15b), the electric field F is evaluated at position $\mathbf{r}^n$ and at position $\mathbf{r}^n + \delta\hbar\mathbf{k}^n/m^* = \mathbf{r}^n + \Delta\mathbf{r}^n$. When we list the four steps in a particle-mesh calculation, we implicitly assume the Poisson equation is solved prior to evaluating particle forces. *This is a critical assumption to maintain zero self force.* Consider the following situation. Given particle positions $\mathbf{r}^n$ at time level n, we spread the charge to the mesh and solve the Poisson equation for ψ^n in order to evaluate forces at mesh locations $\mathcal{F}^n = -e\nabla\psi^n$. Finally we interpolate these forces from mesh locations to particle locations to give $\mathcal{F}^n(\mathbf{r}^n)$. For a single particle ℓ at $\mathbf{r}_\ell^n$, $\mathcal{F}^n(\mathbf{r}_\ell^n)$ gives a numerical *approximation* to the force felt by particle ℓ due to all other charges in the system. This force can formally be written as $\sum_{i,\, i \neq \ell} \mathbf{F}(\mathbf{r}_i;\, \mathbf{r}_\ell)$, where $\mathbf{F}(\mathbf{r}_1;\, \mathbf{r}_2)$ is the force due to charge at $\mathbf{r}_1$ (due to particles and on boundaries) as felt by a particle at $\mathbf{r}_2$. The summation bypasses the $i = \ell$ term because a particle does not feel itself; the numerical procedure bypasses the equivalent of the $i = \ell$ term *only* because we have carefully used a charge spreading and force interpolation scheme which has zero self force. Now, however, independent of the scheme used for spreading and interpolation, if we move the ℓ-th particle from $\mathbf{r}_\ell^n$ to a new position $\mathbf{r}_\ell^n + \Delta\mathbf{r}_\ell^n$ and attempt to evaluate the force on that particle $\mathcal{F}^n(\mathbf{r}_\ell^n + \Delta\mathbf{r}_\ell^n)$, the self force contribution is no longer zero! The ℓ-th particle at $\mathbf{r}_\ell^n + \Delta\mathbf{r}_\ell^n$ will experience a *nonzero* force from the ℓ-th particle at $\mathbf{r}_\ell^n$ implicit in the force field $\mathcal{F}^n$. This self force will have an upper bound of $\mathbf{F}(\mathbf{r}_\ell^n;\, \mathbf{r}_\ell^n + \Delta\mathbf{r}_\ell^n)$, with lower values expected as charge spreading to the mesh decreases this self force.

In general, whenever forces are interpolated to particle positions which are *inconsistent* with the particle positions used to establish the charge for the Poisson solution, a nonphysical self force is included, *independent* of whether the charge spreading and force interpolation scheme has zero self force. The size of this self force can vary widely, becoming significant when there are few particles per mesh element.

There are two solutions to the problem of self force in a multistep integration scheme: (i) *always* solve the Poisson equation prior to force evaluation using the most recent particle positions, or (ii) delay force updating for the particle until after the Poisson equation has again been solved. The former solution is computationally expensive; the latter is adopted here in cases where the multistep self force problem is judged important. In this case, adopting the second solution causes scheme RK to degenerate identically into scheme SF. Thus, we have employed both RK and SF in our device simulations as appropriate.

Boundaries

Boundaries enter into any realistic device calculation. Boundaries can either represent physical entities (interfaces between materials, or true contact locations) or artificial constructs introduced in an effort to limit the extent of the region to be simulated; this distinction is discussed by Selberherr [3]. The treatment we adopt for potential boundary conditions along dielectric interfaces, contacts and artificial domain-edge boundaries follows Selberherr's discussion precisely. The treatment of boundaries for Monte Carlo particles merits some discussion, however. With respect to particle boundary conditions, consider two categories: boundaries that serve to contain particles and boundaries that may permit particle transmission. We consider each of these in turn.

Examples of boundaries which serve to contain particles are artificial domain-edge boundaries and semiconductor-insulator boundaries. In a drift-diffusion simulator, such boundaries would introduce a zero-gradient density condition normal to the boundary, *i.e.*, $\nabla n \bullet \hat{\mathbf{n}} = 0$ where $\hat{\mathbf{n}}$ is the unit vector normal to the interface. Such a boundary condition serves to preclude any particle flux from leaving the device. An analogue to this boundary condition is to simply reflect back into the device domain any particle that impinges on the boundary. There are three types of reflections implemented in DAMOCLES: (i) elastic with specular reflection, (ii) elastic with diffuse reflection, and (iii) thermalizing with diffuse reflection. Elastic, specular reflection is defined by equating the angle of incidence to the angle of reflection in both **r** and **k**-space; energy is conserved by setting $\mathbf{k}' = \mathbf{k} - 2(\mathbf{k} \bullet \hat{\mathbf{n}})\hat{\mathbf{n}}$ which insures $|\mathbf{k}'| = |\mathbf{k}|$, where $\mathbf{k}'$ is the **k** state after reflection. Since our timesteps are so small (~ 0.2 psec), the actual distance traveled per timestep is also very small (worst case of 10^8cm/sec × 0.2 fsec = 2 Å) and the loss of energy conservation introduced by reversing a particle's motion without explicitly accounting for the local field direction can be neglected (a field of 2×10^5 V/cm in this case would give a 4 meV energy imbalance maximum). Elastic, diffuse reflection scatters a particle into a **k**-state chosen from among all final states which conserve final energy and yield a velocity directed away from the boundary, with a probability proportional to the density-of-states at **k**. This is slightly non-standard, and will not necessarily result in particle reflection uniformly into a half plane. Thermalizing, diffuse scattering selects states randomly from **k**-states in local thermal equilibrium with a velocity directed away from the boundary. This type of boundary scattering decouples past particle behavior from future behavior: it is as if the incident particle was transmitted through the boundary, and at the same instant, a particle in thermal equilibrium enters at the same boundary position. We have used this boundary condition only infrequently, generally in connection with artificial boundaries along the domain edge.

Boundaries which permit particle transmission are heterointerfaces and contacts. Our model for behavior at heterointerfaces is conceptually "standard", with new complications due to the use of the full band structure. Whenever a particle hits a heterointerface, conservation of total energy and parallel momentum to the interface are used to determine whether the particle can be transmitted. If transmission is allowed by this first test, a $\mathbf{k} \bullet \mathbf{p}$ approximation is employed to evaluate the overlap integrals between the incident and transmitted wavefunctions, for all possible valleys and bands for which transmission is allowed. Finally, a Monte Carlo technique is employed to select a final transmitted or reflected state, using the squared overlap integrals as probability. Should particle reflection ultimately be selected, the discussion of the preceding paragraph applies.

Our treatment of contacts is somewhat nonstandard. By way of definition, contacts are unions of mesh nodes where our device domain touches an ideal source/sink of carriers. Contacts represent a challenge to Monte Carlo models.

In drift-diffusion models, contacts are trivially modeled by Dirichlet boundary conditions which fix the density value, generally as dictated by local charge neutrality. In hydrodynamic models, we further assume local thermal equilibrium, yielding temperature and/or kinetic energy Dirichlet boundary conditions. We seek an appropriate analogue for Monte Carlo modeling, where particle behavior at the boundary is driven by assumptions of local charge neutrality and thermal equilibrium. Often, this problem is circumvented by either (i) reinjecting particles which escape from other contacts (often employed in one-dimensional modeling, *i.e.*, a periodic condition), or (ii) explicitly including a heavily doped semiconductor layer adjacent to the contact, with particle injection governed by the requirement to maintain charge neutrality in this layer. The former alternative is not sufficiently general so as to be useful; the later can be computationally expensive and introduces an (often unknown) contact potential. We have explored many alternative contact injection schemes, and the following approach has produced the best results, particularly in maintaining charge neutrality over long simulation times without the use of a heavily doped layer. In addition to a particle injection model, we also must account for particles which escape at contacts. These particles are simply tallied as current, and then discarded from the simulation.

Our contact injection scheme proceeds as follows: at the beginning of a run, a target carrier density is calculated at each node along the contact as given by local charge neutrality consistent with Fermi-Dirac statistics. Note no distinction is made between "injecting" or "noninjecting" contacts; all contacts inject as required to maintain charge neutrality as the local conditions require. After each timestep, charge neutrality is examined along each contact. *The contact is considered as a whole*, *i.e.*, a line integral along the contact of actual carrier density is compared to a line integral of target density. If the contact has a net accumulation of charge, the injection algorithm terminates; if not, a search is made along the contact to determine the mesh node where the local carrier density is lowest compared to the local target density. If there are multiple nodes where carrier depletion is equal, one node is randomly selected, weighted by the mesh interval along the contact associated with the node. The amount of charge that is carried by one particle is computed, and the particle is added *only* if this addition will reduce the net deviation from charge neutrality. In the case where statistical enhancements are present at the contact (see next subsection), all injected particles are assigned the smallest charge weight currently in force at the contact. Having decided to add a particle, the **k** coordinates of this particle are computed consistent with two requirements: (i) the particle velocity is directed into the device, (ii) the particle is randomly selected from a Fermi-Dirac distribution function consistent with the local target density and temperature *weighted* by the velocity component perpendicular to the contact. The perpendicular velocity weight is essential, and accounts for the higher probability of a "fast" particle to enter the device from the conceptual thermal carrier gas considered to touch the contact. After successfully injecting a particle, the line integral of actual carrier density along the contact is updated, and the injection algorithm repeats until injection is no longer possible.

Statistical Enhancements

A final ingredient required to produce a bona fide device simulator is some means of enhancing rare events. We adapt the approach of Phillips and Price [41] to include enhancements in both real space and kinetic energy. Our algorithm is designed to populate a given region of space and kinetic energy with a prespecified number of particles, with automatic adjustment of the charge weighting factor S which scales two dimensional particle density into physical volume carrier density (with units of super-particles per electron per distance). A larger S implies fewer particles for a given volume density, and vice versa. Thus, S directly impacts the size of the potential fluctuations due to plasma oscillations present on the Poisson mesh. While S can be chosen as required by statistical estimators in cases where potential fluctuations are negligible and Coulomb scattering is unimportant (*i.e.*, low densities or high internal fields), it is becoming increasingly obvious to us that S is essentially constrained when the amplitude of potential fluctuations must be represented quantitatively.

A simple qualitative estimate concerning an appropriate value for S in a *non-degenerate* situation proceeds as follows. We choose a priori a uniform 2D mesh spacing equal to the screening length (consider the case of uniform density for simplicity). This mesh spacing sets the spatial resolution which defines the boundary between "long-range" and "short-range" forces. This occurs because forces calculated from the Poisson solution on the mesh represent the "long-range" force component, and this force component shrinks as the particles enter the same mesh element. The Monte-Carlo procedure to calculate short-range forces becomes active at this point, for the particles are now within a screening circle of each other. Therefore, we are interested in properly resolving inter-particle forces derived from the Poisson solution at distances greater than a screening length; forces at a closer range are treated separately as previously discussed. To determine the best way to represent long-range forces attributable to point charges, but necessarily represented as line charges in the two-dimensional cross-section, requires consideration of the way charge is screened in two and three dimensions. It is fortunate that in devices we have a situation where mobile charge can move in response to a charge perturbation, *i.e.*, screening is present, for the unscreened case is a disaster. If we were required to represent unscreened potentials due to point charges (potential $\propto |\mathbf{r}|^{-1}$) with line charges (potential $\propto \ln \mathbf{r}$), we could never hope for an accurate representation of long-range forces. Since we have screened potentials, however, the situation is far less problematic. Instead of solving the equation $-\nabla_{\mathbf{r}} \bullet (\varepsilon \nabla_{\mathbf{r}} \psi) = Q\delta(\mathbf{r})$ in two and three dimensions (which yields the unscreened potentials referred to in the previous sentence), we must solve instead $-\nabla_{\mathbf{r}} \bullet (\varepsilon \nabla_{\mathbf{r}} \psi) + \beta^2 \psi = Q\delta(\mathbf{r})$, where β is the screening parameter. This equation is a linearized version of the Poisson equation; see [42] for details of the derivation. Denote the solution in three dimensions as ψ_{3D} and in two dimensions as ψ_{2D}; these solution are

$$\psi_{3D}(\mathbf{r}) = \frac{Q_{point}}{4\pi\varepsilon\,|\,\mathbf{r}\,|} e^{-\beta\,|\,\mathbf{r}\,|} \quad , \qquad (20a)$$

$$\psi_{2D}(\mathbf{r}) = \frac{Q_{line}}{2\pi\varepsilon} K_0(\beta \mid \mathbf{r} \mid) \sim \frac{Q_{line}}{2\pi\varepsilon} \sqrt{\frac{\pi}{2}} \frac{e^{-\beta \mid \mathbf{r} \mid}}{\sqrt{\beta \mid \mathbf{r} \mid}} \quad \text{as} \mid \mathbf{r} \mid \rightarrow \infty, \tag{20b}$$

where K_0 is the modified Bessel of the second kind. The exponential decay present in both ψ_{3D} and ψ_{2D} makes these two behaviors far more similar than the unscreened potentials; for this reason we can hope to represent long-range forces due to point charges by long-range forces due to line charges. The charge weighting factor S in this case gives the scaling between point and line charge in the model, *i.e.*, $SQ_{point} = Q_{line}$. The question becomes: how can we optimally represent three-dimensional screening behavior with two-dimensional screening behavior? We choose to minimize the mean-square error between the force due to the potential perturbation in 2D and 3D, *i.e.*, find the S which minimizes $\int_{\beta^{-1}}^{\infty} (\psi'_{3D} - \psi'_{2D})^2 dr$, where $'$ denotes differentiation with respect to $\mid \mathbf{r} \mid$ as required by our desire to minimize the error in force rather than potential. The solution is strikingly simple: $S_{opt} = 0.5\beta = (2L_S)^{-1}$.

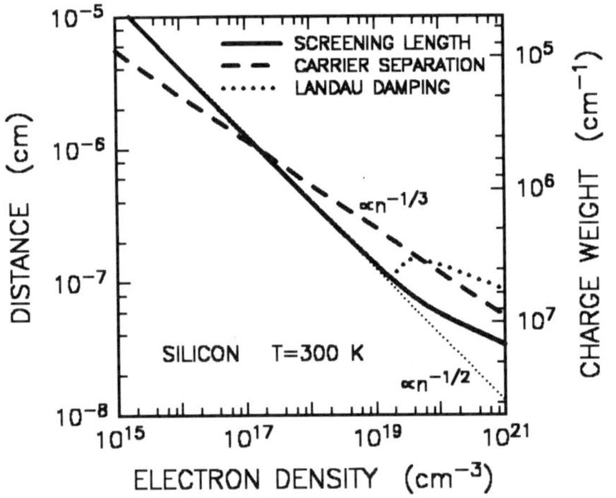

Fig. 4 Equilibrium screening length, average electron separation and characteristic distance derived from Landau damping versus electron density in Si at 300 K. The optimum charge weight S_{opt} at a given distance is read from the right-hand ordinate of the plot.

Fig. 4 plots the values of the screening length L_S at equilibrium, average electron separation and the characteristic distance $(\bar{q})^{-1}$ defined by Landau damping (as defined in [5]) versus electron concentration. The optimum charge weight $S_{opt} = (2L_S)^{-1}$ is also indicated. In nondegenerate situations, we seek to use a mesh spacing equal to the screening length and a charge weight given by S_{opt}. In degenerate situations, there are not enough electrons within a screening sphere to believe our screening model is valid (as inferred by the average electron separation plotted in Fig. 4); however, at these high concentrations ($\gtrsim 10^{19}$ cm^{-3}) the mesh spacing should follow the Landau damping curve[5]. We suggest, then, that mesh design and particle charge weights go hand-in-hand: in nondegenerate cases, follow the curve in Fig. 4 derived from the screening length to determine mesh spacing and charge weight; in degenerate cases, follow the curve due to Landau damping. This should provide the best possible trade-off between conflicting requirements on mesh spacing and charge weight S.

The preceding relationship between screening length L_S and optimum charge weight S can be obtained from a simpler argument as well. We seek to maintain the same number of carriers in the 2D screening "area" in the model as are physically present in the 3D screening "volume". In so doing we have the best possible chance of reproducing screening behavior in the model consistent with the true three-dimensional physics. In the 2D model, the areal density of electrons is n/S, where n is the electron density. The screening area can be approximated by a square with sides of length $2L_S$. In three dimensions, the density is n and the screening volume is approximated by a cube with sides of length $2L_S$. Equating the number of electrons in their 2D and 3D screening environments, we obtain:

$$\frac{n}{S}(2L_S)^2 = n(2L_S)^3 \quad \rightarrow \quad S = \frac{1}{2L_S} \quad , \tag{21}$$

in agreement with the analysis of the preceding paragraph.

PROGRAM IMPLEMENTATION

DAMOCLES has been in use for over three years as of this writing. Program function has continued to evolve, as described in [15]. Fig. 5 summarizes the program structure. Beginning at the top of the circle, execution begins with an initial guess consisting of particle positions and wavevectors. These may derive from a previous solution, or may simply reflect the local doping density and a thermal energy distribution. Proceeding in a clockwise fashion, the Poisson equation is solved to determine the instantaneous potential $\psi(t)$. From the potential, the equations of motion are integrated, as indicated in the box labeled "free-flight". The resultant phase-space coordinates are thus advanced by the timestep Δt. Next, carrier interaction with the device boundaries are processed (labeled "geometry scattering"), revising the phase-space coordinates of a portion of the particle ensemble ($\mathbf{r}$, $\mathbf{k} \rightarrow \mathbf{r}^*, \mathbf{k}^*$). Finally, probabilities for physical scatterings are calculated using carrier average energy, density, Fermi energy, ionized dopant concentrations and information about other particles within a screening circle's distance. These physical scatterings again change the phase-space coordinates for some of the particles; however, the particle positions are unaltered ($\mathbf{r}$, $\mathbf{k} \rightarrow \mathbf{r}^*, \underline{k}^*$). As a last step, statistics are gathered and can be viewed on a graphics workstation. Time is advanced and the cycle begins again. There are two dotted paths indicated in the figure as well, which are conditionally followed. The first path permits the Poisson equation to be skipped, which reduces the compute time at the expense of introducing particle self-forces and possible temporal instability and inaccuracy. This path is rarely taken. The second path conditionally skips the calculation of physical scatterings. Since we do not employ self-scattering in our Monte Carlo scheme to select operative scattering mechanisms, we simply require that the time between scattering evaluation be small compared to the reciprocal of the maximum scattering rate $1/\tau_{max}$, *i.e.*, $\Delta t_{scat} \lesssim 0.1\tau_{max}$. This second path permits us to evaluate the physical scattering consistent with this requirement, independent of the timestep used to integrate the equations of motion.

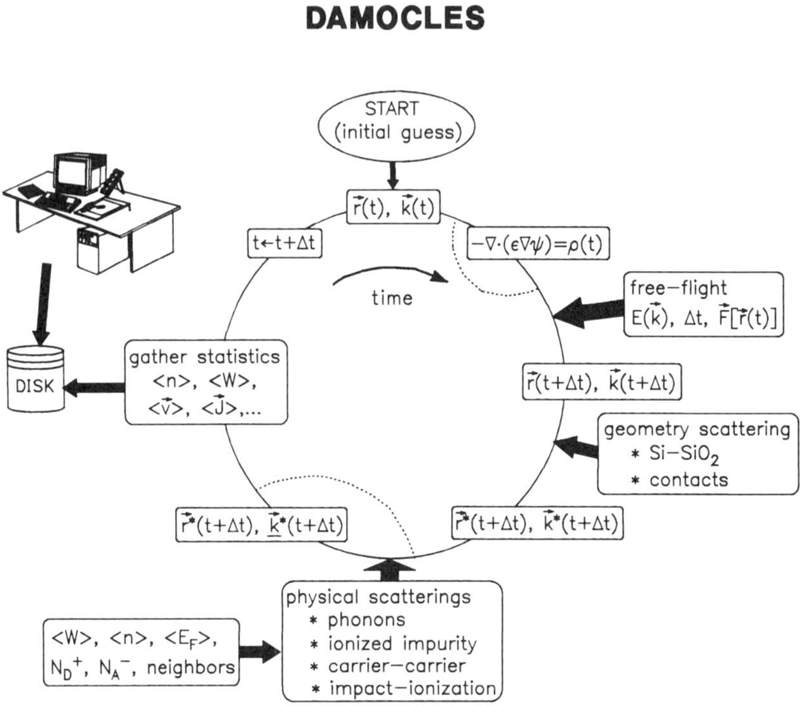

Fig. 5 Flowchart of the DAMOCLES program. Program execution is given by a clockwise path around the circle. The two dotted paths indicate the possibility of skipping the Poisson solution or physical scatterings, effectively introducing different timestep values for the field update and scattering calculations.

As a true litmus test of the program, the code is in active use by five colleagues who are not code developers. We are indebted to them, for they have allowed us to explore many more device scenarios then otherwise possible, and they have demanded a program which delivers results. The (essentially!) singular complaint is that DAMOCLES consumes too much cpu time, failing to meet the often invoked criterion that jobs submitted in the evening should be finished by the next morning. The program executes in 10-100 normalized time units, where this time unit is defined as the cpu time in seconds to follow one particle for one picosecond of physical time. In simpler terms, the code can run for a few cpu days before a useful result is obtained. This is significantly slower than "conventional" band structure Monte Carlo device simulation codes; however, this is the cost of additional physics. For this reason, DAMOCLES practices a "checkpoint-restart" algorithm, *i.e.*, every 15 cpu minutes DAMOCLES writes to disk enough information to restart the calculation should a system failure occur. In addition, after a system failure, the job is resubmitted for execution automatically, without intervention from the user. As a final convenience, users can execute a dedicated graphics package against an ongoing calculation and examine job progress without actually stopping the job. This is the crudest form of "computation steering", but it has served us very well to date.

RESULTS

In this section, we will highlight results obtained with DAMOCLES for Si (and III-V!) MOSFETs, GaAs MESFETs, Si homojunction bipolar transistors and GaAs, InP, $Ga_{0.47}In_{0.53}As$ and InAs homojunction bipolar transistors.

The first devices ever simulated with DAMOCLES were silicon n-MOSFETs, with metallurgical channel lengths nominally 0.1 μm long. The doping and device geometry were selected to correspond to exploratory Si n-MOSFETs fabricated by Sai-Halasz, *et al.* [43]. Silicon MOSFETs are particularly challenging because of the extreme doping dynamic range present in the device, as well as the very difficult question concerning how to treat the Si-SiO_2 interface. The former question was addressed with statistical enhancements, and by judicious choices aimed at including as little of the source and drain as possible in the modeling cross-section. The latter question was addressed by adopting a simple model of interface scattering, consisting of a prespecified mixture of diffuse and specular scattering at the interface. If carriers near the interface are heated sufficiently by the drain field so as to not be confined in subbands, but rather act more "bulk-like" in the channel, this assumption is reasonable. For lack of better guidance, a equal mixture was assumed for these two surface scattering mechanisms.

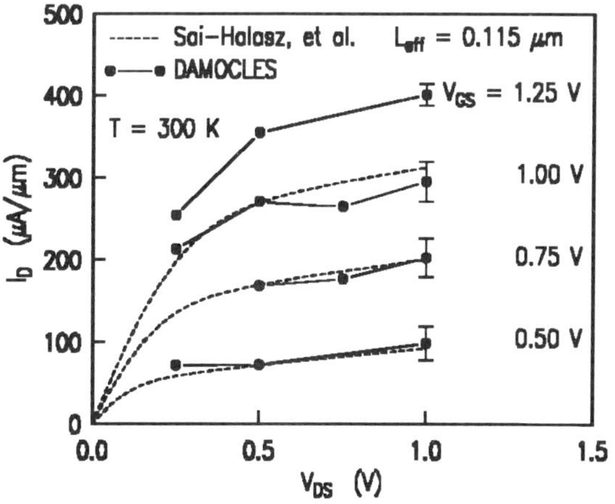

Fig. 6 Output characteristic generated by DAMOCLES for a 0.1 μm Si n-MOSFET at 300 K, together with an overlay of the experimental data.

The first output characteristic ever generated by DAMOCLES is shown in Fig. 6, together with an overlay of the n-MOSFET experimental data. This agreement on the very first try was very gratifying. Although a different mixture of diffuse and specular scattering does have a first-order effect on the drain current at a given bias [12], improving upon this heuristic interface scattering model is very difficult. It is preposterous to attempt to calculate surface roughness scattering in the channel, for the distance to the interface from an electron is quite uncertain: semiclassical models yield a peak density directly at the interface, while quantum effects push the peak away from the interface. Thus, a better interface scattering model must be preceded by a better estimate of the position of the charge in the channel. DAMOCLES has a two-dimensional electron gas transport model, with subband energies calculated self-consistently with the fluctuating potential *and* with the full band structure [8]. This portion of the code needs to be further exercised before conclusions can be made about a more rigorous channel transport approach. Only one aspect is clear, and that is the program runs even slower still with subband quantization included.

The initial comparisons between DAMOCLES and n-MOSFET hardware were very satisfying [6], giving us confidence that the basic physical ingredients necessary to predict device transconductance to very small channel lengths were present in the model. Ultimately, this study of n-MOSFETs and p-MOSFETs [7], together with queries from a co-worker [44], led us to consider a model-based comparison of many different semiconductors in an invariant device structure. The semiconductors to be considered were Si (with both holes and electrons as current carriers), and GaAs, Ge, InP and $In_{0.53}Ga_{0.47}As$ with electrons as current carriers. Only through careful calibration of the bulk transport model for each material could we hope to produce a useful result; this calibration procedure has been painstakingly documented [16]. The device structure to be selected is a MOSFET cross-section, analogous to the nominal 0.1 μm channel-length design used in the early silicon simulations. The same interface scattering model is used throughout these simulations as provided good results for Si MOSFET simulations. This is rationalized by considering the group of MOSFETs, all with different substrate material, to have a SiO_2 deposited insulator film in common. We feel this is a proper approach, for it forces differences between the devices to be solely attributable to the substrate material type. Perhaps more controversial than a common insulator film in this study is the notion that a single device structure can exploit the strengths of the different semiconductors. We respect this criticism, noting silicon and GaAs device designers can argue bitterly over the superiority of their respective, and very different, device design points. Nevertheless, [17] summarizes the findings of our computational study. As device dimensions fall below 0.1 μm, and especially at 300 K, the similar transconductance of the Si p-MOSFET, Si n-MOSFET, GaAs n-MOSFET and the Ge n-MOSFET is striking. To first order, device transconductance can be taken as a direct measure of device switching performance; thus, the similarity of the results is particularly profound. The In-containing compounds InP and $In_{0.53}Ga_{0.47}As$ are markedly superior, mainly due to a lower effective Γ-valley mass together with a large separation in energy between Γ- and satellite-valleys. GaAs is so similar to Si and Ge because of upper valley transfer, which proves difficult to avoid and still obtain acceptable short-channel effects in the MOSFET. Because the effective behavior of GaAs, Si and Ge becomes so similar for situations where the BZ is populated away from valley minima (due to carrier heating in the field), a "universal" behavior is obtained. A detailed account of this model-based comparison of MOSFETs can be found in [11,17].

GaAs MESFETs have been simulated with DAMOCLES, with results discussed in [12]. Velocities greatly exceeding the scattering-limited velocity in GaAs ($\sim 10^7$ cm/sec) fill a progressively greater fraction of the entire channel length as the channel shrinks from 0.75 μ to 0.15 μm. Such high velocities in the GaAs bulk demonstrate the GaAs MESFET will continue to be a viable competitor to GaAs surface-channel devices (*e.g.*, HEMT, SISFET, TEGFET), a conclusion recently underscored from an experimental perspective [45].

In the realm of Si bipolar devices, both npn and pnp homojunction transistors have been simulated. In addition to quantifying the overshoot in the base-collector depletion region, we have also focused on base transport [10]. For transport via diffusion through the base, the transit time should vary as $\tau_B \sim W_B^2$, where W_B is the neutral base thickness assumed to have uniform doping. For ballistic transport through the field-free base, we expect $\tau_B \sim W_B$. The results from DAMOCLES shown in [10] predict a cross-over between diffusion and ballistic transport around 50 nm, or roughly a 5 times larger distance scale that the mean-free path in silicon. This larger distance scale comes about because a single collision experienced in the base is not adequate to randomize the electron's path, for Coulomb scattering dominates over phonon scattering in the device simulated.

Finally, recently results have been obtained for a model-based comparison of GaAs, InP, $Ga_{0.47}In_{0.53}As$ and InAs homojunction transistors [13]. Large signal switching simulations indicate the importance of displacement current effects, corroborating a previous simple theory concerning signal delay [46]. This study indicated that $Ga_{0.47}In_{0.53}As$ and InP transistors have broader and higher velocity profiles than GaAs transistors, resulting in improved switching performance. Indeed, if the low breakdown voltage of InAs is tolerable, this material may be even better suited for high-speed transistors.

CONCLUSIONS

In this chapter, we have attempted to provide detail concerning aspects of the DAMOCLES program which have not yet appeared in print, focusing on physics and especially numerical aspects and program implementation. While there have been many applications of DAMOCLES to device problems, these are well documented in the literature; in this case, we have simply summarized the relevant references. Overall, our conclusion is that physics does indeed matter in the context of device modeling, making a program like DAMOCLES an indispensable part of device research and development.

ACKNOWLEDGEMENTS

We gratefully acknowledge the users of the DAMOCLES program, W. Lee, S. Tiwari, E. Crabbé, D. Frank and D. Cole for their criticism and patience, as well as management support throughout the course of this work.

REFERENCES

[1] R. W. Hockney and J. W. Eastwood, "Computer Simulation Using Particles", Maidenhead: McGraw-Hill, 1981.

[2] C. Jacoboni and P. Lugli, "The Monte Carlo Method for Semiconductor Devices Simulation", New York: Springer, 1989.

[3] S. Selberherr, "Analysis and Simulation of Semiconductor Devices", New York: Springer, 1984.

[4] For example, W. Hänsch and S. Selberherr, "MINIMOS 3: A MOSFET Simulation that Includes Energy Balance", *IEEE Trans. Electron Devices.*, vol. ED-34, p. 1074-1078, May 1987, or A. Forghieri, R. Guerrieri, P. Ciampolini, A. Gnudi, M. Rudan and G. Baccarani, "A New Discretization Strategy of the Semiconductor Equations Comprising Momentum and Energy Balance", *IEEE Trans. Computer-Aided Design.*, vol. 7, p. 231-242, February 1988. A longer sampling of groups working on these approaches can be found in the references contained in [12].

[5] M. V. Fischetti and S. E. Laux, "Monte Carlo Analysis of Electron Transport in Small Semiconductor Devices Including Band-Structure and Space-Charge Effects", *Phys. Rev. B*, vol. 38, p. 9721-8745, 1988.

[6] S. E. Laux and M. V. Fischetti, "Monte Carlo Simulation of Submicrometer Si n-MOSFETs at 77 and 300 K", *IEEE Electron Device Lett.*, vol. EDL-9, p. 467-469, 1988.

[7] M. V. Fischetti and S. E. Laux, "Monte Carlo Simulation of Submicron Si MOSFETs", *Simulation of Semiconductor Devices and Processes, vol. 3*, G. Baccarani and M. Rudan (ed.), Technoprint, Bologna, Italy, p. 349-368, 1988.

[8] M. V. Fischetti, S. E. Laux and D. J. DiMaria, "The Physics of Hot Electron Degradation of Si MOSFETs: Can We Understand It?", *Applied Surface Science.*, vol. 39, p. 578-596, 1989.

[9] M. V. Fischetti, S. E. Laux and W. Lee, "Monte Carlo Simulation of Hot-Carrier Transport in Real Semiconductor Devices", *Solid-State Electronics*, vol. 32, p. 1723-1729, 1989.

[10] W. Lee, S. E. Laux, M. V. Fischetti and D. D. Tang, "Monte Carlo Simulation of Non-Equilibrium Transport in Ultra-Thin Base Si Bipolar Transistors", *Technical Digest, 1989 International Electron Devices Meeting*, p. 473-476, 1989.

[11] M. V. Fischetti and S. E. Laux, "Are GaAs MOSFETs Worth Building? A Model-Based Comparison of Si and GaAs n-MOSFETs", *Technical Digest, 1989 International Electron Devices Meeting*, p. 481-484, 1989.

[12] S. E. Laux, M. V. Fischetti and D. J. Frank, "Monte Carlo Analysis of Semiconductor Devices: The DAMOCLES Program", *IBM J. Res. Develop.*, vol. 34, p. 466-494, July 1990.

[13] S. Tiwari, M. V. Fischetti and S. E. Laux, "Transient and Steady-State Overshoot in GaAs, InP, $Ga_{0.47}In_{0.53}As$ and InAs Bipolar Transistors", *Technical Digest, 1990 International Electron Devices Meeting*, 1990.

[14] E. F. Crabbé, J. M. C. Stork, G. Baccarani, M. V. Fischetti and S. E. Laux, "The Impact of Non-Equilibrium Transport on Breakdown and Transit Time in Bipolar Transistors", *Technical Digest, 1990 International Electron Devices Meeting*, 1990.

[15] S. E. Laux and M. V. Fischetti, "The DAMOCLES Monte Carlo Device Simulation Program", *Computational Electronics: Semiconductor Transport and Device Simulation*, K. Hess, J. P. Leburton and U. Ravaioli (ed.), Kluwer Academic Publishers, Boston, p. 87-92, 1991.

[16] M. V. Fischetti, "Monte Carlo Simulation of Transport in Technologically Significant Semiconductors of the Diamond and Zinc-blende Structures. Part I: Homogeneous Transport", to appear in *IEEE Trans. Electron Devices*, March 1991.

[17] M. V. Fischetti and S. E. Laux, "Monte Carlo Simulation of Transport in Technologically Significant Semiconductors of the Diamond and Zinc-blende Structures. Part II: Submicron MOSFETs", to appear in *IEEE Trans. Electron Devices*, March 1991.

[18] W. Fawcett, A. D. Boardman and S. Swain, "Monte Carlo Determination of Electron Transport Properties of Gallium Arsenide", *J. Phys. Chem. Solids*, vol. 31, p. 1963-1990, 1970.

[19] C. Canali, C. Jacoboni, F. Nava, G. Ottaviani and A. Algerigi-Quaranta, "Electron Drift Velocity in Silicon", *Phys. Rev. B*, vol. 12, p. 2265-2284, 1975.

[20] G. Ottaviani, L. Reggiani, C. Canali, F. Nava and A. Alberigi-Quaranta, "Hole Drift Velocity in Silicon", *Phys. Rev. B*, vol. 12, p. 3318-3329, 1975.

[21] R. Brunetti, C. Jacoboni, F. Nava and L. Reggiani, "Diffusion Coefficients in Electrons in Silicon", *J. Appl. Phys.*, vol. 52, p. 6713-6722, 1981.

[22] C. Jacoboni, F. Nava, C. Canali and G. Ottaviani, "Electron Drift Velocity and Diffusivity in Germanium", *Phys. Rev. B*, vol. 24, p. 1014-1026, 1981.

[23] C. Jacoboni, R. Minder and G. Majni, "Effects of Band Non-Parabolicity on Electron Drift Velocity in Silicon above Room Temperature", *J. Phys. Chem. Solids*, vol. 36, p. 1129-1133, 1975.

[24] C. Jacoboni and L. Reggiani, "The Monte Carlo Method for the Solution of Charge Transport in Semiconductors with Application to Covalent Materials", *Rev. Mod. Phys.*, vol. 55, p. 645-705, 1983.

[25] M. A. Littlejohn, J. R. Hauser and T. H. Glisson, "Velocity-Field Characteristics of GaAs with Γ_6^c-L_6^c-X_6^c Conduction Band Ordering", *J. Appl. Phys.*, vol. 48, p. 4587-4590, 1977.

[26] W. Fawcett and D. C. Herbert, "High-Field Transport in Gallium Arsenide and Indium Phosphide", *J. Phys. C: Solid State Phys.*, vol. 7, p. 1641-1654, 1974.

[27] D. C. Herbert, W. Fawcett and C. Hilsum, "High-Field Transport in Indium Phosphide", *J. Phys. C: Solid State Phys.*, vol. 9, p. 3969-3975, 1976.

[28] J. R. Hauser, M. A. Littlejohn and T. H. Glisson, "Velocity-Field Relationship of InAs-InP Alloys Including the Effect of Alloy Scattering", *Appl. Phys. Lett.*, vol. 28, p. 458-461, 1976.

[29] P. J. Price, "Monte Carlo Calculations of Electron Transport in Solids", *Semiconductors and Semimetals*, vol. 14, p. 249-308, 1979.

[30] M. L. Cohen and T. K. Bergstresser, "Band Structures and Pseudopotential Form Factors for Fourteen Semiconductors of the Diamond and Zinc-blende Structures", *Phys. Rev.*, vol. 141, p. 789-796, 1966.

[31] H. Shichijo and K. Hess, "Band-Structure-Dependent Transport and Impact Ionization in GaAs", *Phys. Rev. B*, vol. 23, p. 4197-4207, 1981.

[32] J. Y. Tang and K. Hess, "Impact Ionization of Electrons in Silicon", *J. Appl. Phys.*, vol. 54, p. 5139-5144, 1983.

[33] J. Y. Tang and K. Hess, "Theory of Hot Electron Emission from Silicon into Silicon Dioxide *J. Appl. Phys.*, vol. 54, p. 5145-5151, 1983.

[34] K. Brennan, K. Hess, J. Y. Tang and G. J. Iafrate, "Transient Electronic Transport in InP under the Conditions of High-Field Electron Injection", *IEEE Trans. Electron Devices*, vol. ED-30, p. 1750-1754, 1983.

[35] K. Brennan and K. Hess, "High-Field Transport in GaAs, InP and InAs", *Solid-State Electron.*, vol. 27, p. 347-357, 1984.

[36] M. V. Fischetti and J. M. Higman, "Theory and Calculation of the Deformation Potential Electron-Phonon Scattering Rates in Semiconductors," in Monte Carlo Simulations of Semiconductors and Semiconductor Devices, ed. by K. Hess, Kluwer Academic Press, Norwell, MA, 1991.

[37] F. Venturi, E. Sangiorgi, R. Brunetti, W. Quade, C. Jacoboni and B. Ricco, "An Efficient Monte Carlo Simulator for High-Energy Electrons and Holes in MOSFETs", *Workshop on Numerical Modeling of Processes and Devices for Integrated Circuits: NUPAD III*", Honolulu, Hawaii, June 3-4, 1990

[38] J. Bude, K. Hess and G. J. Iafrate, "Field Assisted Impact Ionization in Semiconductors", *Computational Electronics: Semiconductor Transport and Device Simulation*, K. Hess, J. P. Leburton and U. Ravaioli (ed.), Kluwer Academic Publishers, Boston, p. 131-136, 1991.

[39] R. E. Bank and D. J. Rose, "Parameter Selection for Newton-like Methods Applicable to Nonlinear Partial Differential Equations," *SIAM J. Numer. Anal.* vol. 17, p. 806-822, 1980.

[40] O. G. Johnson, C. A. Micchelli and G. Paul, "Polynomial Preconditioners for Conjugate Gradient Calculations", *SIAM J. Numer. Anal.* vol. 20, p. 362-376, 1983.

[41] A. Phillips, Jr. and P. J. Price, "Monte Carlo Calculations on Hot Electron Tails", *Appl. Phys. Lett.*, vol. 30, p. 528-530, 1977.

[42] B. K. Ridley, "Quantum Processes in Semiconductors", Oxford: Clarendon Press, 1988.

[43] G. A. Sai-Halasz, M. R. Wordeman, D. P Kern, S. Rishton and E. Ganin, "High Transconductance and Velocity Overshoot in NMOS Devices at the 0.1-μm Gate-Length Level", *IEEE Electron Device Lett.*, vol. 9, p. 464-466, 1988.

[44] A. B. Fowler, private communication.

[45] M. Feng, C. L. Lau, V. Eu and C. Ito, "Does the Two-Dimensional Electron Gas Effect Contribute to High-Frequency and High-Speed Performance of Field-Effect Transistors", *Appl. Phys. Lett.*, vol. 57, p. 1233-1235, 1990.

[46] S. E. Laux and W. Lee, "Collector Signal Delay in the Presence of Velocity Overshoot", *IEEE Electron Device Lett.*, vol. 11, p. 174-176, 1990.

2

SCATTERING MECHANISMS FOR SEMICONDUCTOR TRANSPORT CALCULATIONS

J. Bude

Beckman Institute for Advanced Science and Technology and ECE
University of Illinois at Urbana-Champaign, Illinois 61801

Monte Carlo simulations for transport in semiconductors numerically solve the Boltzmann equation while offering the physically intuitive picture of free flights and carrier scatterings on a microscopic level. The aim of a good Monte Carlo simulator is to give the most physically correct realization of the scattering and free flight processes in an efficient manner. This chapter treats the scattering mechanisms and their implementation in detail. The basic scattering mechanisms treated in this chapter fall into the categories of phonons, static impurities, and scattering due to the coulomb potential of other carriers (carrier-carrier interactions). Photon scattering is important in specific cases (radiative recombination rates for instance), but occurs on time scales much longer than the scattering processes mentioned above, and so is not usually a factor in determining basic transport parameters.

In almost all approaches to electron transport in crystals, the electrons are decoupled from the ions, impurities, and from each other, and familiar Bloch states are assumed for a complete basis set for the electrons. Then, the effect of each of these interactions is reintroduced as a perturbation to the simple, one-electron states forcing transitions between them, ie. scattering. In order to accomplish the decoupling, one makes use of the adiabatic approximation (Born-Oppenheimer approximation) which is the topic of the first section. However, the Boltzmann equation, which is the basis for all Monte Carlo simulations, is a fully classical equation, originally derived for transport of gasses whose constituents interact weakly. The generalization of this equation to the semiclassical regime in which collisions are seen as localized events in space and time but calculated quantum mechanically is a simple extension of the classical Boltzmann equation. The steps leading to this extension which are relevant to the treatment of scattering are discussed in the second section of this chapter. It is then easy to obtain the proper form for the scattering rates. Then, a short discussion of dielectric screening (an important consideration for the calculation of scattering) is given. The remainder of the chapter is devoted to the calculation of specific scattering rates and their implementation in Monte Carlo simulations.

1 The Adiabatic Approximation

To begin our discussion of electron transport in semiconductors, we define electrons to be in one of three classes: *core electrons*, which are tightly bound to the nuclei, *valence electrons*, which are loosely bound and form the covalent bonds between ions, and *conduction electrons*. Valence electron states lie beneath the band gap and are completely filled at zero temperature. Electrons which occupy the excited states above the band gap are termed conduction electrons. In the following discussion of the adiabatic approximation, the crystal is decomposed into *ions* (the nuclei and the core electrons treated as a unit moving rigidly together) and the *outer electrons* which are either valence electrons or conduction electrons.

The effect of the lattice (ions) on electron transport is typically evaluated by making use of the *adiabatic approximation* which separates the electronic (outer electrons) and ionic motion so that their interaction may be treated in perturbation theory. The crystal Hamiltonian in the adiabatic approximation can be derived from the general Hamiltonian as follows. The full Hamiltonian is [1],

$$H_{tot} = T_{ion}(\{R\}) + U_{ion}(\{R\}) + T_{el}(\{r\}) + U_{ee}(\{r\}) + U_{ei}(\{r\},\{R\}) \quad (1)$$

where the ionic and electronic kinetic energies are T_{ion} and T_{el}, and the potential energies U_{ion} and U_{ee}. (U_{ee} is the electron-electron interaction.) U_{ei} is the electron-ion interaction, and the ion and electron coordinates are denoted by $\{R\}$ and $\{r\}$. Because the ions are more massive than the electrons, they will move much more slowly. The key step in formulating the adiabatic approximation is the assumption that the electrons respond *adiabatically* to the motion of the ions — the ion motion does not force transitions between electronic states, but the electron eigen-states evolve adiabatically as the ion positions change. Then, for a fixed set of ion coordinates $\{R\}$, there corresponds a full set of electronic eigen-functions:

$$[T_{el}(\{r\}) + U_{ee}(\{r\}) \quad (2)$$
$$+U_{ei}(\{r\},\{R\})]\psi_n(\{r\},\{R\}) = E_n(\{R\})\psi_n(\{r\},\{R\}) , \quad (3)$$

and the eigen-function of the full Hamiltonian $H_{tot}\Phi = E_{tot}\Phi$, is given by,

$$\Phi = \sum_n \chi_n(\{R\})\Psi_n(\{r\},\{R\}) \quad (4)$$

$$[T_{ion}(\{R\}) + U_{ion}(\{R\}) + E_n]\,\chi_n(\{R\}) = \omega_n\,\chi_n(\{R\}) . \quad (5)$$

In (5) several small terms have been neglected which involve the action of T_{ion} on the electronic wave-function. This is a justifiable approximation since these terms have been shown to be of order $(m/M)^{3/2}$ where m is the electron mass and M is the ion mass (see [1]). The ionic wavefunctions must be calculated self-consistently with the electronic wavefunctions as shown in (5).

In order to untangle the electron states from each other, the Hartree or

Hartree-Fock approximation can be employed which decouples the many-electron state into simple, one electron states by the introduction of exchange and correlation energies. The electron-electron interaction U_{ee} introduces another effect which dresses the remaining coulomb interactions, such as U_{ei}, with a screening cloud resulting in screened interactions (see section 3). We shall designate this screened electron-ion interaction as $\tilde{U}_{ei}$. Similarly, the electron-electron polarization screens other coulomb interactions such as the interaction of conduction electrons (holes) with ionized impurities, V_i, and the electron-electron (carrier-carrier) interaction between conduction electrons or between conduction electrons and valence electrons, V_{cc}. V_i and V_{cc} are assumed weak and treated as the perturbations responsible for ionized impurity scattering and impact ionization (the inverse Auger process) discussed in later sections of this chapter.

The many electron state is retrieved by placing electrons into the one electron states obeying the Pauli exclusion principle. This is the essence of the one-electron approximation for which,

$$\left[T_{el}(r) + \tilde{U}_{ei}(r, \{R\}) + U_{EX}\right] \psi_n(r, \{R\}) = E_n(\{R\})\psi_n(r, \{R\}) \,, \tag{6}$$

where U_{EX} is the exchange energy.

The electron mobilities are generally large in most semiconductors, and it is appropriate to treat the electron-ion interaction $\tilde{U}_{ei}$ as a perturbation [2]. Furthermore, because ionic vibrations involved are relatively small, a good choice for a perturbation approach to decouple the electrons from the ions is an expansion in the displacements of the ions from their equilibrium positions $u^\alpha(R_l)$. If these equilibrium positions are designated as $\{R^0\}$, the solution of (6) for the one-electron wave-functions yields the familiar Bloch wave-functions of band index n and wave-vector k:

$$\begin{aligned} \left[T_{el}(r) + \tilde{U}_{ei}(r, \{R^0\}) + U_{EX}\right] \psi_{nk}(r) &= E_n(k)\psi_{nk}(r) \qquad (7) \\ H_e(r)\psi_{nk}(r) = E_n(k)\psi_{nk}(r), \quad \psi_{nk}(r) &= u_{nk}(r)e^{ik\cdot r} \,. \end{aligned}$$

Labeling the equilibrium lattice sites by R_l^0 for integers l, the basis vectors α, and keeping only the first two terms in the expansion of $\tilde{U}_{ei}(r - R_l)$,

$$\tilde{U}_{ei}(r - R_l) \approx \tilde{U}_{ei}(r - R_l^0 - \alpha) \; + \; u^\alpha(R_l) \cdot \nabla_{R_l} \tilde{U}_{ei}(r - R_l)\Big]_{R_l = R_l^0 + \alpha} \,, \tag{8}$$

the one electron crystal Hamiltonian in the adiabatic approximation becomes:

$$\begin{aligned} H_{tot} &= T_{el} + V_c(r) + V_p(r, R) + V_i(r) + V_{cc}(r, r') \qquad (9) \\ V_p(r) &= \sum_{l\alpha} u^\alpha(R_l) \cdot \nabla_{R_l} \tilde{U}_{ei}(r - R_l - \alpha) \end{aligned}$$

Here, $V_c(r)$ is the equilibrium periodic crystal potential, $\tilde{U}_{ei}(r, \{R^0\})$. $V_p(r)$ represents the phonon scattering perturbation.

The ion eigen-states are, (to second order in the ion-ion interaction) harmonic oscillator states or phonons. Specifically, the ion Hamiltonian is diagonalized by appealing to the raising and lowering operators for the normal modes of vibration, the phonon modes of the crystal. In the phonon occupation number basis, [3]

$$H_i \;=\; \frac{1}{2}\sum_{qj} \hbar\omega_q^j (a_j(q)a_j^+(q) \;+\; a_j^+(q)a_j(q)), \tag{10}$$

where $a_j^+(q)$ and $a_j(q)$ are the raising and lowering operators for the phonon mode of wavevector q and branch j. In this notation, $u^\alpha(R_l)$ can be expressed by an expansion over the normal modes of wavevector q of the lattice as [4]

$$u^\alpha(R_l) \;=\; \sum_{qj}\left[\frac{\hbar}{2Nm_\alpha\omega_q^j}\right]^{\frac{1}{2}} e^{iq\cdot(R_l+\alpha)}[a_j(q)+a_j^+(-q)]\;\xi_{qj}^1 \tag{11}$$

where N is the number of primative cells in the lattice, m_α is the mass of the ion at basis location α, and ξ_{qj}^1 is the phonon wavevector.

2 Scattering in the Semi-classical Boltzmann Equation

To derive the expressions for scattering in the semi-classical Boltzmann equation (SCBE), we start with the equation of motion for the density matrix ρ for the adiabatic crystal. As discussed in section 1, the crystal system can be separated into a carrier part (conduction electrons or holes) and a phonon part. The phonons are considered to be a thermodynamic heat bath in equilibrium at some temperature T. Then, weakly coupled to it are the carriers which we can enumerate with the Bloch wavevector and band index quantum numbers. Furthermore, the carriers are weakly coupled to each other, V_{cc}, and to impurities, V_I. As noted above, these interactions are weak, and we can approximate the full density matrix as a product of the carrier density matrix and the equilibrium phonon density matrix at all times. This derivation ignores V_{cc} so that simple one-particle states can be used; however, to include V_{cc} to first order, two particle antisymmeterized states can be chosen instead. For one-particle states, the carrier density matrix is labeled only by the wavevector k ($\rho_{kk'}^e$) (suppressing n) and the phonon bath density matrix by $\rho_B = e^{-\beta H_B}$. Here, $\beta = 1/k_BT$, where k_B is Boltzmann's constant, and $H_B = H_i$, the bath (ionic) Hamiltonian given above. In general, the quantum analogue of the phase space distribution function is the Wigner distribution which is a transformation of ρ. To demonstrate the form of the scattering kernel ("collision integral") in the SCBE, the field can be ignored and it is permissible to identify the diagonal of $\rho_{kk'}$ as the distribution function $f(k)$ which occurs in the classical Boltzmann equation.

First, the standard quantum Liouville equation neglecting the electric field is:

$$i\hbar\frac{\partial\rho}{\partial t} = H_{tot}^{\times}\rho, \tag{12}$$

where the hyper-operator notation

$$A^{\times}B \equiv [A, B] = AB - BA \tag{13}$$

has been employed and A and B are operators. H_{tot} is the adiabatic Hamiltonian as derived in section 1.

The projection operator method (see for example Kubo [5]) separates the diagonal of the system density matrix so that an equation for the time evolution of $\rho^{e}_{kk} = f(k)$ is obtained. Then, the two projection operators

$$P = \frac{e^{-\beta H_B} tr_B}{tr_B e^{-\beta H_B}} \otimes \delta_{k',k} \tag{14}$$

$$Q = 1 - P \tag{15}$$

with the properties $PP = P$, $QQ = Q$, and $PQ = QP = 0$, can be used in order to derive equations of motion for the diagonal and off-diagonal parts of ρ in terms of each other. Here, tr_B is the many body trace over the phonon bath states. The projection operator P projects the portion of ρ which is diagonal in the carrier quantum numbers (containing the probability of occupancy of the carrier stationary states) and leaves the phonon bath in its equilibrium configuration.

By substitution, we arrive at an equation for only the diagonal part in the form,

$$i\hbar(P\frac{\partial\rho}{\partial t}) = -\frac{i}{\hbar}P(H_p)^{\times}\int_0^t dt' e^{-\frac{i}{\hbar}Q[H_{tot}]^{\times}}Q(H_p)^{\times}(P\rho(t')) \tag{16}$$

where the density matrix has been assumed to be diagonal at $t = 0$. Here, all the perturbations discussed earlier have been included in H_p.

Then, noting the hyper-operator identity for operators A and B,

$$e^{A^{\times}}B = e^{A}Be^{-A} \tag{17}$$

and that

$$e^{\frac{i}{\hbar}H_{tot}(t-t')} = U_s(t, t') \tag{18}$$

where $U_s(t, t')$ is the Schroedinger picture propagator for the full Hamiltonian from t to t', we can write equation (16) in terms of these propagators [11],

$$\frac{\partial}{\partial t}P\rho(t) = -\frac{i}{\hbar}P(H_p)^{\times}\int_0^t dt' U_s(t, t')(H_p)^{\times}P\rho(t')U_s(t', t). \tag{19}$$

Since H_p is considered weak, it can be neglected in the exponential of (18) compared to H_B and $H_0 = T_{el} + V_c$. This allows an expression of U_s in terms of products of the free propagators for the bath and the carriers,

$$U_s(t,t') \approx e^{-\frac{i}{\hbar}H_0(t-t')}\ e^{-\frac{i}{\hbar}H_B(t-t')}. \tag{20}$$

If the matrix element for the electron phonon interaction given in (10) is defined as,

$$(V_p)_{kk'} \equiv < n'k' \mid V_p(r) \mid nk >, \tag{21}$$

where the phonon operators in $u^\alpha(R_l)$ have produced to N_{jq} as discussed above, and the matrix element for the non-phonon scattering terms is defined as

$$(V_i)_{kk'} \equiv < n'k' \mid V_i(r) \mid nk > \tag{22}$$

for the ith perturbation, then the equation of motion for $\rho^e_{kk} = f(k)$ is given by,

$$\frac{\partial}{\partial t} f(k,t) = \tag{23}$$

$$\frac{2Re}{\hbar^2}\sum_{k'i} \mid (V_i)_{k,k'} \mid^2 \int_0^t [T_i(k',k;t')f(k',t') - T_i(k,k';t')f(k,t)]dt' +$$

$$\frac{2Re}{\hbar^2}\sum_{k'} \mid (V_p)_{k,k'} \mid^2 \int_0^t [T_p(k',k;t')f(k',t') - T_p(k,k';t')f(k,t')]dt'$$

with

$$T_i(k,k') = e^{-i(\omega(k)-\omega(k'))(t-t')} \tag{24}$$

$$T_p(k,k') = (N_{k-k'}+1)e^{-i(\omega(k)-\omega(k')-\omega_{k-k'})(t-t')} + (N_{k-k'})e^{-i(\omega(k)-\omega(k')+\omega_{k-k'})(t-t')}, \tag{25}$$

with $\omega(k) \equiv E(k)/\hbar$. $T_i(k,k')$ and $T_p(k,k')$ are functions associated with non-phonon and phonon scattering from k to k' respectively. Furthermore, the trace over the phonons and the action of the raising and lowering operators in the electron-phonon perturbation have produced the factors $N_{jq}+1$ and N_{jq} for emission and absorption respectively with N_{jq} being the Bose-Einstein average occupation number for the phonon of wavevector q and phonon branch j:

$$N_{jq} = \frac{1}{e^{\hbar\beta\omega_{jq}} - 1}. \tag{26}$$

If $f(t)$ were outside of the time integrals, the integral over t of the $T(k,k')$ becomes the familiar delta functions of energy conservation. For example,

$$\lim_{t\to\infty} \int_0^t dt' Re\,(T_i(k,k';t')) \to \pi\hbar\delta(E(k) - E(k')). \tag{27}$$

Since $f(t)$ typically changes very little in the time it takes to establish a delta function, $f(t)$ can indeed be taken outside of the time integral. The result is the standard Boltzmann equation with the scattering rate

$S(k, k')$ from k to k' given by the familiar Golden Rule expression for energy conservation:

$$S_i(k, k') = \frac{2\pi}{\hbar} | (V_i)_{kk'} |^2 \delta(E(k) - E(k')) \tag{28}$$

for elastic scattering processes, and

$$S_p(k, k') = \frac{2\pi}{\hbar} (N_q + \frac{1}{2} \pm \frac{1}{2}) \mid (V_p)_{kk'} |^2 \delta(E(k) - E(k') \mp \omega_q) \tag{29}$$

for phonon scattering processes where the upper sign is taken for emission and the lower sign for absorption and $q = k - k'$. Thus, the scattering rates in the SCBE are simply given by the Fermi Golden Rule from quantum mechanics.

However, the scattering events have been treated as if they happened instantaneously as a result of the approximation in (27). This approximation breaks down in three cases. The first case is obvious. If transients are to be resolved on a time scale of the order of the time it takes to establish (27) and the scattering rates are high, $f(k, t)$ cannot be pulled out of the time integral and the simple forms in (28) or (29) will not apply. If the perturbations are sufficiently weak, then $f(k, t)$ can still be removed from the time integral, but (28 and (29) must be replaced with partially completed delta functions, which relax energy conservation. This is just a manifestation of the energy and time uncertainty principle.

Second, if scattering rates are high, the approximation of a weak interaction leading to equation (18) breaks down, and in the time it takes to establish the delta function in (27), the original state can decay appreciably. To account for the depletion of the initial state in the finite time it takes for the collision to become "complete" the approximation in (18) must be improved to include the effect of H_p in $U_s(t, t')$. One way to account for H_p is to calculate the full propagator $U_s(t, t')$ from field theory. This leads to two additions to (23). First, the simple propagators in the scattering terms are *dressed* by virtual transitions in the self-energy $\Sigma(k, E)$ [6]. The second addition unfortunately leads to the inclusion of terms which do not resemble the Boltzmann equation. Typically these terms are assumed small and are ignored. At present, it is unclear what effect these terms have. Following the general approach that these terms are small, we can augment the SCBE by including $\Sigma(k, E)$ in equation (18).

Assuming that the self-energy can be calculated, (see for instance reference [7] for a self-consistent calculation of the self-energy for realistic band structures) quantum field theory states that for $t > t'$ [6],

$$U_s(k; t, t') = e^{-\frac{i}{\hbar}(E(k)+\Delta(k))(t-t')} e^{-\frac{i}{\hbar}(\Gamma(k))(t-t')}, \tag{30}$$

with

$$\begin{aligned} \Delta(k) &= Re\,(\Sigma(k)) \\ \Gamma(k) &= Im\,(\Sigma(k)) \ . \end{aligned} \tag{31}$$

Physically, $\Delta(k)$ corresponds to a shift in the energy level $E(k)$, and $\Gamma(k)$ corresponds to the finite lifetime of the state k. For example, the probability that a particle which starts at time t' in state k is in state k at time t $(t \geq t')$ is

$$| U_s(k;t,t') |^2 = e^{-\frac{2}{\hbar}\Gamma(k)(t-t')}. \tag{32}$$

The total scattering rate out of state k, $S_{tot}(k)$, can be identified as

$$S_{tot}(k) = \frac{2}{\hbar}\Gamma(k), \tag{33}$$

by interpreting the lifetime of the state as the inverse of the total scattering rate. (This can also be shown through application of the Optical Theorem of quantum mechanics [8].) This clearly demonstrates the connection between high scattering rates and the finite lifetime of the state.

When (32) is substituted into (23), the limit in equation (27) becomes,

$$\begin{aligned} A(E,E') \equiv & \qquad\qquad\qquad\qquad\qquad\qquad\qquad\qquad (34) \\ & \lim_{t\to\infty} Re \int_0^t dt' e^{-\frac{i}{\hbar}(E(k)+\Delta(k)-E(k')+\Delta(k'))(t-t')} e^{-\frac{1}{\hbar}(\Gamma(k)+\Gamma(k'))(t-t')} \\ & = \frac{\hbar}{\pi} \frac{\Gamma(E)+\Gamma(E')}{(E+\Delta(E)-E'-\Delta(E'))^2+(\Gamma(E)+\Gamma(E'))^2}. \end{aligned}$$

Therefore, one way to add the effects of high scattering rates to the SCBE is to replace $\delta(E-E')$ with $A(E,E')$ which is the familiar Lorentzian line shape. The effect of broadening the delta function through high scattering rate is termed *collision broadening.*

Lastly, if a large electric field is present, the carriers can be accelerated appreciably during the time of the collision. This is referred to the *intra-collisional field effect.* In general, this effect can also broaden the energy conserving delta function, however, it is more difficult to deal with in a compact way. The interested reader is referred to the following references for a detailed treatment of this effect and collision broadening: [9]-[11]. Also, for an overview of attempts to include collision broadening and the intracollisional field effect in Monte Carlo Simulations see: [12]-[13].

3 Dielectric Screening

As discussed in section 1, the scattering potentials V_{cc}, V_i and V_p are self-consistent, screened potentials which result from bare perturbations. Because in most cases the bare potentials are simple Coulomb potentials, $v(q) = 1/(4\pi\epsilon_0 q^2)$, the easiest way to calculate the self-consistent perturbations is to screen the bare perturbations with the dielectric function which connects the two.

Suppose the potential, $V_{ex}(r,t)$ is introduced as a bare, external potential to the crystal. The resulting (true) potential felt at time t' and position r' is $V_{tr}(r',t')$ and the induced charge is $\rho_{in}(r,t)$. If V_{ex} is weak enough that first order perturbation theory is adequate, then we can calculate the redistribution of charge, $\rho_{in}(r,t)$ as a linear functional of the self-consistent potential $V_{tr}(r,t)$. In general [18]

$$\rho_{in}(r,t) = \int d^3r' \int dt' \chi(r,r';t-t') V_{tr}(r',t') \tag{35}$$

where $\chi(r,r';t-t')$, the electronic susceptibility, is calculated from first order perturbation theory in which V_{tr} is the perturbation. A related quantity called the dielectric function can similarly be defined:

$$V_{ex}(r,t) = \int d^3r' \int dt' \epsilon(r,r';t-t') V_{tr}(r',t'). \tag{36}$$

The "inverse" of the dielectric function is the quantity we want since it expresses the true potential in terms of the bare (external) potential.

The Fourier transform $\rho(q+G,\omega)$ of equation (35) can be written,

$$\rho_{in}(q+G,\omega) = \int_{Vol} d(q'+G')\chi(q+G,q'+G';\omega)\; V_{ex}(q'+G',\omega)\;, \tag{37}$$

where G is a reciprocal lattice vector, q is a wavevector in the first Brillouin zone (BZ), and Vol is the crystal volume. Because χ is lattice translationally invariant, $\chi(r,r') = \chi(r+R_j, r'+R'_j)$, and the integral in equation (35) becomes a summation: [14]

$$\rho_{in}(q+G,\omega) = \sum_{G'} \chi(q+G,q+G',\omega)\; V_{tr}(q+G',\omega)\;. \tag{38}$$

A similar result holds for $\epsilon(q+G,q+G',\omega)$:

$$V_{ex}(q+G,\omega) = \sum_{G'} \epsilon(q+G,q+G',\omega)\; V_{tr}(q+G',\omega)\;. \tag{39}$$

Application of Poisson's equation connects V_{tr} with V_{ex} [16] Fourier transforming the Poisson's equations for the total charge $\rho = \rho_{in} + \rho_{ex}$ and the external charge ρ_{ex} we have,

$$(q+G)^2 V_{tr}(q+G,\omega) = -\frac{1}{\epsilon_0}\rho(q+G,\omega) \tag{40}$$

$$(q+G)^2 V_{ex}(q+G,\omega) = -\frac{1}{\epsilon_0}\rho_{ex}(q+G,\omega)\;. \tag{41}$$

Solving (40) and (41) for ρ_{in} and substituting this into (38) identifies the dielectric function in (39) as:

$$\epsilon(q+G,q+G',\omega) = \delta_{GG'} - \frac{1}{\epsilon_0(q+G)^2}\,\chi(q+G,q+G',\omega). \tag{42}$$

The random phase approximation (RPA), a type of first order perturbation calculation including the temperature through the distribution function

$f_n(k)$, defines χ as:

$$\chi^e(q+G, q+G', \omega) = \sum_{nn'k} \frac{f_{n'}(k+q) - f_n(k)}{E_{n'}(k+q) - E_n(k) + \hbar\omega}$$
$$< nk \mid e^{-i(q+G)\cdot r} \mid n'(k+q) >< n'(k+q) \mid e^{i(q+G')\cdot r} \mid nk > . \quad (43)$$

This together with (42) expresses the effects of screening on a bare, external potential. Then, the inverse of the matrix $\epsilon(q+G, q+G')$, where the rows and columns are labeled by G and G', expresses the true perturbations V_{cc}, V_i and V_p in terms of the bare Coulomb potentials which give rise to the perturbations. A plot of $\epsilon(q,q)$ is shown in figure 1 as a function of ω for silicon at zero temperature calculated using equation (43).

Several simplifications of this result are applicable for scattering in semiconductors. First, the off-diagonal terms can usually be neglected for non-phonon scattering since they are usually smaller than the diagonal. They must be kept for phonon scattering since they are necessary to fulfill certain sum rules ([2]). Furthermore, since there are usually many more electrons in the valence bands than there are electrons (holes) in the conduction (valence) bands, $f_n(k) \approx 1$ for valence bands and $f_n(k) \approx 0$ for conduction bands. This is the zero temperature approximation.

One simple way to include the effect of the free conduction band electrons in the $T = 0$ model is to add the susceptibilities for the $T = 0$ case, χ^0 and the susceptibility for a free electron gas χ^c with density equal to the density of conduction band electrons. Two expressions for χ^c are given by the Thomas-Fermi screening theory and the Lindhard screening theory for a free electron gas [17]. For example, in the Thomas-Fermi theory

$$\chi^c(q) = -e^2 \frac{\partial f_0(E, E_f)}{\partial E_f} \quad (44)$$

where f_0 is an equilibrium Fermi distribution in the conduction bands, and E_f is the Fermi energy. Then, a natural screening length can be defined as

$$k_{TF}^2 = \frac{e^2}{\epsilon_0} \frac{\partial f_0(E, E_f)}{\partial E_f}, \quad (45)$$

so that

$$\epsilon(q) = 1 + \frac{k_{TF}^2}{q^2}. \quad (46)$$

Having derived relations between the bare perturbing potentials and the screened potentials which result, we can explicitly calculate important scattering matrix elements for use in the SCBE of section 2.

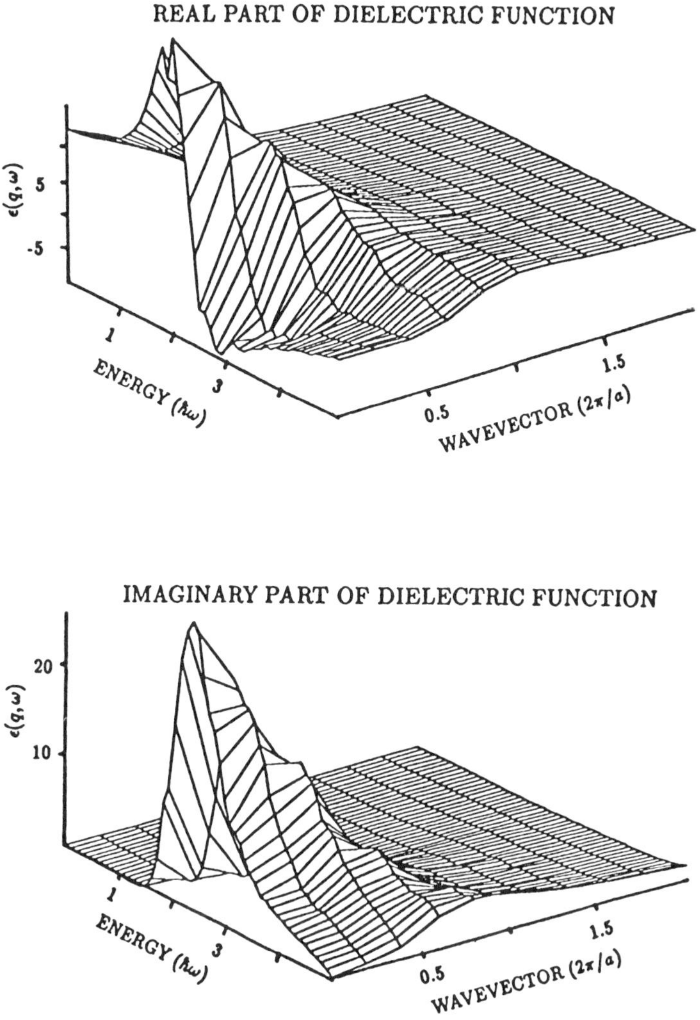

Figure 1: Real and Imaginary parts of the frequency and wavevector dependent dielectric function in silicon calculated using the random phase approximation (RPA).

4 Phonon Scattering

4.1 Phonon Perturbation Potential

In this section, the phonon perturbation is rewritten in terms of the bare potential and the dielectric function, and the matrix elements $(V_p)_{n'k',nk}$ from (21) are calculated. From section 1, the perturbing potential was found to be

$$V_p(r) = \sum_{l\alpha} u^\alpha(R_l) \cdot \nabla_{R_l} \tilde{U}_{ei}(r - R_l - \alpha) \tag{47}$$

where $\tilde{U}_{ei}(r - R_j - \alpha)$ is the self-consistent pseudopotential (effective potential outside of the ion itself) felt by an electron at r due to the ion at site R_l and basis vector α, and $\nabla_{R_l} \cdot \tilde{U}_{ei}(r - R_l)$ is the local gradient of that pseudopotential taken with respect to the ionic location. Proceeding as in Vogl [2], we Fourier transform $V_p(r)$:

$$V_p(r) \;=\; \sum_{qG} e^{i(q+G)\cdot r} \partial V(q+G) \;, \tag{48}$$

where, q is a vector in the first Brillouin zone and G is a reciprocal lattice vector. The summation above must contain all vectors of reciprocal space because $V_p(r)$ depends on the displacements $u^\alpha(R_l)$ which are not in general periodic in a primitive lattice translation. Since the displacements are small, we assume that $\tilde{U}_{ei}$ is related to the unscreened ionic potential v_α by the inverse dielectric function $\epsilon^{-1}(r, r')$ as described in section 3. In this context the unscreened ionic pseudopotential, v_α, is the potential of the nucleus and the core electrons associated with the basis ion α. If the crystal is monoatomic, v_α is independent of α. v_α behaves like a Coulomb potential at long distances, but has a repulsion for short distances due to the core electrons (see [17]). A more exact expression for this potential is given in the chapter by Fischetti and Higman.

In terms of the bare pseudopotential and the dielectric function,

$$\begin{aligned} \partial V(q+G) \;=\; & \frac{1}{Vol} \int_{Vol} d^3r \; e^{-i(q+G)\cdot r} \sum_{l\alpha} u^\alpha(R_l) \cdot \\ & \int_{Vol} d^3r' \epsilon^{-1}(r, r') \nabla_{R_l} v_\alpha(r' - R_l - \alpha) \;, \end{aligned} \tag{49}$$

where Vol is the crystal volume. The resulting Fourier components, $\partial V(q+G)$, of the screened perturbation are referred to as pseudopotential perturbations.

From section 3 we can write equation (50) in reciprocal lattice space as,

$$\begin{aligned} \partial V(q+G) \;=\; & -i \sum_{l\alpha} u^\alpha(R_l) \cdot \\ \sum_{G'} \epsilon^{-1}(q+G, q+G')(q+G') & e^{i(R_l+\alpha)\cdot(q+G')} v_\alpha(q+G') \;. \end{aligned} \tag{50}$$

The result of equation (50) comes from the recognization that $\nabla_{R_l} v_\alpha(r - R_l - \alpha) = -\nabla_r v_\alpha(r - R_l - \alpha)$, and an application of the shifting and derivative properties of the Fourier transform. We do not need to consider the frequency dependence in (50) since in the adiabatic approximation, the ions are assumed to be stationary.

Substituting the expression in equation (11) for $u^\alpha(R_l)$ into equation (50), we can write,

$$\partial V(q+G) = \frac{-i}{Vol} \sum_{\alpha j q' G'} \left[\frac{\hbar}{2N m_\alpha \omega_q^j} \right]^{\frac{1}{2}} [a_j(q') + a_j^+(-q')] \, \xi_{qj}^\alpha \cdot \tag{51}$$
$$(q+G')\epsilon^{-1}(q+G, q+G') v_\alpha(q+G') e^{-i(q+G')\cdot\alpha} \sum_l e^{i(q'-q-G')\cdot R_l} .$$

Concentrating on the sum over l, we can make several simplifications. First, $G' \cdot R_l$ is always 2π, so the G' in the first exponential can be ignored. Second, if $q' \neq q + G''$ with G'' being any reciprocal lattice vector, the sum vanishes because

$$\sum_l e^{iR_l \cdot (q-q')} = N \delta_{q, q'+G''} \, . \tag{52}$$

Hence, we can considerably simplify equation (51) and write it as

$$\partial V(q+G) = \frac{-i}{\Omega} \sum_{\alpha G'} \left[\frac{\hbar}{2N m_\alpha \omega_q^j} \right]^{\frac{1}{2}} [a_j(q) + a_j^+(-q)] \, \xi_{qj}^1 \cdot$$
$$(q+G')\epsilon^{-1}(q+G, q+G') v_\alpha(q+G') e^{-i\alpha\cdot q}, \tag{53}$$

where the fact that $\Omega = Vol/N$, the unit cell volume, has been used. Equation (53) is substituted into equation (48) to give the full electron-phonon perturbation $V_p(r)$ in terms of the bare ion pseudopotential and the dielectric function.

4.2 General Electron-Phonon Matrix Elements

To calculate the matrix elements of the electron-phonon interaction as required in the scattering kernel (see section 2, equation (29)), we need only calculate the matrix element of $e^{i(q+G)\cdot r}$ between Bloch states. For example,

$$< n'k' \mid V_p(r) \mid nk > = \sum_{qG} \partial V(q+G) < n'k' \mid e^{i(q+G)\cdot r} \mid nk > . \tag{54}$$

where $\mid nk >$ are the normalized Bloch states $N^{-\frac{1}{2}} u_{nk}(r) e^{ik\cdot r}$ of band index n and wave vector k. The phonon raising and lowering operators in equation (53) have already acted on the phonon occupation states to give N_{qj} as in section 2, equation (26). The electronic part of the matrix element provides conservation of the total crystal momentum as follows:

$$< n'k' \mid e^{i(q+G)\cdot r} \mid nk >=$$

$$\frac{1}{N}\int_{Vol} u^*_{n'k'}(r)u_{nk}(r)\; e^{i(k'-k)\cdot r}\; e^{\mp i(q+G)\cdot r} d^3r \tag{55}$$

$$= \delta_{k',k\pm q+G'} \int_{\Omega} u^*_{n'k'}(r)u_{nk}(r)\; e^{i(\mp G-G')\cdot r} d^3r \tag{56}$$

Here, the integral in (55) was factored into a sum of integrals over the primative cell. The result of (56) follows from the identity given in (52). The G' in the Kronecker delta refers to a general reciprocal lattice vector, not necessarily the particular G in the integral. When $G' \neq 0$, the scattering process is called an Umklapp process (assisted by a reciprocal lattice vector). In the following sections, $\delta_{k',k\pm q}$ always implies that Umklapp processes are allowed ($\delta_{k',k\pm q} \rightarrow \delta_{k',k\pm q\pm G'}$), although the G' will be suppressed in the notation. If an Umklapp process occurs, then an extra term, $e^{iG'\cdot r}$, appears in each integral.

4.3 Phenomenological Phonon Scattering Processes

The form of the electron-phonon matrix element as given in (54), written in terms of microscopic quantities is exact within the rigid-pseudo-ion approximation (see the chapter in this book by Fischetti and Higman) [19]. However, because it requires a detailed knowledge of both the dielectric matrix and the bloch wavefunctions, it is very difficult to calculate. With some approximations, simple forms for these interaction matrix elements which correspond to well known phenomenological scattering processes in semiconductors can be demonstrated [2]. In particular, the acoustic and optical deformation potentials and the polar optic interactions can be extracted from (54), and the approximations leading to these familiar processes can be examined.

From equation (53) it is apparent that $V_p(r)$ can contain both *long range* and *short range* components. Long-range components vary negligibly within the unit cell and produce fields, whose average over many cell lengths does not vanish. The long-range components are therefore responsible for the macroscopic fields produced in the crystal, which have been identified with the phenomenological polar optic interaction in polar materials, and the piezoelectric interaction in both polar and nonpolar materials ([2]). In contrast, the short-range components of the perturbing potential involve rapid oscillations within the unit cell, and consequently, their average on any macroscopic length scale vanishes. They do, however, contribute to scattering through the phenomenological deformation potential interaction.

The separation of $V_p(r)$ into short and long-range components has been shown by Vogl by considering the behavior of $\partial V(q+G)$ as $q \rightarrow 0$ [2]. We quote only the result,

$$V_p(r)_{lr} = e^{iq\cdot r}\left[1+\sum_{G\neq 0} e^{iG\cdot r}\frac{\epsilon^{-1}(q+G,q)}{\epsilon^{-1}(q,q)}\right]\partial V(q) \tag{57}$$

$$V_p(r)_{sr} = \sum_{G\neq 0} e^{i(q+G)\cdot r}\partial V(q+G), \tag{58}$$

where $V_p(r)_{lr}$ and $V_p(r)_{sr}$ are the long and short range components respectively.

4.3.1 Phenomenological Deformation Potential

The electron-phonon matrix element for the short-range interaction is obtained by combining equations (58), (53) and (55). The resulting matrix element can be recast into the form of the deformation potential interaction, and will be shown to involve, for small q, dilation for optical modes and elastic strain for acoustic modes. For large q, the form of the phenomenological intervalley scattering matrix element will be discussed.

We start by putting $< n'k' \mid V_p(r)_{sr} \mid nk >$ into a more manageable form for a crystal with a basis of two ($\alpha = 1$ and $\alpha = 2$). For a crystal with a basis there are two phonon types, *optical* and *acoustic*, and they behave quite differently. For acoustic modes, $\lim_{q\to 0}\omega(q) \to 0$, corresponding to a rigid displacement of the entire lattice, whereas, for optic modes, $\lim_{q\to 0}\omega(q) \to \omega_{op} \neq 0$ corresponding to rigid displacement of the two sublattices. The latter case can cause transitions between the adiabatic Bloch states leading to scattering, but the former case cannot lead to scattering [20]. For small q it is possible to write the phonon polarization vectors, ξ^{α}_{qj}, as $\xi^{1}_{qj} = K{m_1}^{1/2}\hat{\xi}_{jq}$ and $\xi^{2}_{qj} = \pm K{m_2}^{1/2}\hat{\xi}_{jq}$ where $\hat{\xi}_{jq}$ is a unit vector, K is a constant, and the plus (minus) sign is chosen for acoustic (optical) modes. Furthermore, in this limit, there are three acoustic and three optical branches corresponding to two modes polarized transversely to the direction of propagation and one poplarized longitudinally.

$$< n'k' \mid V_p(r)_{sr} \mid nk > = \left(N_{jq}(T)+\frac{1}{2}\pm\frac{1}{2}\right)\left[\frac{\hbar}{2N\mu\omega^j_q}\right]^{\frac{1}{2}} \tag{59}$$
$$-i\,\delta_{k',k\pm q} \;< n'k' \mid e^{-iq\cdot r}V^{sr}_{jq}(r) \mid nk > \cdot\hat{\xi}_{jq}$$

with V^{sr} given by comparison with (58) and (53). μ can be chosen to be the reduced mass. Since, the final form for (59) is phenomenological, the constant can be chosen to accommodate the choice of μ.

To compare (59) with the phenomenological forms for deformation potential scattering, we expand $e^{iq\cdot r}$ in a Taylor series in q and look for the higest order non-vanishing terms.

For small q, the exponential in equation (59) can be expanded as $e^{iq\cdot r} \approx 1 + iq\cdot r$. For the acoustic phonon case, the highest order term in (59) which does not vanish is the $q \cdot r$ term, since as discussed above, the constant term corresponds to a rigid displacement of the entire crystal which cannot scatter [2]. For a particular mode j, (59) then becomes

$$< n'k' \mid V_p(r)_{sr} \mid nk >_{ac} = \left(N_{jq}(T) + \frac{1}{2} \pm \frac{1}{2} \right)^{\frac{1}{2}}$$
$$\delta_{k',k\pm q} \; < n'k' \mid (q_x r_x + q_y r_y + q_z r_z) V_{sr}(r) \mid nk > \cdot \hat{\xi}_{jq} \qquad (60)$$

Equation (60) can be put in the form of the familiar deformation potential interaction as first expressed by Shockley, which for a given phonon branch is [21]-[23],

$$< n'k' \mid V_p(r)_{sr} \mid nk >_{ac} = \qquad (61)$$
$$\delta_{k',k\pm q} \left[\frac{\hbar}{2N\mu\omega_q^j} \right]^{\frac{1}{2}} \left(N_{qj}(T) + \frac{1}{2} \pm \frac{1}{2} \right)^{\frac{1}{2}} \sum_{il} \Xi_{il} \, S_{il}$$

The terms in (60) correspond to the following terms from (61):

$$\Xi_{il} \;=\; < n'k' \mid \left[(V_j^{sr})_i r_l + (V_j^{sr})_l r_i \right] \mid nk > \qquad (62)$$

where $(V_j^{sr})_l$ is the lth component of V_j^{sr}, and

$$S_{ij} \;=\; \frac{1}{2}(\xi_i q_j + \xi_j q_i) \;=\; \frac{1}{2}\left(\frac{\partial u_i}{\partial x_j} + \frac{\partial u_j}{\partial x_i} \right) \qquad (63)$$

Since the displacement, u, is proportional to $\hat{\xi}_q e^{iq\cdot r}$, S_{il} is the elastic strain tensor [25]. Thus, Ξ_{il} is the deformation potential tensor which couples the local strain set up by the acoustic phonon to the scattering matrix element as given in the phenomenological theories of scattering by acoustic phonons. Equation (61) is valid for small q, which corresponds to intravalley scattering for low energies and represents the anisotropy of the coupling constant. For larger q, higher order terms in $iq \cdot r$ come into the integral in (60), and hence the coupling to the phonon wavevector involves higher rank tensors.

Now, returning to (59) for the case of optical phonons, the highest order term which can be non-zero is the zero-order term. In contrast to the acoustic mode case , the zero-order term can be nonzero because when $q \to 0$ for an optic mode, the two sublattices are rigidly displaced with respect to one another, and this can scatter carriers. Thus, for the case of optical phonons, we look to highest order at the zero-order term and write

$$< n'k' \mid V_p(r)_{sr} \mid nk >_{op} = \qquad (64)$$
$$-i \left(N_{qj}(T) + \frac{1}{2} \pm \frac{1}{2} \right)^{\frac{1}{2}} \delta_{k',k\pm q} \; < n'k' \mid V_{jq}^{sr}(r) \mid nk > \cdot \hat{\xi}_{jq}$$

This is valid for small q (intravalley scattering). Equation (64) can also be written in terms of an optical deformation potential vector D_{op} as,

$$< n'k' \mid V_p(r)_{sr} \mid nk >_{op} = \tag{65}$$

$$\left(N_{qj}(T) + \frac{1}{2} \pm \frac{1}{2}\right)^{\frac{1}{2}} \left[\frac{\hbar}{2N\mu\omega_q^j}\right]^{\frac{1}{2}} D_{op} \cdot \xi_{qj} \delta_{k',k\pm q} \, . \tag{66}$$

The form of the electron-phonon matrix element for small q and optical modes involves a direct dilation of the local crystal structure as seen from the dot product coupling $D_{op} \cdot \xi$. Thus, 66 yields the phenomenological optical deformation potential matrix element ([25]). It has been shown that for symmetry reasons the intravalley optical deformation potential scattering vanishes for X and Γ_1 valleys for the zero-order matrix element discussed above [26]-[27]. In general, the higher order tensor coupling (acoustic-like) can be non-zero.

While for small q the phenomenological form of the deformation potential interaction for acoustic and optic modes is quite different, for large q, the coupling contains many complicated higher order tensor modes. The standard treatment for these situations assumes that, for a particular intervalley transition (for example, X-X, Γ - X), q is confined to a small cone of allowed directions. Because the angular orientation is fairly constant within this cone, the tensor coupling can change only a small amount for any scattering into the cone. Then, it may be approximately correct to treat the coupling for intervalley transitions by a constant. Usually the intervalley deformation potential matrix element is written phenomenologically as [24]

$$< n'k' \mid V_p(r)_{sr} \mid nk >_{iv} = D_{iv} \left[\frac{\hbar}{2N\mu\omega_q^j}\right]^{\frac{1}{2}} \left(N_{qj}(T) + \frac{1}{2} \pm \frac{1}{2}\right)^{\frac{1}{2}} \delta_{k',k\pm q} \, . \tag{67}$$

Both acoustic and optical modes have been shown to participate in intervalley scattering with this type of scattering rate. It has been shown, for instance, that to fit experimental values of conductivity in silicon, it is necessary to include acoustic phonons in intervalley scattering [28]. It is evident from the nature of the integral in (59) that the coupling is widely determined by the orientation of the final valleys and the phonon branch. Thus, for each phonon branch and each set of initial and final valleys, a new intervalley coupling constant, D_{iv}, must be chosen. Typically, the values for D_{iv} are selected to best fit experimental data for a given Monte Carlo simulation [29]-[31].

Finally, for the phenomenological deformation potential interaction, some authors have factored out the overlap integral,

$$I(n'k', nk) = \int_\Omega d^3r \; u_{nk}(r) u^*_{n'k'}(r) \, , \tag{68}$$

from the matrix elements. This is only valid for the long range interaction, as discussed in the next section, and can readily be seen from (56) and (57).

4.3.2 The Phenomenological Polar Optical Interaction

In this section, the phenomenological polar optic phonon matrix element will be derived from the long-range matrix element in (57), using the equations of macroscopic electrostatics [2]. The first step is to factor out of (57) all terms proportional to the Coulomb potential $v(q) \propto 1/q^2$ so that the behavior of the matrix element for small q is clear. These terms can be factored out giving,

$$< n'k' \mid V_p(r)_{lr} \mid nk >_{po} \;=\; -iV_m(q)I(n'k', nk) \qquad (69)$$

$$V_m(q) \;\equiv\; \frac{1}{\Omega}\frac{N_{eff}(T)}{\epsilon(q)\,q}\sum_{\alpha} Z_{\alpha}^{q} \cdot \xi_{qj}^{\alpha} \,,$$

with,

$$N_{eff}(T) \equiv \left(N_{qj}(T) + \frac{1}{2} \pm \frac{1}{2}\right)^{\frac{1}{2}} \left[\frac{\hbar}{2N\mu\omega_q^j}\right]^{\frac{1}{2}} . \qquad (70)$$

The quantity Z_{α}^{q} is the remainder of (57) after the singular $1/q$ terms are removed (see Vogl [2]). The whole matrix element is proportional to $1/q$ and so, acts quite differently than the deformation potential matrix element. Here, $\epsilon(q,q) = \epsilon(q)$, the macroscopic dielectric function. Equation (69) is a fairly intractable formula, since the function $Z_{\alpha}^{q}$is generally difficult to calculate. However, their exists a simple expression for Z_{α}^{q} which can be seen by using macroscopic electrostatics. This is only possible because of the long-range nature of the perturbation discussed earlier.

The term containing the product $\xi_{qj}^{\alpha} N_{eff}(T)$, can be treated as the effective phonon polarization vector, u. In addition, $V_m(q)$ is the true macroscopic potential perturbation set up by this phonon. As such, it can be examined using macroscopic electrostatics and related to the polarization wave of the phonon.

Two equations from macroscopic electrostatics directly apply: first,

$$\nabla \cdot D(r) \;=\; \nabla \cdot (\epsilon_0 E(r) \;+\; P(r)) \;=\; 0 \;\Longleftrightarrow\; E(q) \;=\; -P(q)/\epsilon_0 \qquad (71)$$

in the absence of excess charge (charge neutrality is not necessary for the proof but it is a convenient assumption), and $P(q)$ is the polarization. Second,

$$\nabla \cdot V(r) \;=\; -eE(r) \;\Longleftrightarrow\; iqV(q) \;=\; -eE(q) \cdot \hat{q} \qquad (72)$$

where $V(r)$ is the potential energy felt by an electron. Together these equations give

$$iqV_m(q) \;=\; e\hat{q} \cdot P(q)/\epsilon_0 \qquad (73)$$

which can be substituted into (3.25) to give

$$< n'k' \mid V_p(r)_{lr} \mid nk >_{po} = \frac{I(n'k', nk)\hat{q} \cdot eP(q)}{\epsilon_0 q} . \qquad (74)$$

Comparing 74 with 69 yields the polarization P(q) in terms of microscopic quantities.

For small q, $P(q)$ is the dipole polarization set up by the optical phonon mode in the crystal. For larger q $P(q)$ acquires higher order poles (quadrupole polarization, etc.). We are only interested in the small q limit for polar optic phonon scattering, so for these purposes, $P(q) \approx P_{dipole}$. Fortunately, the dipole polarization associated with a long wavelength mode is easily calculated from a self-consistent, first order lattice dynamics theory. The result is that [32],

$$\hat{q} \cdot P_{dipole} = \frac{e_L^* u \cdot \hat{q}}{\Omega}, \tag{75}$$

where e_L^* is the effective dipole charge associated with a longitudinal optical phonon and u is the ionic displacement. Furthermore, an application of the well known Lyddane-Sachs-Teller relation relates the effective charge, e_L^*, to the experimentally known quantities ϵ_r, ϵ_∞ and the longitudinal optical phonon frequency at zone center ω_{LO} [32]:

$$(e_L^*)^2 = \Omega \epsilon_0 \mu \omega_{LO}^2 \left(\frac{1}{\epsilon_r} - \frac{1}{\epsilon_\infty} \right). \tag{76}$$

This allows us to rewrite (69) for small q in a more transparent form involving experimentally known quantities:

$$< n'k' \mid V_p(r)_{lr} \mid nk >_{po} = \frac{e N_{eff}(T) e_L^* \; \hat{q} \cdot \xi_{qj}^{\alpha}}{\Omega \epsilon_0 q} \; I(n'k', nk) \tag{77}$$

or, in the familiar Frohlich interaction form,

$$\begin{aligned} |< n'k' \mid V_p(r)_{lr} \mid nk >_{po}|_{LO}^2 &= I^2(n'k', nk) \\ \delta_{k',k\pm q} \left(N_{qj}(T) + \frac{1}{2} \pm \frac{1}{2} \right) & \frac{e^2 \hbar \omega_{LO}}{2\epsilon_0 Vol q^2} \left(\frac{1}{\epsilon_r} - \frac{1}{\epsilon_\infty} \right) \end{aligned} \tag{78}$$

Pure transverse modes do not scatter because for them, $\hat{q} \cdot \xi_{qj}^{\alpha} = 0$. So, only longitudinal modes participate in polar optical phonon scattering. Also, although not stated explicitly, the Frohlich interaction (78) vanishes in non-polar crystals. The reason for this is obvious on physical grounds– in a non-polar crystal, no dipole polarization can arise from optical mode displacements. Mathematically,

$$Z_1^q = Z_2^q, \quad \xi_{qj}^1 = -\xi_{qj}^2 \tag{79}$$

so the sum over α vanishes.

The lowest order (in q) long range acoustic mode process which doesn't vanish is the scattering due to quadrupole polarization. Piezoelectric scattering results from quadrupole polarization and can be present in both polar and non-polar crystals [25].

4.3.3 Phonon Scattering Rates

In the first part of this section the general electron-phonon matrix element, and the standard phenomenological electron-phonon matrix elements have been presented. To make use of these rates in a Monte Carlo simulation, we need to calculate from these matrix elements, the total scattering rates out of a particular state, and the differential scattering probability between a given final and initial state for use in the SCBE.

In section 2, it was stated that for weak perturbations and low electron energies, the Fermi Golden Rule (28) adequately described the scattering rate into a particular set of final states from a given initial state. The general form for the total scattering rate, $S_{tot}(nk)$, is the Golden Rule rate summed over all final states:

$$S_{tot}(nk) \;=\; \sum_{n'k'} \frac{2\pi}{\hbar} \left|< n'k' \mid V_p(r) \mid nk >\right|^2 \delta(E_{n'}(k) - E_n(k) \mp \hbar\omega_q) \,. \quad (80)$$

where ω_{qj} is the phonon associated with the particular scatterer in $V_p(r)$. This expression can be calculated exactly using the full electron-phonon matrix element in (53) as discussed in the chapter by Fischetti and Higman. However, it is standard practice for Monte Carlo simulations to take the much simpler route and calculate the scattering rates for the phenomenological matrix elements of the previous sections.

In fact, in order to obtain closed form expressions for these scattering rates, it is necessary to assume simple analytic forms for the band structure, the most general of which is the non-parabolic, ellipsoidal band structure describing the bands near the minimum of a "valley". The analytic form most often used is,

$$\gamma(E) \;=\; E(1+\alpha E) \;=\; \frac{\hbar^2}{2}\left(\frac{(k_x - k_x^0)^2}{m_x} + \frac{(k_y - k_y^0)^2}{m_y} + \frac{(k_z - k_z^0)^2}{m_z}\right) , \quad (81)$$

where k^0 is the minimum of the valley. Accordingly, this description is only valid near the minimum of a particular valley. In general, the band structure is very complicated [33], and in the case of hot electron transport, the electrons are far these minima. One way to deal with this problem is to calculate the matrix elements discussed above for a full band structure. Another, much simpler way is to calculate the various scattering mechanisms valid near the minima and normalize the scattering rates to the density of states for higher energies [30]. This is a reasonable approach since most scattering rates are proportional to the density of states.

In this section the standard scattering rates valid near valley minima are discussed. Since the derivations leading to the scattering rates from the phenomenological matrix elements are well known and given in many texts (see for instance [34] and [36]), we quote only the results here. Typical values for the phenomenological parameters and material parameters for silicon are given in the appendix of this chapter.

However, as a prelimeninary, we list facts useful for deriving them. First, to transform a summation over k' in (80) into a more manageable integral over k' we use the transformation [34]

$$\sum_{k'} \longrightarrow \frac{Vol}{(2\pi)^3} \int dk' \,. \tag{82}$$

Note that spin is conserved in phonon collisions. Furthermore, if the m_i are equal, these integrals are normally done in spherical coordinates. Then the magnitude of k can be transformed into an integral over final energy E_f by use of the transformation,

$$dk' \;=\; dE_f \, \frac{(1+2\alpha E_f)m^*}{\hbar^2 k'} \,. \tag{83}$$

If the masses are unequal, the Herring-Vogt transformation is useful because it maps the problem into a "starred-space" which has a spherical $E(k)$ relation [35]:

$$k_i^* \longrightarrow k_i\sqrt{m_i} \,. \tag{84}$$

The problem is then solved in the "starred-space" and then transformed back to the physical space. As far as total scattering rates are concerned, this only has the effect of replacing m^* in the result by $m_D \equiv (m_x m_y m_z)^{1/3}$.

For **intra-valley optical deformation potential** scattering:

$$S_{tot}(E) \;=\; \frac{D_{op}^2}{8\pi^2 \rho\omega_{op}} \left[N_{\omega_{op}} + \frac{1}{2} \pm \frac{1}{2}\right] \, g(E \pm \hbar\omega_{op}) \,, \tag{85}$$

where $g(E)$ is the density of states given by,

$$g(E) \;=\; \frac{(m_D)^{3/2}}{2^{1/2}\hbar^3\pi^2} \, (1 \,+\, 2\alpha E)[E(1 \,+\, \alpha E)]^{1/2} \;, \quad E > 0 \tag{86}$$

and D_{op} is the optical deformation potential, ρ is the crystal density and ω_{op} is the relevant optical phonon energy. To determine the final state after scattering, we only have to enforce energy conservation, $E(k') \;=\; E(k) \pm \hbar w_{op}$. Any state on this energy conserving surface is equally probable.

For **inter-valley deformation potential** scattering:

$$S_{tot}(E) \;=\; \frac{D_{op}^2}{8\pi^2 \rho\omega_{iv}} \, Z_e \left[N_{\omega_{op}} + \frac{1}{2} \pm \frac{1}{2}\right] \, g(E \pm \hbar\omega_{iv} - \Delta_{iv}) \,, \tag{87}$$

where ω_{iv} is the intervalley phonon energy for a particular phonon branch, and Z_e is the number of equivalent final valleys. For instance, in Silicon, there are six X-mimima, and two "different" types of intervalley scattering. For scattering across to the X-mimimum on the same axis (g-scattering), $Z_e = 1$; for scattering to one of the minima on the plane perpendicular to the initial state axis (f-scattering), $Z_e = 4$.

Δ_{iv} is the energy difference between the initial and finial state minima. For X-X scattering in silicon, $\Delta_{iv} = 0$. For Γ−L scattering in GaAs, $\Delta_{iv} \approx 0.2$eV. As in the intra-valley optical deformation potential scattering

$E \leq E_s$	Absorption	$x_{min} = 4E_s^{*1/2}(E_s^{*1/2} - E^{1/2})/K_BT$ $x_{max} = 4E_s^{*1/2}(E_s^{*1/2} + E^{1/2})/K_BT$
	Emission	None
$E \leq E_s$	Absorption	$x_{min} = 0$ $x_{max} = 4E_s^{*1/2}(E_s^{*1/2} + E^{1/2})/K_BT$
	Emission	$x_{max} = 4E_s^{*1/2}(E_s^{1/2} - E^{*1/2})/K_BT$ $x_{min} = 0$

rate, $g(E)$ is the density of states in the final valley. The density of states plays an important role in all scattering rate determination, but it is most apparent in the expressions for the optical and intervalley deformation potential cases. The final state is selected randomly from all states conserving energy, and the final valley is picked randomly from the equivalent valleys. A list of typical intervalley phonons for Si is given in the appendix to this chapter.

For **intra-valley acoustic deformation potential** scattering: Because of the complicated tensor form for acoustic coupling, there is an intrinsic anisotropy in the coupling. However, as shown by Conwell, the effect of this anisotropy is small [24] and can be removed by a suitable averaging. This amounts to defining an average sound velocity $v_s = (2v_s^t + v_s^l)/3$ where v_s^t and v_s^l are the velocity of sound propagated by transverse and longitudinal modes respectively and an effective coupling strength Ξ. Ξ is usually chosen to fit experimental data. The acoustic phonon scattering rate is[36]:

$$S_{tot} = \frac{\Xi^2 (m_D)^{1/2} (K_B T)^3}{2^{5/2} \pi \rho \hbar^4 v_s^4} \gamma^{-1/2}(E) \int_{x_{min}}^{x_{max}} dx N^{\pm}(x)(1 + 2\alpha E \mp 2\alpha K_B T x)\, x^2 \tag{88}$$

with

$$N^{\pm}(x) = \left(\frac{1}{e^x - 1} + \frac{1}{2} \pm \frac{1}{2} \right) \tag{89}$$

and the dimensionless limits of integration: where $E_s^* \equiv \frac{1}{2}(m_D v_s^2)$.

For **polar optical** scattering: The total scattering rate for polar optical phonon scattering has been calculated by Boardman-Fawcett-Swain which takes into account the overlap integrals in (78) [37]. These integrals can be evaluated for k and k' near the bottom of non-parabolic bands using $k \cdot p$ perturbation theory. The result is:

$$\begin{aligned} I^2(k', k) &= (a_k a_{k'} + c_k c_{k'} \cos\theta)^2 && (90) \\ a_k &= [(1 + \alpha E(k))/(1 + 2\alpha E(k)]^{1/2} \\ c_k &= [\alpha E(k)/(1 + 2\alpha E(k))]^{1/2} \end{aligned}$$

(91)

where θ is the angle between k and k'. Using this expression for the overlap integral, the total scattering rate for polar optical phonon scattering can be written:

$$S_{tot}(E) = \frac{e^2\omega_{LO}(m^*)^{1/2}}{2^{5/2}\pi\epsilon_0\hbar}\frac{1+2\alpha E'}{\gamma(E)}\left(\frac{1}{\epsilon_\infty} - \frac{1}{\epsilon_0}\right)\left[N_{\omega_{op}} + \frac{1}{2} \pm \frac{1}{2}\right]$$
$$A\left(B\ \ln\left[\frac{\gamma^{1/2}(E')+\gamma^{1/2}(E)}{\gamma^{1/2}(E')-\gamma^{1/2}(E)}\right] - C\right),\ E \geq \hbar\omega_{LO} \qquad (92)$$

where

$$\begin{aligned} A &= [4(1+\alpha E)(1+\alpha E')(1+2\alpha E)(1+2\alpha E')]^{-1} \\ B &= [2(1+\alpha E)(1+\alpha E') + \alpha[\gamma(E')+\gamma(E)])^2 \\ C &= 2\alpha\gamma^{1/2}(E')\gamma^{1/2}(E)[4(1+\alpha E)(1+\alpha E') + \alpha[\gamma(E')+\gamma(E)]] \\ E' &= E - \hbar\omega_{LO} \end{aligned}$$

Because of the $1/q^2$ dependence of the matrix element and the angular dependence of the overlap integrals, polar optical phonon scattering is a anisotropic scattering mechanism. To choose the final state we can use the von Neumann rejection method to pick the angle θ between k and k', which is given by the probability density,

$$P(\theta)d\theta = \frac{(a_k a_{k'} + c_k c_{k'}\cos\theta)^2 \sin\theta\ d\theta}{\gamma(E') + \gamma(E) - 2\gamma^{1/2}(E')\gamma^{1/2}(E)\cos\theta}. \qquad (93)$$

The azimuthal angle is completely random since the scattering probability density is independent of ϕ. Thus, we can choose ϕ with a uniform random number $r : r \in [0,1]$ by $\phi = 2\pi r$. The magnitude of the final state wavevector, k', is selected by energy conservation for the given scattering event (phonon emission, or absorption), and thus, the final state vector is completely determined.

5 Impact Ionization

The multiplication of carriers by impact ionization is of central importance in the theory of semiconductor devices both as a limiting mechanism and as a basis of device functionality. Impact ionization is a two electron process, corresponding to the exact inverse of the Auger process: a highly energetic conduction band electron collides with a valence band electron which is ionized over the band gap, leaving two conduction electrons and a hole. The process can also occur for holes, in which a highly energetic hole creates two holes and an electron. Impact ionization for holes can be seen as a "mirror image" of the impact ionization for electrons, so we will treat only the electron impact ionization.

As in section 1, the electron-electron interaction is designated as $V_{cc}(r, r')$, and following section 3, we screen the bare electron-electron interaction with the dielectric screening function. We will neglect the off-diagonal terms of

$\epsilon^{-1}(q+G, q+G'; \omega)$ and allow q to take on all values in the reciprocal lattice.

The bare electron-electron interaction is of course just the coulomb interaction between two electrons. Figure 2 shows the impact ionization process schematically. States three and four are final conduction band electron states. State one is the initial electron state, and state two is the initial valence band electron state. The crystal momenta and band indices are designated as k_i and n_i, with $i \in [1,4]$. Figure 2 shows the electron in states 1 and 2 interacting via the screened Coulomb potential to generate electrons in states 3 and 4. The final hole state corresponds to the missing valence electron in state two.

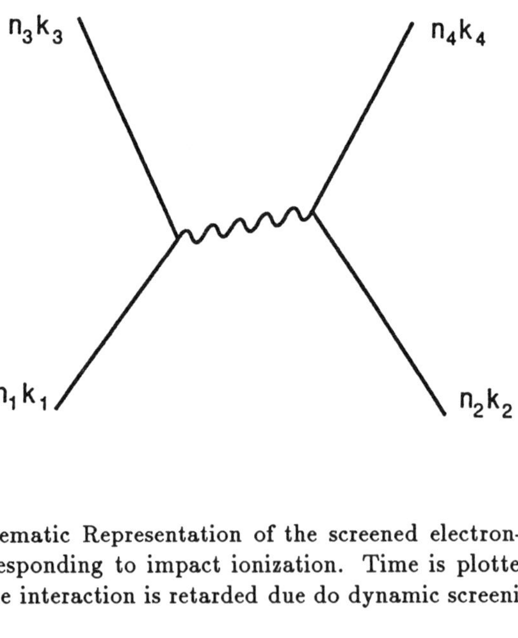

Figure 2: Schematic Representation of the screened electron-electron interaction corresponding to impact ionization. Time is plotted vertically. Notice that the interaction is retarded due do dynamic screening effects.

Since $V_{cc}(r, r')$ is a two particle interaction explicitly carrying two spatial coordinates, its matrix elements are between two-particle states. Since the electrons are fermions, the two-particle wave-functions must be anti-symmetric linear combinations of two-particle states with the Pauli spin matrices. We write these two-particle states as,

$$\begin{aligned} \mid n_1k_1\sigma_1; n_2k_2\sigma_2 >_A \; \equiv \; & \frac{1}{\sqrt{2}}[\psi_{n_1k_1}(r_1)\psi_{n_2k_2}(r_2)\sigma_1(s_1)\sigma_2(s_2) \; - \\ & \psi_{n_1k_1}(r_2)\psi_{n_2k_2}(r_1)\sigma_1(s_2)\sigma_2(s_1)] \; , \end{aligned} \quad (94)$$

where the σ are the Pauli spin matrices, and s_1 and s_2 are the spin coordinates. The coordinate wave-functions ψ are normalized Bloch wave-functions. Also, we write the subscript A to signify the anti-symmetrized state.

In this language, the event shown in figure 2 corresponds to the matrix element,

$$M(12;34) \equiv < n_1 k_1 \sigma_1; n_2 k_2 \sigma_2 \mid_A V_{cc} \mid n_3 k_3 \sigma_3; n_4 k_4 \sigma_4 >_A \ . \qquad (95)$$

The matrix element contains four terms with different arrangements of the coordinates and the wavefunctions. For simplicity of notation we write the simple product states as $\mid 12 >= \psi_{n_1 k_1}(r_1)\psi_{n_2 k_2}(r_2)$. Then, equation (95) expands as

$$M(12;34) = \qquad (96)$$

$$\frac{1}{2}\left[\delta_{\sigma_1\sigma_3}\delta_{\sigma_4\sigma_2} < 34 \mid V_{cc} \mid 12 > -\delta_{\sigma_1\sigma_4}\delta_{\sigma_2\sigma_3} < 43 \mid V_{cc} \mid 12 > - \right.$$
$$\left. \delta_{\sigma_1\sigma_4}\delta_{\sigma_2\sigma_3} < 34 \mid V_{cc} \mid 21 > +\delta_{\sigma_2\sigma_4}\delta_{\sigma_1\sigma_3} < 43 \mid V_{cc} \mid 21 >\right] \ . \qquad (97)$$

For a given initial state spin σ_1, there are three distinct physical situations, with equal proability of occurence corresponding to the different configurations of the remaining spin indices. They are:

1. $\sigma_1 = \sigma_2 = \sigma_3 = \sigma_4$
2. $\sigma_1 \neq \sigma_2 : \quad \sigma_1 = \sigma_3, \ \sigma_2 = \sigma_4$
3. $\sigma_1 \neq \sigma_2 : \quad \sigma_1 = \sigma_4, \ \sigma_2 = \sigma_3$

The rate for each configuration must be calculated separately, and then summed to give the total rate independent of spin. For instance, if we define,

$$\begin{aligned} M_1 &= < 34 \mid V_{cc} \mid 12 >, \\ M_2 &= < 43 \mid V_{cc} \mid 12 >, \end{aligned} \qquad (98)$$

then the squares of matrix elements corresponding to the spin configurations in the list above (the probabilities) can be written in terms of M_1 and M_2 as:

1. $\mid M_1 - M_2 \mid^2$
2. $\mid M_1 \mid^2$
3. $\mid M_2 \mid^2$.

The sum of these probabilites gives the square of the total effective matrix element, summed over all internal spins for a given initial spin which we designate as M_{tot}^2 and write as:

$$M_{tot}^2 = 2 \mid M_1 \mid^2 + 2 \mid M_2 \mid^2 - (M_1^* M_2 + M_2^* M_1). \qquad (99)$$

Thus, we need only to calculate M_1 and M_2 to find the total rate. Furthermore, if we find an expression for M_1, M_2 is easily found by exchanging the final state indices, so we only calculate M_1.

The simplest way to include frequency dependent screening in calculating M_1 above and the associated scattering rate $S(12;34)$ is to calculate the two-particle propagator from $|\ 12>$ to $|\ 34>$ in time t, square it to obtain the probability, and take the time derivative for long times (time rate of change of the proability of going from $|\ 12>$ to $|\ 34>$):

$$S(12;34) \ = \ \lim_{t\to\infty} \frac{\partial}{\partial t} |< 34 \mid V_{cc} \mid 12 >|^2 \tag{100}$$

This is wholly equivalent to the expression in section 2. In figure 2, the electron in state 1 feels the screened coulomb potential of state 2 at time t' and state 2 feels state 1 at time t''. The dielectric function retards the effects of states 1 and 2 on each other; for instance, even if electron 2 has passed electron 1, the valence electrons may still be readjusting to its passage and this can effect electron 1.

We write $M_1 \ = \ < 34 \mid V_{cc} \mid 12 >$ by letting states 1 and 2 propagate freely until times t' and t'', and then scatter into states 3 and 4 which propogate until time t. The potential felt at each time is weighted by $\epsilon^{-1}(r'' - r'; t'' - t')$. Then we must integrate over all t' and t''. This is basically a restatement of the Feynman rules for a first order, two-particle, time-dependent interaction which give

$$M_1 = \int d^3r \int d^3r' \psi_3^*(r)\psi_4^*(r')\psi_1(r)\psi_2(r') \sum_q \frac{e^2}{Vol\ q^2} e^{-iq\cdot(r-r')} I_q(t) \tag{101}$$

with

$$I_q(t) \equiv \int_0^t dt' e^{-i\omega_1 t'} e^{-i\omega_3(t-t')} \int_0^t dt'' e^{-i\omega_2 t''} e^{-i\omega_4(t-t'')} \epsilon^{-1}(q, t' - t''). \tag{102}$$

In equation (101) we have written the Coulomb potential as a Fourier series over vectors in the reciprocal lattice q. $I_q(t)$ is easily evaluated by making a change of variables in the t'' integral to $t' - t''$. Then,

$$I_q(t) = e^{-i(\omega_3+\omega_4)t} \int_0^t dt' e^{-i(\omega_1+\omega_2-\omega_3-\omega_4)t'} \int_{t'-t}^t dt'' e^{i(\omega_2-\omega_4)t''} \epsilon^{-1}(q, t''). \tag{103}$$

Since we need only the large t limit, the inner integral becomes $\epsilon^{-1}(q, \omega_2 - \omega_4)$. In (101) we can evaluate the spatial coordinate integrals by using the trick in equation (56) section 4.

$$M_1 = \frac{1}{Vol} \sum_q \delta_{k_4, k_2+q+G_1} \delta_{k_3, k_4-q+G_2} \int_\Omega d^3r \int_\Omega d^3r' \tag{104}$$
$$\frac{e^{i(k_4-k_2-q)\cdot r} e^{i(k_3-k_1+q)\cdot r'} u_3^*(r) u_4^*(r') u_1(r) u_2(r')}{q^2\ \epsilon(q, \omega_2 - \omega_4)} I_q(t)$$

In the above equation, G_1 and G_2 are arbitrary reciprocal lattice vectors. For a given set of k_i, q is determined up to a reciprocal lattice vector. We have the following crystal momentum conservation laws:

1. $k_4 - k_2 + q + G_1 \;=\; 0$
2. $k_3 - k_1 - q + G_2 \;=\; 0$
3. $k_1 + k_2 - k_3 - k_4 \;=\; G_0 \;=\; G_1 + G_2$

The last equation, which follows from the first two, asserts the conservation of crystal momentum for the entire process. Furthermore, we can write each of the Bloch wavefunctions as sums over reciprocal lattice vectors (since they are lattice translation invariant):

$$u_{nk}(r) = \sum_G z_{nk}(G)e^{iG\cdot r}. \tag{105}$$

Then,

$$M_1 = \delta_{k_1+k_2,k_3+k_4+G_0} \sum_{G_1G_3G_4} I_{k_1-k_3+G_1-G_3}(t) \tag{106}$$

$$\frac{z_1(G)z_2(G_3+G_4-G_1-G_0)z_3^*(G_3)z_4^*(G_4)}{Vol \mid k_1-k_3+G_1-G_3 \mid^2} \tag{107}$$

This form is particularly well suited for numerical calculations since the expansions in (105) can be calculated by pseudopotential band structure calculations. To calculate the scattering rate given in (100), we must square (106) and take the time derivative. The only time dependence in the magnitude squared of $I_q(t)$ is in the t' integral in equation (103). The time derivative of this gives the Golden rule rate expression as in equation (28).

Now, we can combine (106), (99) and (100) to obtain the total impact ionization scattering rate $S(12;34)$ summed over secondary particle spins:

$$\begin{aligned}
S(12;34) &= \frac{2\pi}{\hbar} \mid M_{tot} \mid^2 \delta(E_1+E_2-E_3-E_4)\delta_{k_1+k_2,k_3+k_4+G_0} \quad (108)\\
M_{tot} &= \mid M_1 \mid^2 + \mid M_2 \mid^2 - \frac{1}{2}(M_1^*M_2 + M_2^*M_1)\\
M_i &= \sum_{G_1G_3G_4} \frac{z_1(G)z_2(G_3+G_4-G_1-G_0)z_3^*(G_3)z_4^*(G_4)}{Vol\; q_i^2\epsilon(q,\omega_3-\omega_4)}\\
q_1 &= k_1-k_3+G_1-G_3\\
q_2 &= k_1-k_4+G_1-G_4\\
G_0 &= k_1+k_2-k_3-k_4
\end{aligned}$$

This expression was first obtained by Kane [38].

To get the total scattering rate we need only sum over the two independent k vectors and the secondary particle band indices in in $S(12;34)$. Given n_1 and k_1, the total impact ionization scattering rate, S_{ii}, from that state is simply,

$$S_{ii}(n_1k_1) = \sum_{n_1n_2n_3} \sum_{k_3k_4} S(12;34). \tag{109}$$

The wavevector sums (109) run over the first BZ. G_0 is then the Umklapp wavevector necessary to ensure that k_2 is in the first BZ for a given k_1.

To gain some qualitative insight into equation (109), we look at the two conservation laws it implies–conservation of crystal momentum and conservation of energy. First, energy conservation requires the initial electron to be at least E_g (gap energy) above the bottom of the conduction band in order to excite a valence electron across the gap and into some conduction band state. Therefore, E_g is a minimum energy for impact ionization. Additionally, the conservation of crystal momentum further shifts this minimum energy. The smallest energy necessary to initiate an impact ionization event in a particular band structure is known as the *impact ionization threshold energy* E_{th}.

Some authors have defined a wavevector dependent E_{th} by minimizing E_i (the initial electron energy) for a given k_i with the energy and momentum conservation constraints. This minimization procedure requires that [39]

$$\nabla_k E_2(k) = \nabla_k E_3(k) = \nabla_k E_4(k). \tag{110}$$

We shall re-evaluate the usefulness of this result later. Now we will look at the first attempt to evaluate the impact ionization scattering rate known as the Keldysh formula. The Keldysh formula makes two rather extreme approximations. The first is to approximate the matrix element as a constant, and the second, to assume simple parabolic bands. Of these, the second is very poor, yet the resulting formula, known as the Keldysh formula, has received considerable attention in the past. The result given by Keldysh involving the two adjustable parameters P and B is [40]:

$$S_{ii}(n_1k_1) = \frac{B}{S_{ph}(E_{th})} \left(\frac{E - E_{th}}{E_{th}}\right)^P . \tag{111}$$

Typically, $1 \leq P \leq 2$ and S_{ph} is the total phonon scattering rate. Although this phenomenological form is simple to use, the approximations leading to it disregard the important features of the band structure, which at the high energies involved in impact ionization, has no resemblance to simple parabolic bands.

A much better solution for the impact ionization rate in silicon was given by E. O. Kane [38] in 1967. He numerically calculated the rate in equation (109) using the expressions developed in equation (108). First he calculated the pseudopotential band structure and wavefunctions (see the chapter by Fischetti and Higman) and used them to calculate the dielectric function in the RPA approximation (see equation (43)). The summations in equation (109) were evaluated by employing a Monte Carlo integration algorithm. The results of this calculation for $S_{ii}(E)$, ($S_{ii}(k)$ averaged over all k with $E(k) = E$) are shown in figure 3.

Figure 3 also plots the impact ionization scattering rate in silicon calcu-

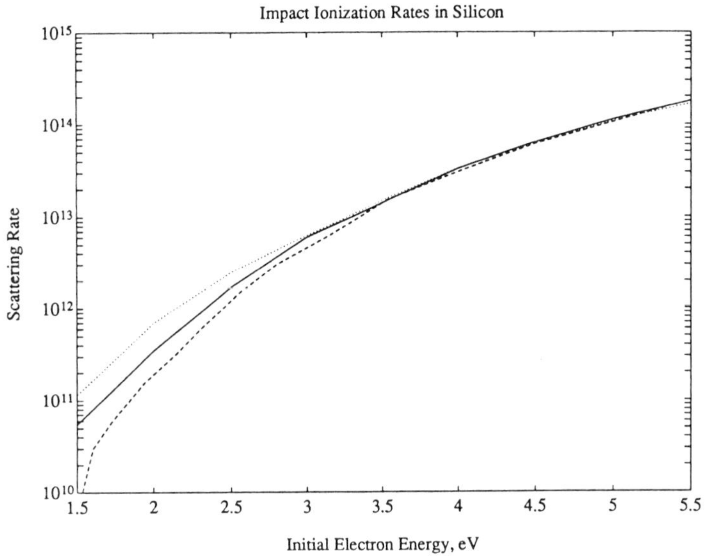

Figure 3: Impact ionization rates in s^{-1} for silicon averaged over initial electron energy measured from the bottom of the conduction band. Dashed curve, Kane's result; solid curve, no collision broadening; dotted curve, collision broadening.

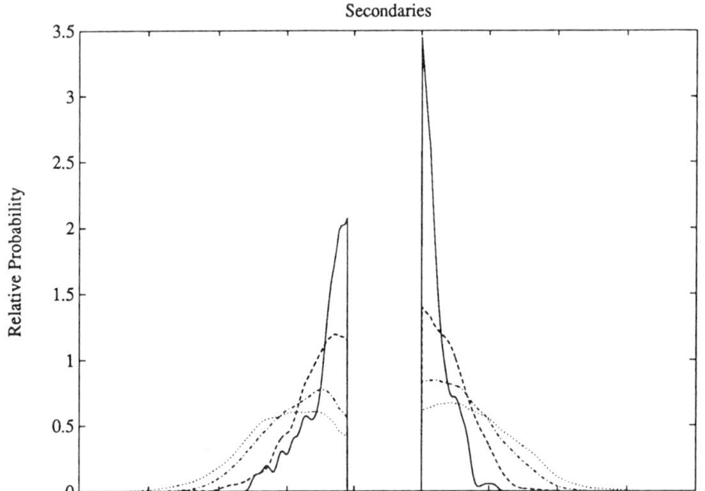

Figure 4: Secondaries produced by impact ionizing electrons in silicon. Solid curve, initial electron of 1.5 eV; dashed curve, 2.5eV; dashed-dot, 3.5eV; dotted, 4.5 eV. The two sets of curves correspond to holes for the $E < -E_G$, and electrons, for $E > 0$.

lated to include the effects of collision broadening and the intra-collisional field effect (see section 2) [41]. The effects of high fields and high phonon scattering rates on the impact ionization process shift the threshold down from that obtained by Kane. Figure 4 shows the distribution of secondaries (two final conduction electrons and a hole) produced initiating electrons of various energies. These distributions include collision broadening, which has a large effect near threshold.

Figure 5 addresses the question of k-space anisotropy for the impact ionization scattering rate. As discussed above, it is possible to calculate a wave-vector dependent threshold which would seem to be relevant to the anisotropy of the scattering rate. It is, however, difficult to guess an appropriate form for the scattering rate as a function of this threshold which is consistent with the true physics in equation (109). For instance, the wave-vector dependent scattering rate may not even be directly related to these thresholds for energies greater than $E_{th}(k)$. Furthermore, if collision broadening is included (an important effect for high energy electrons for which the phonon scattering rate is high) the threshold condition is greatly relaxed. Figure 5 shows the impact ionization rate in silicon for electrons on the equi-energy surfaces $E(k) = 2.5$eV and $E(k) = 3.0$eV for electrons in the second conduction band in the $k_z = 0$ plane [41]. As can be seen the scattering rate shows little anisotropy. In addition this energy range is important for impact ionization in transport calculations [42], so it is probably a good approximation to use the average, energy dependent scattering rates given in figure 3, for most Monte Carlo simulations.

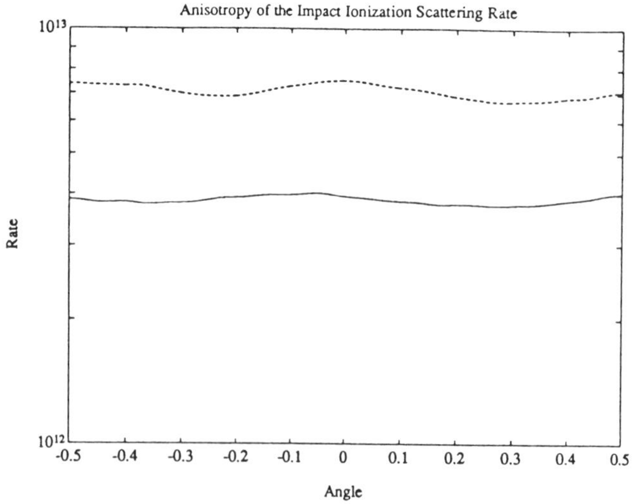

Figure 5: k-space anisotropy of the impact ionization scattering rate in silicon for electrons with $k_z = 0$ on the equi-energy surfaces $E = 2.5$eV (solid line) and $E = 3.0eV$ (dotted line). The angle is in units of π from the k_y axis.

6 Ionized Impurity Scattering

For even moderately doped semiconductors, ionized impurity scattering plays an important role and can dominate the total scattering rate for high doping concentrations. At 300K when doping densities reach levels above $10^{17}cm^3$, scattering rates for ionized donors (acceptors) become comparable to low energy phonon scattering rates, while at low temperatures ionized impurity scattering becomes even more important since equilibrium phonon populations disappear exponentially as $T \rightarrow 0$. Thus, in cases of high doping or low temperatures, an accurate Monte Carlo model must include ionized impurity scattering.

Ionized impurities are usually assumed to be simple Coulomb potentials with a charge Ze (e the electron's charge). Typically they are associated with ionized acceptors or donors. At first glance one may propose that the ionized impurity perturbation $V_i(r)$ is just this Coulomb potential, calculate the scattering rate for one impurity and multiply by the total number of impurities. Unfortunately, the calculation of ionized impurity scattering rates in this way is complicated by the long range nature of the Coulomb potential associated with the ions. Unscreened, the interaction of a single carrier and an ion leads to a diverging scattering rate which is of course an unphysical result. Therefore, the view of the ionized impurity interaction as one electron interacting with a single Coulombic potential cannot adequately describe the situation. Although there are many problems with this simplified model, the two most flagrant are the neglect of mobile charge screening, and the correlation of the other charges present. Each of these limits the effective length of the Coulomb interaction and removes the singularity. The first is treated by the Brooks-Herring Model [43], and the second by the Conwell-Weisskopf model [44]. A third model proposed by Ridley, called, *Third Body Exclusion*, reconciles these two approaches [45]. We will treat each in turn.

6.1 The Brooks-Herring Model

The Brooks-Herring model assumes that the electron (or hole) interacts solely with one ionized impurity site and deals with the many-body effects by introducing mobile charge screening. If mobile charge densities are high, it is necessary to include the effects of electronic screening, and these effects will limit the scattering rate to finite values.

Because the Brook-Herring model assumes that a carrier sees each charge one at a time, in a semiconductor sample with N_D^+ ionized donors and N_A^- ionized acceptors we must add the scattering rate due to each separately instead of using the net charge present. Thus, compensated ions can contribute to the total scattering rate as well. For this model, we will assume N_{tot} total ionized impurities of charge Ze.

Next we turn to Thomas-Fermi screening theory of section 3, equation (46). If the distribution function is Maxwellian, (46) reduces to the Debye-Huckel screening formula, introducing a screening length proportional to the square root of the mobile carrier density n:

$$k_{TF}^2 = \frac{e^2 n}{\epsilon_0 \epsilon_r K_B T} . \tag{112}$$

The Fourier transform of the screened ionized impurity potential $V_i(q)$ is:

$$V_i(q) = \frac{Ze}{e_r \epsilon_0 q^2} \frac{1}{e_{TF}(q)} = \frac{Ze}{e_r \epsilon_0 (q^2 + k_{FT}^2)} \tag{113}$$

For $K_{FT} \neq 0$ in equation (113) there is no divergence for small q, and therefore, there is no problem with infinite scattering rates. For degenerately doped semiconductors, the situation is more complicated, however an effective screening length can still be calculated from (46).

In real space, the Coulomb potential acquires an exponential tail as:

$$V_i(r) = \frac{Ze^2}{4\pi e_r \epsilon_0 \mid r - R \mid} e^{-k_{FT}|r-R|} \tag{114}$$

where R is the location of the ion. It should be emphasized that the screened potential approximation is not necessary to limit the cross section to finite values. In fact, at low mobile charge density, k_{FT} may not be sufficient to screen the potential and a model similar to the Conwell-Weisskopf method must be used.

In this section we will assume simple plane wave states for the electrons which is a good approximation for low energies. Ionized impurity scattering is essentially important only for low energy electrons, because, the phonon scattering rate dominates the impurity scattering rate for high electron energies.

We write the perturbation matrix element between plane wave states as $M_{kk'}$ where k is the incident wave vector and k' is the scattered wave vector as

$$M_{kk'} = \frac{1}{V} \int e^{-ik \cdot r} \frac{Ze^2}{4\pi e_r \epsilon_0 \mid r - R \mid} e^{-k_{FT}|r-R|} e^{ik' \cdot r} d^3 r \tag{115}$$

(Vol, the volume of the crystal, enters because of normalization). The matrix element in equation (115) is found by taking the integral over into spherical coordinates with the azimuthal axis being in the direction of $q \equiv k' - k$. The result is:

$$M_{kk'} = \frac{iZe^2 \, e^{iq \cdot R}}{e_r \epsilon_0 q \; Vol} \frac{q}{k_{FT}^2 + q^2} \tag{116}$$

Finally, the scattering rate $S(k, k')$ is given by the Fermi Golden Rule (see equation (28)). From equation (116) we can calculate the total scattering rate S_{tot} by integrating $S(k, k')$ over all k'. Assuming spherical nonparabolic bands,

$$S_{tot} = \frac{2\pi}{\hbar}\frac{(Ze^2)^2}{e_r^2\epsilon_0^2 Vol^2}\frac{\hbar^2}{(1+2\alpha E_f)m^*}\frac{Vol}{8\pi^3}2\pi \int_{-1}^{1} du \int_0^\infty \frac{k' dE_f \delta(E_f - E_i)}{(k^2 + k'^2 - 2k'^2 u + k_{FT}^2)^2} . \tag{117}$$

Here, u is the cosine of the angle between k and k'. Since the ion is strongly coupled to the lattice and is much more massive than the electron, the interaction can be treated as elastic, and we can take the term E_f equal to the energy of the final electronic state with wave vector k' and E_i equal to the energy of the initial electronic state with wave vector k.

The total scattering rate is [43]

$$S_{tot} = \frac{N_I}{32\pi(2m^*)^{\frac{1}{2}}}\left(\frac{Ze^2}{\epsilon_r\epsilon_0}\right)^2 \frac{(1 + 2\alpha E)}{\gamma^{3/2}(E)}\left[\frac{1}{(k_{TF}/2k)^2(1 + (k_{TF}/2k)^2))}\right] , \tag{118}$$

where N_I is the total density of impurities ($N_I = N_{tot}/Vol$). It is important to notice that we have simply summed the interaction of a plane wave and a Coulombic center over all the centers present; we have not calculated any many-body effect other than mobile charge screening–the calculation is a two-body calculation. The probability of a scattering into angular increment $d\theta$ about θ, (θ is the angle between k and k') $P(\theta)d\theta$ is

$$P(\theta)d\theta = \frac{\sin\theta d\theta}{(2k^2(1 - \cos\theta) + k_{TF}^2)^2} \tag{119}$$

To relate $P(\theta)d\theta$ to a uniformly chosen random number, we need only find the normalized probability that θ lies between 0 and θ. The result of this normalization allows us to determine θ:

$$\cos\theta = 1 - \frac{2(1-r)}{1 + 4r(k/k_{FT})^2} . \tag{120}$$

Thus, we can stochastically determine the final state with only two random numbers, one for ϕ and one for θ. k' must equal k because the scattering is elastic and k is predetermined. If the bands are ellipsoidal, non-parabolic bands, we can use the Herring-Vogt transformation mentioned in section 4 equation (84). We need only make two modifications. First, we replace m^* by m_D in equation (118). To choose the final state we pick a k' using (120). Then, we transform it back from the Herring-Vogt space (see equation (84)).

It should however, be emphasized that the screened potential is not necessary to limit the scattering rate to finite values. In fact, for low mobile charge density, k_{FT} may not be sufficient to screen the potential and a model such as the Conwell-Weisskopf model may be more correct.

6.2 Conwell-Weisskopf Model

For comparison with the Brooks-Herring model, we compute the scattering cross section for the Conwell-Weisskopf model which assumes screening by other ions instead of screening by mobile charges. The Conwell-Weisskopf model assumes that the electron is fairly well localized instead of being an infinite plane wave as assumed in the Brooks-Herring model. This assumption remains useful up to moderate energies where the electron is considered a wave packet following a classically defined orbit. By localizing the electron, we can see that in certain positions (midway between ions), the Coulomb forces from each tend to cancel and there is no interaction. In this spirit, Conwell and Weisskopf limit the impact parameter b to half of the mean distance between ions,

$$b_{max} = \frac{1}{2} N_I^{-1/3} \tag{121}$$

where the impact parameter for classical Rutherford scattering is the closest approach of the electron to the ion if its path weren't deflected by the Coulombic force. Thus, this model effectively screens out the Coulomb force to the average distance between ions, and the unscreened Coulomb potential can be used to calculate the scattering rate classically with the Rutherford scattering model [44]. The scattering rate evaluated in this way is:

$$S_{tot}(k) = \frac{\pi v(k)}{4} N_I^{1/3} \tag{122}$$

where $v(k)$ is the electron's velocity. The electron equation of motion states that $v(k) = (1/\hbar)\nabla_k E(k)$. For parabolic bands then, $v(k) = (\hbar k)/m^*$.

Not only is the Conwell-Weisskopf model more applicable for low doping densities than the Brooks-Herring model, but it also behaves more reasonably for high doping densities than the Brooks-Herring model. If we plot mobility versus N_I for both models, we see that the two agree very well for $N_I < 10^{17}$ cm^3.

6.3 Third Body Exculsion

Both models presented above imply two-body, nearest-scatterer processes, but don't expressly prohibit scattering from more distant scattering centers. Ridley argues that, for the sake of consistency, the differential cross-section $\sigma(k, \theta, \phi)$ must be weighted by the probability the the scattering is a nearest-scatterer process. In doing so, he has found that the limiting cases of both the Brooks-Herring and the Conwell-Weisskopf models can be obtained. The method of weighting the cross-section in this manner is called third-body exclusion. (Note: the classical scattering cross-section is related to the scattering rate by $S(k, \phi, \theta) = N_I \; v(k)\sigma(k, \theta, \phi)$).

In order to determine the appropriate factor we again employ the classical notion of the impact parameter b. Ridley has calculated the probability, $P(b)$, that no scattering center exists with impact parameter less than b

from the propability $p = 2\pi N_I ab\, db$ that such a center exists:

$$P(b) = e^{-\pi N_I ab^2} . \tag{123}$$

Here, a is the average distance between ions. Therefore, to prohibit third-body processes, the probability of a scattering event occurring in the solid angle $d\Omega$, must be multiplied by $P(b)$. Thus, we need only calculate b corresponding to the cross-section σ we wish to use. If we take σ from the Brooks-Herring model, we can obtain the limiting cases of both models, so we follow Ridley and calculate the corrected differential cross-section for it [45].

σ_C the corrected cross-section, can be calculated easily from the Brooks-Herring cross-section σ_{BH} by the following:

$$\sigma_C = 2\pi \int_0^\pi \sigma_{BH}(k,\theta) e^{-\pi N_I ab^2} \sin\theta d\theta \tag{124}$$

Roer and Widdershoven give the scattering rate $S_C = N_I v(k) \sigma_C$ as [46]

$$S_C = \frac{v(k)}{a}\left[1 - \exp\left(-\frac{aS_{BH}}{v(k)}\right)\right] \tag{125}$$

This scattering rate incorporates screening of both types (mobile and fixed charges) because it contains both k_{TF} and an ionic screening cutoff from the term in equation (123). Furthermore, it has the advantage that the total scattering rates are ten to a hundred times lower than the peak values of the Brooks-Herring and Conwell-Weisskopf models. This is quite important for Monte Carlo simulations because the higher the scattering rate for all mechanisms, the smaller the time step must be, forcing simulation runs to take much longer times. Roer and Widdershoven also show that at least in the case of GaAs, the low field impurity limited mobilities agree well with experimental results [46].

To find an expression for the angular distribution from the third-body exclusion model, we notice that the probability that an electron has an impact parameter b is $pP(b) = 2\pi N_I a d e^{-\pi N_I ab^2} db$. This is simply the product of the probability that a scattering center exists at a distance b and the probability that no other scattering center is closer. We can relate $P(b)db$ to a uniform random number $r : r \in [0,1]$ by the relation:

$$r = \frac{\int_0^b b' e^{-\pi N_I ab'^2} db'}{\int_0^{b_{max}} b' e^{-\pi N_I ab'^2} db'} = \frac{e^{-\pi N_I ab^2} - 1}{e^{-\pi N_I ab^2_{max}} - 1} . \tag{126}$$

Since the relationship between r and b is transcendental, the Von-Neumann rejection method must be used to select a value of b. Then the scattering angle θ can be determined from b by inverting the relation

$$\pi b^2(\theta) = \int_\theta^\pi \sigma_{BH}(k,\theta') \sin\theta' d\theta' \tag{127}$$

Then, the relation between θ and b is

$$\cos\theta = 1 - \frac{1}{2k^2}\left[\left(\frac{2k^2\pi b^2}{K^2} + \frac{1}{4k^2 + k_{TF}^2}\right)^{-1} - {k_{TF}}^2\right] \tag{128}$$

with

$$K^2 = \frac{(Ze^2)^2(1+2\alpha E)m^*k}{4\pi\hbar^2 e_r^2 e_0^2 v(k)} \tag{129}$$

Although still a rather crude model, third-body exclusion is a compromise including screening of both types which limits scattering rates to manageable levels and fits experimentally obtained low field mobilities fairly well. At high fields, the Born approximation breaks down and other methods must be sought to include collision broadening effects.

7 Acknowledgement

This work has been supported by the Army Research Office, NSF through NCCE, and by the Office of Naval Research. The author would like to thank Jack Higman, Yoshitaka Tanimura, Chris Lee, and Doug Yoder for helpful and insightful conversations, and Karl Hess for his guidance.

Appendix

Table of Physical Constants and Phonon Scattering Parameters for Silicon

The following table gives typical parameter values and material constants for Si which occur in the formulas of section 4. Note that intervalley phonon energies are given in degrees Kelvin ($\hbar\omega = k_B T_{phonon}$). Also, the models shown here are two-valley models– X and L valleys.

	Units	Tang [30]	Canali [47]	Sano [48]	Yoder [42]
a	Å	5.43		5.43	5.43
ρ	g/cm^3	2.329	2.329	2.329	2.329
ϵ	ϵ_0	11.7	11.7		
v_s	10^5 cm/s	9.04	9.037	9.04	9.04
m_{xt}	m_0	0.19	0.1905	0.19	0.19
m_{xl}	m_0	0.9163	0.9163	0.916	0.916
α_x	eV	0.5	0.5	0.5	0.5
Xi_x	eV	9.5	9.0	9.0	9.5
m_{lt}	m_0	0.12		0.12	0.12068
m_{ll}	m_0	1.59		1.59	1.5942
X-X Intervalley Scattering					
f_1	K (10^8 eV/cm)	220 (0.3)	210 (0.15)	210 (0.15)	220 (0.3)
f_2		550 (2.0)	500 (3.4)	500 (3.4)	550 (1.9)
f_3		685 (2.0)	630 (4.0)	630 (4.0)	685 (1.9)
g_1		140 (0.5)	140 (0.5)	140 (0.5)	140 (0.5)
g_2		215 (0.8)	210 (0.8)	210 (0.8)	215 (1.1)
g_3		720 (11.0)	700 (3.0)	700 (3.0)	720 (4.3)
X-L Intervalley Scattering					
1	K (10^8 eV/cm)	672 (2.0)		672 (4.0)	672 (1.8)
2		634 (2.0)		634 (4.0)	634 (1.8)
3		480 (2.0)		480 (1.8)	480 (1.8)
4		197 (2.0)		197 (1.8)	197 (1.8)

References

[1] W. Jones and N. H. March, *Theoretical Solid State Physics*, pp. 611-613, Dover Publications, Inc., New York, (1985).

[2] P. Vogl "The Electron-Phonon Interaction in Semiconductors," pp. 75-116, in *Physics of Nonlinear Transport in Semiconductors* edited by D. K. Ferry, J. R. Barker and C. Jacoboni, Plenum Press, New York, (1979).

[3] J. M. Ziman, *Elements of Advanced Quantum Theory*, pp. 1-12, Cambridge University Press, New York, (1988).

[4] G. D. Mahan, *Many-Particle Physics*, p. 12, Plenum Press, New York, (1981).

[5] R. Kubo, M. Toda, N. Hashitsume, in *Statistical Physics*, Springer-Verlag, New York, (1985).

[6] A. L. Fetter, J. D. Walecka, in *Quantum Theory of Many Particle Systems*, McGraw-Hill Book Company, New York, (1971).

[7] Y. C. Chang, D. Z. Y. Ting, J. Y. Tang, and K. Hess, *Appl. Phys. Lett.*, **42**, 1 (1983).

[8] A. Messiah, *Quantum Mechanics*, John Wiley and Sons, New York, (1958).

[9] J. R. Barker, *J. Phys. C:Solid State Physics*, **6**, pp. 2663-2684, (1973).

[10] J. R. Barker, "Quantum Transport Theory," pp. 126-152, in *Physics of Nonlinear Transport in Semiconductors* edited by D. K. Ferry, J. R. Barker and C. Jaconboni, Plenum Press, New York, (1979).

[11] D. K. Ferry, *Semiconductors*, Macmillan Publishing Company, New York, (1991). See especially the works of D. K. Ferry and J. R. Barker and of I. B. Levinson as well as the other references in chapter 15.

[12] K. Kim, B. Mason, K. Hess, *Phys. Rev. B*, **36** No. 12, p. 6547 (1987).

[13] L. Reggiani, *Physica Scripta*, **T23**, p. 218 (1988).

[14] L. Sham and J. M. Ziman, "Solid State Physics – Advances in Research and Applications," pp. 270-274, in *Solid State Physics* edited by F. Seitz, D. Turnbull, and H. Ehrenreich, Academic Press, New York (1963).

[15] N. W. Ashcroft, N. D. Mermin, *Solid State Physics*, pp. 338-340, Saunders College, Philadelphia, (1976).

[16] H. Ehrenreich, M. H. Cohen, *Phys. Rev.*, **115**, p. 786, (1959).

[17] G. D. Mahan, *Many-Particle Physics*, pp. 405-437, Plenum Press, New York, (1981).

[18] N. W. Ashcroft, N. D. Mermin, *Solid State Physics*, pp. 422-450, Saunders College, Philadelphia, (1976).

[19] J. M. Higman, "Rigid Pseudo-Ion Calculation of the Intervalley Electron-Phonon Interaction in Silicon," pp. 131-136, in *Computational Electronics*, edited by K.Hess, J. P. Leburton, U. Ravaioli, Kluwer Academic Publishers, Boston, (1991).

[20] P. Bruesch, *Phonons: Theory and Experiments I*, p.63, Springer Verlag, New York, (1982).

[21] W. Shockley, Bell System Technical Journal **30**, 990, (1951).

[22] W. Shockley and J. Bardeen, *Phys. Rev.*, **77** pp.407-408 (1950).

[23] J. Bardeen and W. Shockley, *Phys. Rev.*, **80** p.72 (1950).

[24] Conwell, E. M., "High Field Transport in Semiconductors," in *Solid State Physics*, edited by F. Seitz, D. Turnbull, and H. Ehrenreich, Supplement 9, Academic Press, New York, (1967).

[25] B. K. Ridley, *Quantum Processes in Semiconductors*, pp. 106-130, Oxford Press, Oxford, (1988).

[26] W. Harrison, *Phys. Rev. B*, **104**, p. 1281, (1956).

[27] G. L. Bir, G. E. Pikus, *Sov. Physics - Solid State*, **2**, p. 2039, (1961).

[28] D. Long, *Phys. Rev. B*, **120**, No. 6, p. 2077 (1960).

[29] Canali, *Phys. Rev. B*, **12**, No. 4, p. 2275 (1975).

[30] J. Y. Tang, "Theoretical Studies of High Energy Transport in GaAs, Si and Heterostructures," Ph. D. Thesis, University of Illinois, Champaign, Urbana, (1983).

[31] M. V. Fischetti, IEEE Trans. Electron Devices, (1991). to appear

[32] R. Evrard, "The Frohlich Polaron Concept," in *Polarons in Ionic Crystals and Polar Semiconductors*, pp. 37-4, edited by Jozef T. Devreese, American Elsevier Publishing Company, New York (1972).

[33] Cohen, M. L., and T. K. Bergstresser, *Phys. Rev.*, **141**, p. 789, (1966).

[34] K. Hess, *Advanced Theory of Semiconductor Devices*, Prentice Hall, Englewood Cliffs, New Jersey, (1988).

[35] C. Herring and E. Vogt, *Phys. Rev.*, **101**, p. 994, (1956).

[36] C. Jaoboni, L. Regianni, "The Monte Carlo Method for the Solution of Charge Transport in Semiconductors with Application to Covalent Materals," Review of Modern Physics **55**, No. 3 p. 645, (1983).

[37] W. Boardman, A. D. Fawcett, S. Swain, *J. Phys. Chem. of Solids*, **31**, p. 1963, (1970).

[38] E. O. Kane, *Phys. Rev.*, **159**, p. 624, (1967).

[39] C. L. Anderson and C.R. Crowell *Phys. Rev. B*, **5**, p. 2267, (1972).

[40] L. V. Keldysh, *Sov. Phys. JETP*, **21**, p. 1135 (1965).

[41] J. Bude, K. Hess, G. J. Iafrate, "Impact Ionization in Semiconductors: Beyond the Golden Rule," to be published.

[42] D. Yoder, MS Thesis, University of Illinois, May 1991.

[43] H. Brooks, C. Herring, *Phys. Rev.*, **83**, p. 879, (1951).

[44] E. Conwell, V. F. Weisskopf, *Phys. Rev.*, **77**, No. 3, pp. 388-390, (1950).

[45] B. K. Ridley, *J. of Phys. Chem.*, bf 10, pp. 1589-1593, (1977).

[46] T. G. Van de Roer, and F. P. Widdershoven, *J. of Appl. Phys.*, **59**, No. 3, pp. 813-815, (1986).

[47] C. Canali, C. Jaoboni, F. Nava, G. Ottaviani, and A. Alverigi-Quaranta, *Phys. Rev. B*, **12**, pp. 2265-2284, (1974).

[48] N. Sano, T. Aoki, M. Tomizawa, A. Yoshii, *Phys. Rev. B*, **41**, pp. 12122-12128, (1990).

3

Evaluating Photoexcitation Experiments Using Monte Carlo Simulations

C. J. Stanton and *D. W. Bailey*

Department of Physics
University of Florida
Gainesville, FL 32611

This chapter focuses on nonequilibrium carrier dynamics in semiconductors through the evaluation of time dependent optical experiments. Using ensemble Monte Carlo simulations coupled with detailed band structure calculations, we show how the evolution of photoexcited distribution functions are determined. Evaluating photoexcitation experiments with simple models (i.e., without Monte Carlo) poses severe approximations and leads to misunderstandings. We show that Monte Carlo simulations are essential to modeling photoexcitation experiments, but that calculating the distribution functions alone is not enough to fully understand the optical measurements. To relate to the experimental quantities of interest, the distribution functions must weighted by the optical matrix elements and joint density of states and summed over all allowed transitions. With complete calculations including Monte Carlo simulations, the optical experiments are shown to provide reliable details on scattering rates, relaxation channels, and band structure — valuable information for characterizing materials and carrier dynamics.

1. INTRODUCTION

A fundamental problem in understanding the physical processes that occur in modern microelectronic devices is comprehending the relaxation dynamics of highly nonequilibrium carrier distributions. In these devices, carriers are often far from equilibrium and conventional techniques for describing their behavior do not apply. That is to say, theories based on equilibrium distributions and quasi-equilibrium distributions (where the carrier distributions are assumed to have an effective temperature different from the lattice temperature) are invalid. Therefore, to describe the system behavior, one must use a nonequilibrium theory. The typical theory used is the Boltzmann transport equation or some variation based on kinetic theory.[1] As devices move to shorter length and time scales, the validity of these theories is even called into question.

Further complications arise because the carrier scattering rates in these semiconductor systems range from tens to hundreds of femtoseconds (1 fs = 10^{-15} sec). In the past, DC transport measurements, such as drift or diffusion, were central ways of understanding and gathering information about these systems. By comparing theoretical calculated quantities to experimentally measured quantities, one would get information about the scattering processes and band structure of the system under study. Unfortunately, DC transport measurements do not directly measure carrier distribution functions, but instead measure moments of the distribution.[2] As a result, while DC transport measurements can provide some information on the systems under study, they are insensitive to the details of some of the scattering mechanisms, and thus cannot provide a complete picture of the carrier dynamics.

Ideally, because transport parameters are derived from the time dependent carrier distribution functions, one would like to measure the distributions directly. With the development of optical spectroscopy, one can begin to probe the dynamics of photoexcited carriers and gain information about the time evolution of the distribution functions.

Photoexcitation methods have many advantages over drift-diffusion methods. Unlike a DC transport measurement, which integrates over the entire distribution function, a photoexcitation measurement can selectively probe the energy dependence of the carrier density. *Time–dependent* laser spectroscopy has the additional advantage of studying the evolution of the distributions, and thus the carrier relaxation processes. Because of recent advances in femtosecond laser spectroscopy, particularly in tunable and continuum probe techniques,[3–8] distribution functions can be probed on the same time scale on which carrier scattering occurs. By carefully studying these systems, detailed information about band structure, nonequilibrium relaxation processes and scattering rates can be obtained.

Even though optical spectroscopy can probe details of the carrier distributions, quantitative analysis of the data from femtosecond laser spectroscopy, however, is difficult. Nonequilibrium carrier dynamics in GaAs are intricate owing to many scattering mechanisms, multiple conduction–band minima, and degenerate hole bands. The analysis also depends on the band structure as well as the distribution functions and may involve several transitions. These complications have sometimes led to apparently conflicting conclusions between experiments.[9–11]

Correctly and quantitatively interpreting femtosecond spectroscopy in GaAs, however, is possible if both the carrier dynamics and nonlinear absorption are calculated accurately. The Monte Carlo method can be used to solve the carrier dynamics via the Boltzmann transport equation. The nonlinear absorption can then be calculated directly from the distribution functions returned by the Monte Carlo simulation.

In this chapter, we discuss methods of analyzing time dependent optical experiments using Monte Carlo simulations and detailed computer modeling. We focus on the differences between the modeling of photoexcitation experiments and the modeling of more conventional transport phenomena discussed in the other chapters. In the next section, we briefly describe what types of photoexcitation experiments are performed and the corresponding measured quantities. In section 3, we describe our calculational method, and then in section 4 present results illustrating the use of Monte Carlo techniques to understand time dependent optical experiments. Emphasis is placed on transient absorption experiments, but the methods can be generalized to other types of photoexcitation experiments.

2. EXPERIMENTS

When one speaks of investigating the nonequilibrium carrier dynamics in semiconductors, one usually pictures a scenario where the carriers are put into a nonequilibrium state by the application of an external electric field. The experimental investigations then focus on determining the current in the nonequilibrium steady state as a function of the applied electric field. However, there is an alternative method of investigating nonequilibrium effects. One can put the carriers into an initial nonequilibrium distribution with no external fields and then watch the time development of the carrier distributions as they relax to the equilibrium state. The initial nonequilibrium distribution function can be created in several different ways, such as by injection over a barrier or by photoexcitation.

Optical techniques are an alternative way well suited for studying carrier relaxation. While there are many different ways of optically probing carrier dynamics, they fall into two broad classes: absorption and luminescence.

Transient Absorption

In transient absorption experiments, one uses a pump/probe procedure illustrated in fig. 1. The pump pulse creates a nonequilibrium distribution of photoexcited carriers in the material. The probe pulse, delayed by time τ with respect to the pump pulse, is then used to measure the decay of carriers out of their initial states.

The decay is measured through the change in absorption of the probe pulse. This is influenced by the carriers generated by the pump pulse since, by the Pauli exclusion principle, probe pulse photons cannot be absorbed if there are no states available into which new carriers can be photoexcited. As carriers scatter out of their initial states, absorption of the probe pulse increases. By measuring the absorption of the probe pulse as a function of delay τ, one gathers information about scattering rates.

Roughly speaking, the probability of absorption of the probe pulse is proportional to the probability of having an electron in the valence band (to excite to the conduction band) times the probability of *not* having an electron in the conduction band (which would block the transition),

$$P \sim f_e^v(t) \times (1 - f_e^c(t)). \tag{1}$$

In addition, one must subtract the probability for stimulated emission which is proportional to

$$f_e^c(t) \times f_h^v(t). \tag{2}$$

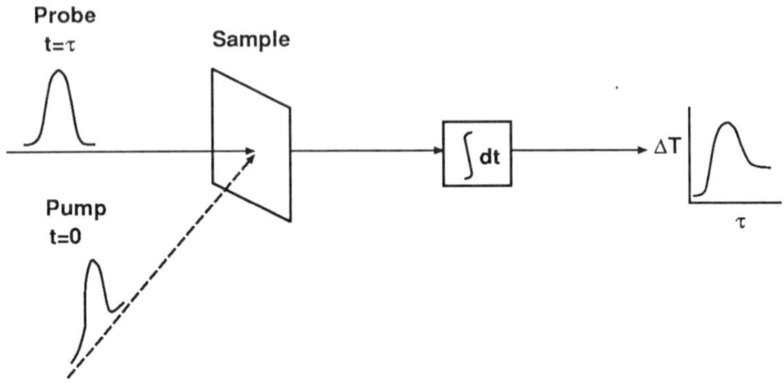

Figure 1. Schematic of a typical transient absorption experiment. At t=0, a pump pulse creates a nonequilibrium distribution of carries in the material. The time dependence of the transmission of a probe pulse, delayed in time by τ, provides information on how carriers are scattered out of their initial states. A nonlinear transmission experiment measures the sum of the carrier distribution functions, $f_e(t) + f_h(t)$, integrated over all possible transitions and weighted by the strength of the transition.

Since the probability of finding an electron in the valence band is given by $f_e^v(t) = (1 - f_h^v(t))$ we see that the probability of absorption of the probe pulse is proportional to :

$$\alpha \sim (1 - f_e^c(t) - f_h^v(t)). \tag{3}$$

Henceforth, we omit the superscripts *c* and *v*; electrons are assumed to be in the conduction band and holes in the valence bands. If the pump pulse is weak, then f_e, $f_h \ll 1$, and the absorption of the probe pulse is given by the linear absorption. If the pump pulse is strong enough to photoexcite a substantial number of carriers, then the change in absorption of the probe pulse measures the sum of the electron and hole distribution functions, $f_e(t)+f_h(t)$. Since scattering rates in semiconductors are on the order of tens to hundreds of femtoseconds, femtosecond lasers could be used to probe the distribution functions on the same time scale as they are changing, and hence provide time dependent dynamical information.

Unfortunately, life is not so simple. The nonlinear absorption does not measure the distribution function in the entire conduction and valence bands, but only in the region where optical transitions are energetically allowed, i.e., in the *optically coupled region.* This is further complicated since there is often more than one allowed transition. In particular, in most diamond/zinc blende semiconductors, there are three valence bands near the energy gap, which means that for a certain range of laser energies, three transitions are possible. In more complicated structures such as quantum wells, many more transitions are possible.[12] The actual nonlinear absorption measures a weighted average of the electron and hole distributions summed over all allowable transitions. The strength of the optical transition is determined by the optical matrix element together with the joint density of states. Thus, the carrier distribution functions are not directly measured.

There are many different types of pump/probe experiments that yield information about the nonlinear transmission. Examples of these are equal energy pump/probe,[11] equal pulse-correlation,[9] pump/continuum probe,[3,6] photon-echo,[13] and time-of-flight measurements.[14]

Figure 2 shows the experimental nonlinear transmission spectra from two different techniques. The first plot is based on the conventional pump/probe technique and is from the work of Lin et al.[5] while the second plot is based on the equal pulse correlation technique of Rosker, Wise and Tang.[9] In principal, the two traces provide the same information.

Luminescence

An alternative to transmission experiments is provided by luminescence experiments. A typical front-surface photoluminescence experiment[12a] is shown in fig. 3. In a luminescence experiment, one looks at the spontaneous recombination of electrons with holes after either CW or pulsed photoexcitation. For an electron in the conduction band to recombine, there must be an empty state (hole) in the valence band at the same wavevector. The probability for luminescence to occur is therefore proportional to the probability of having an electron in the conduction band times the probability of having a hole in the valence band, $f_e \times f_h$. As such, luminescence provides a complimentary measurement to nonlinear transmission.

A variant of luminescence occurs when one measures the recombination of electrons with impurities in the sample, i.e., acceptor luminescence.[16–18] Here one measures only the electron distribution function.

As mentioned above for transmission, the actual luminescence signal is more complicated than the simple picture presented above since it must be summed over the weighted average of all such transitions.

3. CALCULATIONAL METHODS

There are several different procedures for analyzing results from femtosecond laser spectroscopy in GaAs ranging from the simple fitting of experimental curves to detailed models based on determining the carrier dynamics through Monte Carlo simulations.

In the simplest method, the *extracted time constant* method, the experimental differential transmission is fitted to the convolution of a set of one-sided exponential time decays, with different maximum heights, and the probe pulse

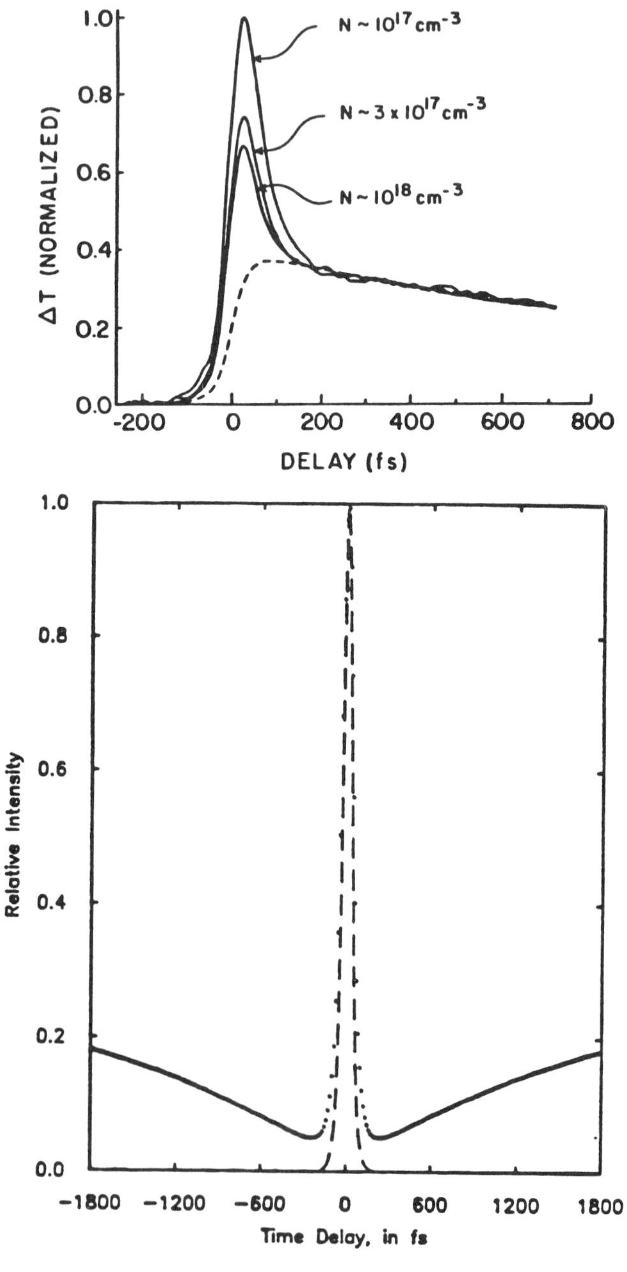

Figure 2. Experimentally determined transient absorption, $\Delta T/T$, in GaAs as a function of probe delay for a conventional pump/probe experiment (from Lin et al.[5]) and of an equal pulse correlation experiment (from Rosker, Wise and Tang[9]). While the traces appear different, they provide the same information.

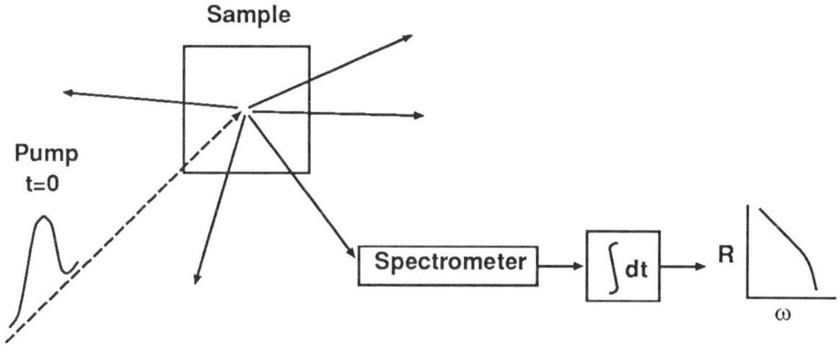

Figure 3. Schematic of a typical photoluminescence experiment. In a photoluminescence experiment, the product of the distribution functions, $f_e \times f_h$, is measured. Luminescence thus provides complimentary information about a system compared to absorption.

autocorrelation. Thus, one to three decay times are extracted from the experimental data and directly attributed to a set of plausible scattering mechanisms. This method is based on the assumption that a multi-level system response model [19,20] can be applied to GaAs to describe carrier relaxation and that the response is well characterized by a sum of exponentials. The extracted time constant method, however, oversimplifies the carrier dynamics and provides misleading results when applied to high-energy photoexcitation in GaAs because of many competing scattering processes and the existence of non-isolated optical transitions. Scattering mechanisms that occur on the same time scale are indistinguishable in the measurement, and carriers that relax out of initially excited levels sometimes re-enter those same or equivalent levels. For these reasons and because the time constants obtained by these methods are often ambiguous, the extracted time constant method can be misleading for explaining the data from femtosecond laser spectroscopy experiments. This ambiguity is best illustrated in figs. 4a and 4b.

In fig. 4a, we show the nonlinear transmission obtained from one of our calculations for a 1.9 eV pump and probe. A 1.5 ps component is fit to match the response at long time and then subtracted from the total response to yield a remaining fast response. For comparison purposes, a 60 fs signal and a 90 fs signal are shown. As can be seen, the remaining fast signal is characterized by a 60 fs response.

In fig. 4b, we show the same 1.9 eV pump/probe signal, but this time subtract off a 5 ps component that is fit to match the response at long time. The remaining fast component is again shown compared to 60 fs and 90 fs responses. As can now be seen, the fast component is 90 fs rather than 60 fs!

We thus see that there is much ambiguity in the extracted time constant method, and the relaxation times obtained from these methods are subjective

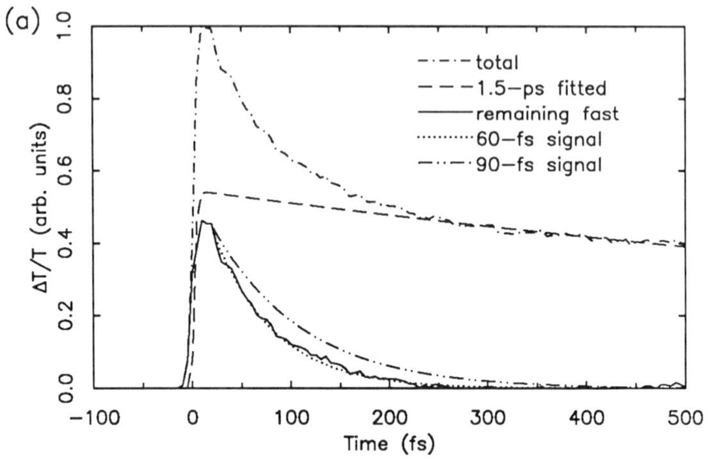

Figure 4a. Extraction of a 60 fs signal using the extracted time constant method. The dash-dotted line is the calculated nonlinear transmission for a 1.9 eV pump and probe. The dashed line is a 1.5 ps convoluted exponential fit to the long time behavior. The solid line is the remaining component after the 1.5 ps response is subtracted. For comparison purposes, 60 fs and 90 fs responses (dotted and dash-double dotted, respectively) are shown. The remaining fast component has a 60 fs decay time.

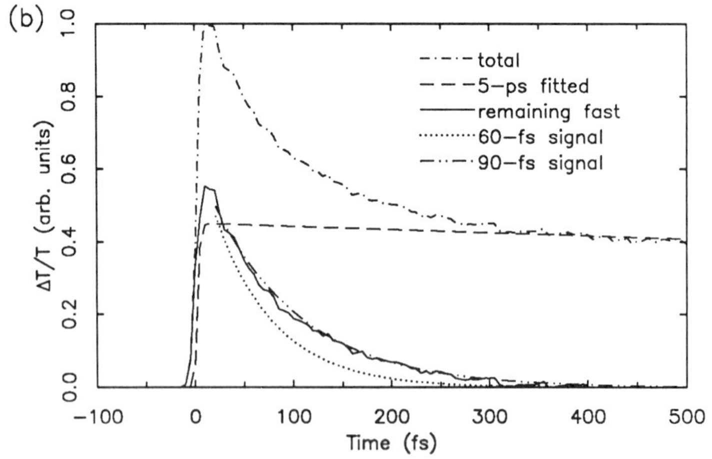

Figure 4b. Extraction of a 90 fs signal using the extracted time constant method. The dash-dotted line is the calculated nonlinear transmission for a 1.9 eV pump and probe, identical to the top line in fig. 4a. The dashed line is a 5 ps convoluted exponential fit to the long time behavior. The solid line is the remaining component after the 5 ps response is subtracted. For comparison purposes, 60 fs and 90 fs responses (dotted and dash-double dotted, respectively) are shown. The remaining fast component has a 90 fs decay time. This shows the ambiguity in extracting useful data from simple exponential fits.

to the way in which they are extracted. There is a great variance in the time constants reported by this method even when the same data is analyzed by different groups.[21]

A more advanced procedure for interpreting high–energy femtosecond laser spectroscopy is to solve the Boltzmann transport equation, either directly[22–25] or using the Monte Carlo method.[26–28] This method used by itself has limited success because the experimental results do not directly measure the distribution functions. Although the carrier dynamics are modeled precisely, nonlinear absorption effects are not treated numerically and conclusions are necessarily qualitative. That is, the nonlinear absorption does not directly measure $f_e(t)$ and $f_h(t)$ solely by themselves. It measures a complicated integral containing the distribution functions together with the joint density of states, optical matrix elements, time convolution, energy integral of the probe pulse, and summation over optical transitions. While trends in the distribution functions may be used to predict why certain features occur in the nonlinear transmission, the agreement can be at best qualitative.

A more complete approach is to use an ensemble Monte Carlo method to model both electron and hole dynamics, and, *in addition*, to couple these simulations to calculations of the nonlinear absorption. In this way femtosecond laser spectroscopy experiments can be *quantitatively* interpreted. Thus, the extracted time constant method is avoided, and instead of just using distribution functions calculated from Monte Carlo simulations, the calculated distributions are used as input in calculating the nonlinear absorption which is then compared directly with the experimental results.

A summary of our method of simulation is shown by the flowchart in fig. 5. The band structure and related variables, such as Bloch overlap factors, are determined from the $\mathbf{k}\cdot\mathbf{p}$ method. The nonlinear absorption is then calculated to determine initial photoexcited electron–hole pair states in the Monte Carlo simulation of the pump pulse. The differential transmission is calculated using the distribution functions from the completed Monte Carlo simulation. The final step is matching the experimental data and thus determining band structure or scattering rate parameters.

Band Structure

The band structure of the material is an extremely important part of the calculation since it determines several key ingredients. It governs the calculation of both the absorption profile (i.e., where the transitions are and their intensity) and the carrier scattering rates. The region where carriers are photoexcited is determined by the energetically allowed optical transitions. In addition, $\varepsilon(\mathbf{k})$ is needed in the calculation of scattering rates and the final state after each scattering event.

For 2–eV photoexcitation in GaAs, the conduction band and heavy, light, and split–off hole bands are involved. The band structure for these four bands

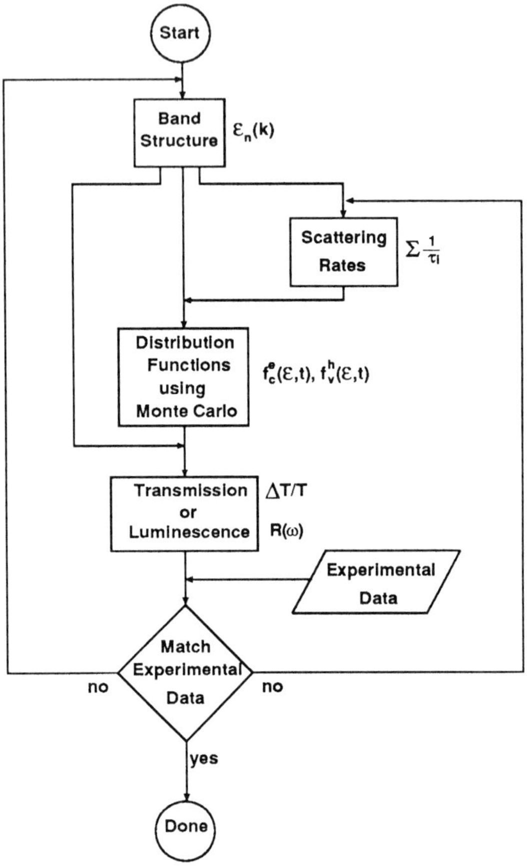

Figure 5. Flowchart of method used to simulate photoexcitation experiment. Band structure is input to scattering rate, Monte Carlo, and optical absorption calculations. A match to experimental data provides information on band structure or scattering rate parameters.

along the Λ and Δ symmetry lines is enlarged in fig. 6. Vertical arrows show the three electron optical transitions allowed for 2–eV photoexcitation. The light hole transition to the conduction band occurs beyond the first inflection point away from the center of the zone, showing it is wrong to use a simple effective mass for the light holes. $\Gamma \rightarrow L$ and $\Gamma \rightarrow X$ intervalley scattering, two of the primary ways electrons are removed from the optically connected region for high-energy photoexcitation, are represented by the horizontal arrows. The electrons excited from the split–off band do not have enough energy to undergo intervalley scattering.

For the range of energies under consideration, for the holes a simple effective mass approximation is invalid because the hole isoenergy surfaces

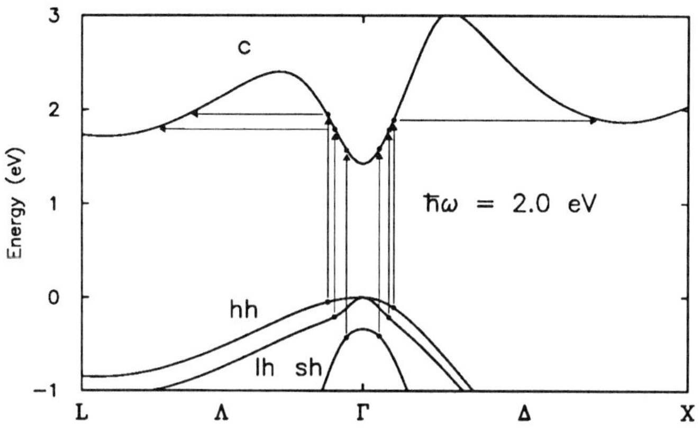

Figure 6. The (top to bottom) conduction, heavy hole, light hole, and split–off hole bands for GaAs in the L and X directions as calculated by the $\mathbf{k} \cdot \mathbf{p}$ method. The 2–eV optical transitions are represented by vertical arrows while horizontal arrows correspond to intervalley scattering that can occur after the photoexcitation of the electrons. Because the X valley lies higher in energy than the L valley, for a 2.0 eV pump photoexcitation, only electrons that come from the transition from the heavy hole band are energetically able to scatter into the satellite X valley.

are warped. Anisotropy and nonparabolicity are large. This is illustrated in fig. 7 which shows the heavy hole band structure close to the Γ point in the $k_z = 0$ plane. Significant anisotropy is displayed even at low energies, showing the inappropriateness of the effective mass approximation for the holes. While one may be able to use an effective mass approximation for the holes in a transport experiment for low electric fields, in most optical experiments the holes are generated away from the zone center and the effective mass approximation cannot be used. In addition, the warping of the hole surface also controls the shape of the electron distribution generated in the conduction band. In our method, therefore, for the heavy, light, and split–off hole bands we diagonalize a full–zone 30×30 $\mathbf{k} \cdot \mathbf{p}$ Hamiltonian following the procedure of Pollak et al.[29,30] to obtain an accurate and reliable band structure, essential for the calculations.

While the anisotropy of the hole bands may not be surprising, what is somewhat surprising is that there is sufficient anisotropy in the conduction bands to warrant using a more complicated expression than the nonparabolic form, $\varepsilon(1+\alpha\varepsilon) = \hbar^2 k^2/2m^*$.

Figure 8 shows the difference between $\mathbf{k} \cdot \mathbf{p}$, non-parabolic, and parabolic conduction bands for GaAs. For 2 eV photoexcitation experiments, carriers may be generated more than 15% away from the zone center to the zone edge. For these wavevectors, there is several meV difference in the energies in the X and L direction. While the X direction closely follows the nonparabolic expression, the L does not.

For CW photoluminescence, where the energy resolution is high, detailed

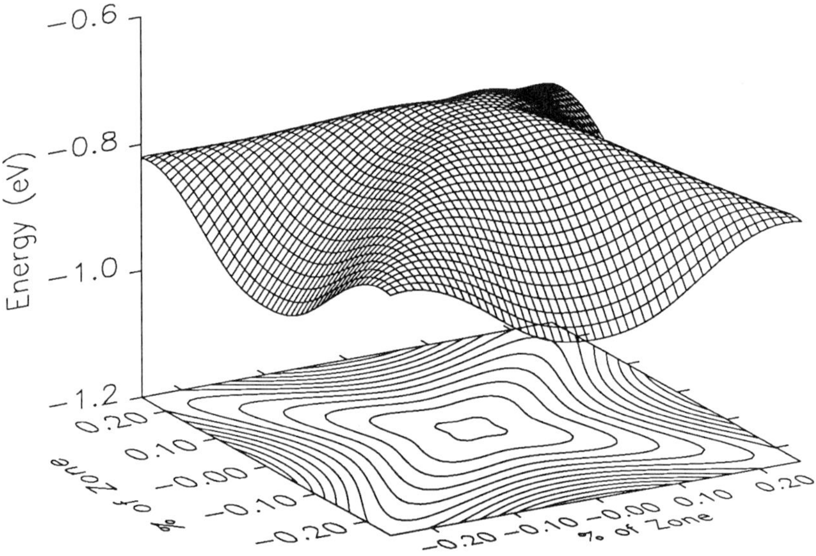

Figure 7. Warping of the heavy hole band. The heavy hole energy surface and isoenergy contours in the $k_z = 0$ plane as calculated by the $\mathbf{k} \cdot \mathbf{p}$ method are shown. For laser energies of 2.0 eV and greater, carriers are generated more than 10% away from the Brillouin zone center toward the zone edge. As can be seen, for high energy photoexcitation experiments, the hole bands are not well characterized by an effective mass model.

band structure calculations such as the $\mathbf{k} \cdot \mathbf{p}$ method are required even for the conduction band. The result of the differences shown in fig. 8 come into play for acceptor-photoluminescence simulations. Figure 9 shows comparisons of experimental acceptor-luminescence[17] at energies of 2.004, 1.938 and 1.732 eV with the calculated initial electron densities with energy relative to the acceptor level. The solid lines are the experimental results of Fasol et al.[17] The remaining three lines are calculations of initial electron densities. Scattering creates the "phonon replicas" seen in the experimental data.

In all the calculated curves, the valence bands are determined from the 30–band $\mathbf{k} \cdot \mathbf{p}$ theory. In the dash-dotted line, the conduction band is calculated using the effective mass approximation, $\varepsilon = \hbar^2 k^2/2m^*$. The dotted line is calculated from 4–band $\mathbf{k} \cdot \mathbf{p}$ theory which gives an isotropic, analytic form for the nonparabolicity, $\varepsilon(1 + \alpha\varepsilon) = \hbar^2 k^2/2m^*$ with $\alpha = 0.67$ for GaAs. The dashed line is calculated from the thirty band $\mathbf{k} \cdot \mathbf{p}$ calculations which includes higher bands and produces anisotropy in the conduction band.

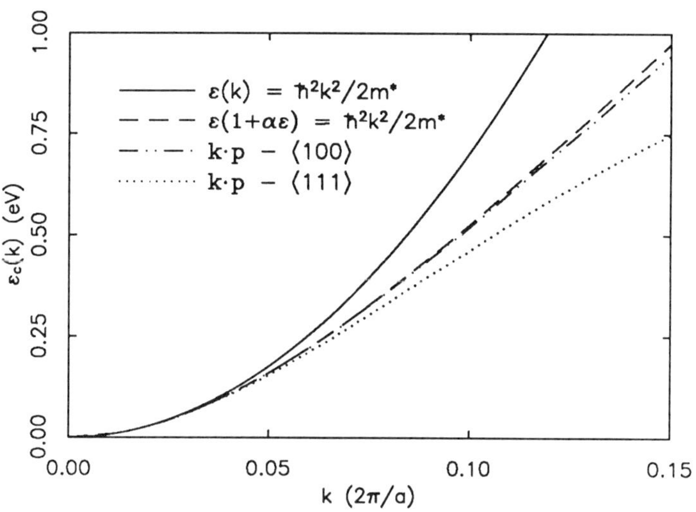

Figure 8. Anisotropy in the conduction energy band in GaAs. The solid line corresponds to the effective mass approximation, $\varepsilon = \hbar^2 k^2/2m^*$, the dashed line to the nonparabolic expression, $\varepsilon(1 + \alpha\varepsilon) = \hbar^2 k^2/2m^*$, and the dash-double dotted and dotted lines correspond to the full $\mathbf{k} \cdot \mathbf{p}$ calculation in the X and L direction, respectively. For 2 eV photoexcitation, carriers may be generated more than 15% away from the zone center to the zone edge. For these wavevectors, there is a several meV difference in the energies in the X and L direction. While the X direction closely follows the nonparabolic expression, the L does not.

At energies close to the band edge, 1.732 eV, all three curves accurately reproduce the experimental data. This concurs with fig. 8 which shows there is little deviation between all three cases at low energies. At higher energies, i.e., 1.938 eV and 2.004 eV, there are discrepancies between the experimental results and the analytic calculations. The parabolic and nonparabolic expressions predict peaks that are too high in energy (for both the heavy and light hole transitions). The full $\mathbf{k}\cdot\mathbf{p}$ calculation gives the best location of the peaks.

This illustrates that acceptor-luminescence experiments are sensitive to small changes in the band structure, almost an order of magnitude more sensitive than photoemission.[17,31] Luminescence techniques should therefore prove to be a valuable tool in studying and characterizing new materials.

There are alternative methods of calculating detailed band structure than the $\mathbf{k}\cdot\mathbf{p}$ method. Most notable is the pseudopotential method used in Monte Carlo calculations by Shichijo and Hess[32] and also Fischetti and Laux.[33] Although the pseudopotential method is slower to calculate, it is generally considered more reliable away from the zone center.

Bloch Overlap Factors

In addition to using the $\mathbf{k}\cdot\mathbf{p}$ program to determine the band structure and hence initial regions of photoexcitation, it is also used to determine the

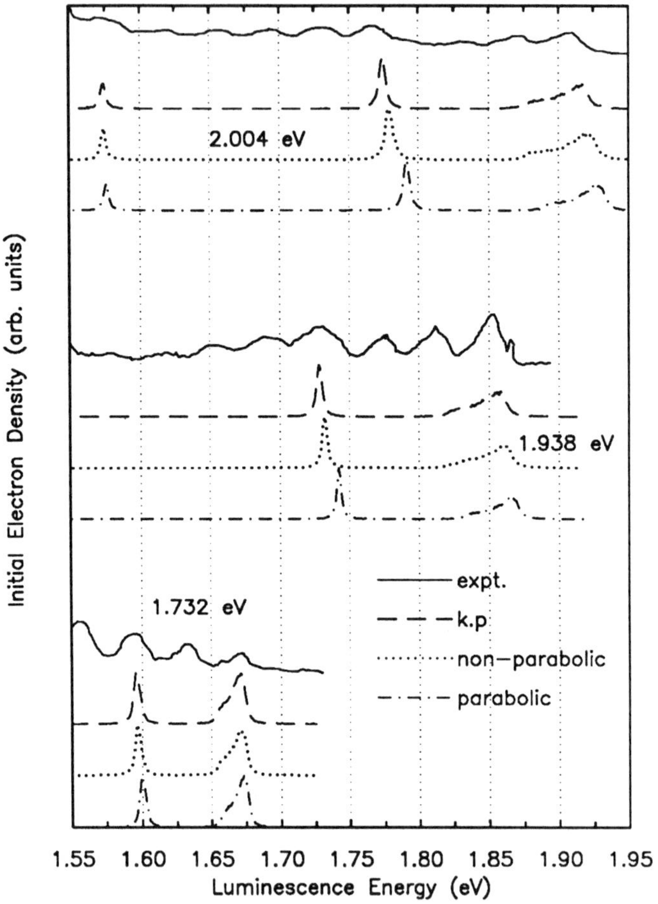

Figure 9. Comparison of the experimental acceptor-luminescence by Fasol et al.[17] (solid lines) at the energies of 2.004, 1.938 and 1.732 eV with the calculated initial electron densities (before scattering occurs). The dashed line corresponds to a full $\mathbf{k} \cdot \mathbf{p}$ calculation used for both the conduction band and hole bands. The dotted line corresponds to a $\mathbf{k} \cdot \mathbf{p}$ used for the holes, but the nonparabolic expression $\varepsilon(1 + \alpha\varepsilon) = \hbar^2 k^2/2m^*$ used for the conduction band. The dashed-dotted line corresponds to a simple effective mass approximation, $\varepsilon = \hbar^2 k^2/2m^*$ for the conduction band. As can be seen, at low energies both the light and heavy hole transitions match the experimental luminescence for all three calculations. At higher energies, the effective mass and nonparabolic expressions predict transitions that are too high in energy. This demonstrates the sensitivity of luminescence experiments to an error of a few meV in band structure.

eigenfunctions and the Bloch overlap factors. The Bloch overlap factors are the inner product of the periodic part of the initial and final Bloch states and must be included in the calculation of the hole scattering rates. Although the Bloch overlap factors can be approximated as unity for electron interactions, they vary between zero and one for holes. The usual approximation, taken by several researchers.[28,34,35] is to use Wiley's approximation[36] for interactions involving light and heavy holes based on a 4–band $\mathbf{k}\cdot\mathbf{p}$ model, near k=0,[37]

$$G_{lh\text{–}lh} = G_{hh\text{–}hh} = \frac{1}{4}\left(1 + 3\cos^2\theta\right) \tag{4a}$$

$$G_{lh\text{–}hh} = G_{hh\text{–}lh} = \frac{3}{4}\sin^2\theta\ , \tag{4b}$$

where θ is the angle between $\mathbf{k}$ and $\mathbf{k}'$. Wiley's expressions, however, are not intended for interactions involving the split–off band and are a poor approximation for high–energy holes in any of the bands.

Bloch overlap factors for small k' heavy hole–heavy hole interactions as a function of k and θ are shown in fig. 10. Significant differences between the calculated results and Wiley's approximation,[36] shown on the front face for comparison, occur away from the band edge.

The rejection method is used to implement the Bloch overlap factors in the Monte Carlo simulation,[35] similar to the usual application of the Pauli exclusion principle.[38] Thus, the overlap factor is included by a Monte Carlo integration of the matrix element. If a hole is selected to scatter, its final state and Bloch overlap factor are determined. If the overlap factor is less than a random number uniformly distributed between 0 and 1, then the scattering event is rejected.[39]

Monte Carlo Simulations

The ensemble Monte Carlo method is described in detail in other chapters in this book. In particular, we use a constant time step and include electron-polar optical phonon, electron-intervalley phonon, Γ-valley electron-electron, electron-hole, hole-polar optical phonon,[34,40] hole-nonpolar optical phonon,[34,41] and hole-hole scattering mechanisms. All types of hole scattering refer to heavy, light, and split-off hole bands, and include both interband and intraband scattering. Other scattering mechanisms which occur on a picosecond scale or slower, such as acoustic scattering, are not included because they do not affect femtosecond calculations.

Differences with Conventional Monte Carlo. This section deals with specific differences between simulations of photoexcitation experiments and drift-diffusion experiments. In particular, band-to-band transitions must be considered in photoexcitation simulations. Energy, time, and spatial resolutions are different; initial distributions are highly nonequilibrium; and photoexcitation techniques have a distinct set of experimental artifacts. These dissimilarities force the programmer to consider a new set of methods.

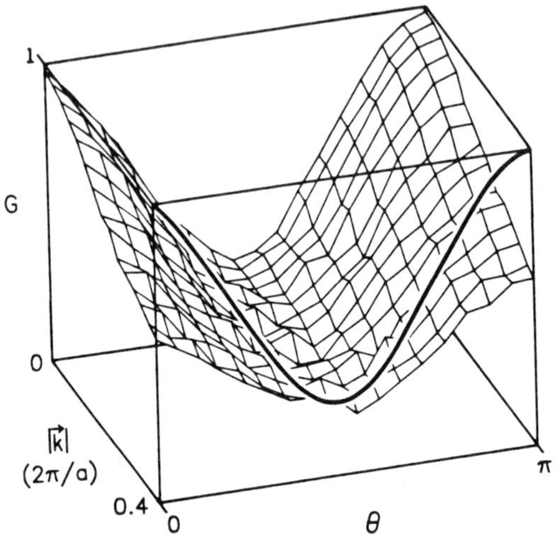

Figure 10. Heavy hole–heavy hole Bloch overlap factors calculated from the k·p method compared with the intraband Wiley approximation (shown on the front face). Factors are plotted for small k as a function of the angle between k and k′ for k' from 0 to 10% of Γ to X

As mentioned before, a great advantage of photoexcitation methods is the fast time resolution. Pulses as short as 6 fs have been obtained and applied in semiconductor laser spectroscopy experiments. However, with the gain in time resolution comes an inescapable loss in energy resolution from the Heisenberg uncertainty principle. For example, a transform-limited hyperbolic secant squared 40–fs pulse has an energy width of approximately 40 meV.

For CW lasers, the spectral linewidth attainable is negligible. However, broadening effects from other sources (i.e., lifetime broadening and band structure anisotropy) dominate the initial carrier distributions in band-to-band transitions in semiconductors.

Another important consideration in these simulations are the multiple transitions during photoexcitation. In cubic semiconductors such as GaAs and Si, the top valence bands are two-fold degenerate: if heavy-hole to conduction band transitions are energetically allowed, then light-hole transitions are allowed also. Therefore, at least two peaks in the electron energy distribution are excited; three if the split-off band to conduction band transition is energetically allowed. This results in an initial distribution that cannot be described by a Maxwell-Boltzmann distribution.

A Monte Carlo simulation of a photoexcitation experiment thus starts with a highly non-equilibrium initial distribution, indescribable by the conventional definition of temperature. Usually to evaluate photoexcitation experiments, one must simulate electrons *and* holes. Also, if interested in phenomena on the time scale of the laser pulse, then one must allow for a variable number of carriers in the simulation.

It is important to be aware of the experimental artifacts possible in photoexcitation measurements. Laser spectroscopy should ideally be applied to thin active regions to achieve saturable absorption independent of thickness. Another critical issue is the magnitude of the coherent artifact, which is a partial measurement of the pump pulse in addition to the probe pulse.[9] Multiple internal reflection is also a possible cause for misleading results, but can be diminished by proper anti-reflection coating of the semiconductor sample. For white–light continuum and femtosecond pulses, chirp should be zero, i.e., all frequencies should arrive at the sample surface at the same time.

Initial Distributions. For a transient simulation for which the time dependence of the pulse is important, the simulation starts with zero or few carriers. A starting time for the simulation can be chosen that is several standard deviations before the center of the pulse without losing much CPU time. This is because few calculations are needed for each time step when there are not many carriers in the simulation. The main challenge is to determine the number of carriers to add at each time step of the simulation.

As an example, the autocorrelation of the CPM copper lasers used to generate femtosecond pulses have a characteristic hyperbolic secant squared time dependence. Therefore, starting from no carriers long before the center of the pump pulse, electron-hole pairs are introduced to the simulation according to the transient photon density

$$N(t) = N_o \mathrm{sech}^2 \left[\frac{2t \, \cosh^{-1}(\sqrt{2})}{\tau_o} \right], \tag{5}$$

where τ_o is the temporal full-width half maximum of the pulse intensity, and $t = 0$ is the center of the pulse. As shown by integrating, the number of carriers to add at each time step from t_0 to t_1 is thus

$$N(t_1) = N(t_0) + \frac{N_{tot}}{2}[\tanh(t_1) - \tanh(t_0)]. \tag{6}$$

Calculating Optical Spectra From Distribution Functions

Transient Nonlinear Absorption. A critical part of any photoexcitation model is the calculation of the nonlinear absorption. It is used not only to determine the differential transmission, it is also needed to determine the initial states for the excited electron-hole pair. The expression for the differential absorption is obtained by summing the number of photons over allowed optical transitions assuming Fermi's Golden Rule,

$$\alpha(t) \propto \frac{1}{\omega} \int d\omega' \int dt' \, N(\omega', t-t') \sum_v \int d\mathbf{k} \, |H_{\mathbf{kk}'}|^2 \delta[E_c(\mathbf{k}) - E_v(\mathbf{k}) - \omega'] \times \left[1 - f_c^e(\mathbf{k}, t') - f_v^h(\mathbf{k}, t')\right] . \tag{7}$$

Therefore,

$$\Delta\alpha(t) \propto -\frac{1}{\omega}\int d\omega' \int dt'\; N(\omega', t-t') \sum_{v} |H_v(\omega')|^2 \rho_v(\omega') \times \left[f_c^e(\omega', t') + f_v^h(\omega', t')\right] , \tag{8}$$

where $N(\omega,t)$ is the transient photon energy density. By $f(\omega,t)$ we mean the time-dependent distribution function in the optically connected region, which should not be confused with $f(E,t)$. That is, ω is the laser energy, whereas E is the energy with respect to the bottom of the band. To simplify the calculation we approximate the optical matrix element $|H_{\mathbf{kk}'}|^2$ in eq. 7 by $|H_v(\omega)|^2$ in eq. 8. To calculate the optical matrix elements we average over polarizations. As expected, $|H_v(\omega)|^2$ decreases with increasing energy as the coupling between the bands decreases. To evaluate $\Delta\alpha(t)$, the rejection method is used for each term in the integrand of eq. 8.

In the thin sample limit, the expression for the differential transmission is derived from eq. 8 as

$$\frac{\Delta T(t)}{T_0} \propto \frac{1}{\omega}\int d\omega' \int dt'\; N(\omega', t-t') \sum_{v} |H_v(\omega')|^2 \left[n_c^e(\omega', t') + n_v^h(\omega', t')\right] , \tag{9}$$

where $n(\omega,t)$ is the time-dependent density of carriers in the optically connected region as given by the Monte Carlo simulations. Equation 9 is calculated more efficiently than eq. 8 since the rejection technique is used only to generate $N(\omega, t-t')$.

Implicit in the evaluation of eq. 9 is the assumption that the probe pulse does not appreciably disturb the carrier populations. This is a reasonable approximation since for these experiments the intensity of the pump pulse is several orders of magnitude larger than the intensity of the probe pulse.

In pump/continuum probe experiments, the probe energy is varied from the pump energy. Therefore, the region in energy space sampled by the probe, i.e., the optically connected region, also varies from the pump pulse. Figures 11a and 11b show the optically connected regions, or energy distributions being sampled, of the continuum probe in conduction and heavy hole bands, respectively. This is where the conduction band electrons or heavy holes block absorption of the probe pulse photons. Not shown are the optically connected regions for the light and split-off transitions. At energies above 1.73 eV, the split–off transition is not allowed. Both parts of fig. 11 show the degree of anisotropy of the heavy hole band. Although the optically connected region in the heavy hole band is narrow in **k** space, it is broad in energy.

Thus, in GaAs at each probe energy, the nonlinear absorption has three to four components. These are from conduction band electrons, heavy and light holes, and, when energetically allowed, split-off holes. In addition, the contribution in the conduction band can be divided up into two or three components depending on whether the electrons are probed by a transition from the heavy, light, or split-off hole bands.

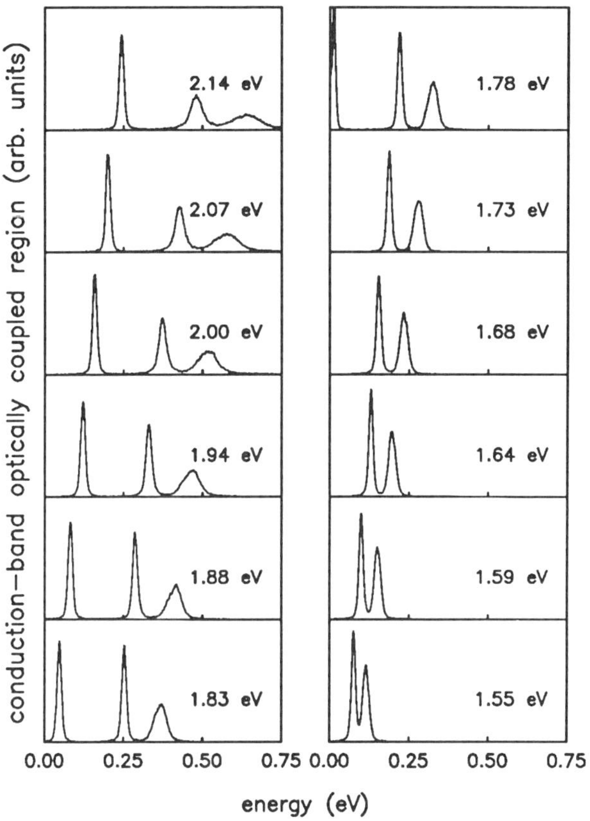

Figure 11a. Conduction band optically coupled region in GaAs for 75–fs photoexcitation. The split–off transition is not allowed for laser energies of 1.73 eV and lower. Broadening is caused by the laser linewidth, by collision broadening, and by band structure anisotropy.

Luminescence. Spontaneous recombination can be written in a form similar to the transient absorption:

$$R(t) \propto -\frac{1}{\omega}\int d\omega' \int dt'\, W(\omega', t-t') \sum_v |H_v(\omega')|^2 \rho_v(\omega') \times f_c^e(\omega', t') f_v^h(\omega', t') \,, \tag{10}$$

$N(\omega, t)$ in eq. 9, the transient photon energy density, is replaced by $W(\omega, t)$, the transient energy weight. $W(\omega, t)$ defines the measurement characteristics of the

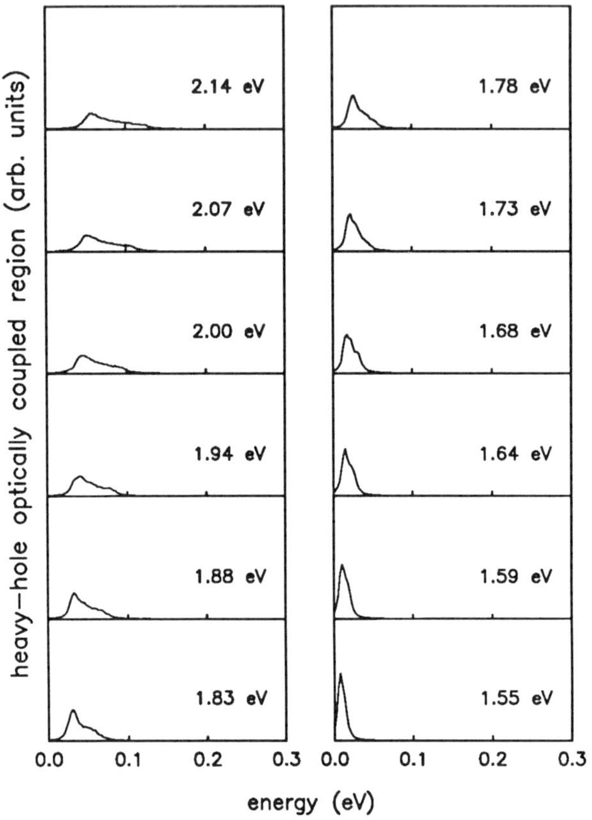

Figure 11b. Heavy hole band optically coupled region in GaAs for 75–fs photoexcitation. band structure anisotropy causes the optically coupled region to widen with increasing energy.

spectrometry equipment. It serves the same purpose as $N(\omega, t)$, but is usually spectrally much narrower and temporally much wider.

Developing this further into terms easily dealt with by the Monte Carlo simulation gives

$$R(t) \propto \frac{1}{\omega} \int d\omega' \int dt' \, N(\omega', t-t') \sum_{v} |H_v(\omega')|^2 \frac{n_c^e(\omega', t') n_v^h(\omega', t')}{g_c(\omega') g_v(\omega')} , \qquad (11)$$

where $g(\omega)$ is the density of states. Caution must be observed if different energy meshes are chosen for the different bands and variables.

Collision broadening. In addition to the laser linewidth, there is energy uncertainty in the initially excited states resulting from collisional broadening.[42,43] This is accounted for by modifying the expression for nonlinear absorption (eq. 7, which is derived for $t \rightarrow \infty$) by broadening the delta function consistent with the Fermi's Golden Rule for finite times. A good approximation for collision broadening is to randomly distribute each carrier about its unbroadened initial state according to a Lorentzian distribution. The spread of the Lorentzian is given by $\Delta E = \hbar/\tau(E)$ where $\tau(E)$ is the energy–dependent total scattering rate which must be determined self–consistently from the Monte Carlo simulation. As the carrier density changes during photoexcitation, so do the scattering rates and the broadening. Typically, the broadening is about 10 meV, but can vary significantly with energy and band depending on the scattering mechanisms.

Broadening is needed to accurately reproduce the initial hole distributions. Most of the width in energy of the laser pulse occurs in the conduction band. Thus, collision broadening only slightly modifies the initial shape of the electron distributions. The holes, because of their larger effective mass, only have a small width in energy due to the energy width of the laser pulse. Without the additional broadening of the hole distributions resulting from lifetime effects, the initial hole distributions are too sharp and the corresponding initial transient absorption calculated from these distributions is too strong.[30] It is essential to include this broadening in any semiconductor nonlinear absorption calculation.

4. RESULTS

To illustrate the above method, we apply it to a specific experiment: pump/continuum probe transient absorption. These types of experiments provide more information than single energy pump/probe experiments and hence give a more complete picture of the carrier dynamics.

Experimental results from pump/continuum probe experiments of Schoenlein et al.[3] are shown in fig. 12. In the experiment, the carriers are generated with a 75 fs, 2.0 eV pump pulse and are then probed with 100 fs laser pulses varying in energy from 1.55 eV to 2.14 eV. Of interest in this figure is that at 1.55 eV there is a rapid change in the transmission. Initially, it was thought that the electrons were scattering rapidly to the bottom of the band, but results from our calculations show that it is primarily the holes quickly scattering to the band edge which causes the change in transmission.

Distribution Functions

First, we show the density of carriers as a function of energy. The density of carriers is distinguished from the distribution function $f(\epsilon)$ by containing the extra factor of the density of states $g(\epsilon)$ which is determined from the full band structure calculation. The carrier densities are calculated from the Monte Carlo simulations and provide a basis for understanding the carrier dynamics and interpreting features in the nonlinear absorption.

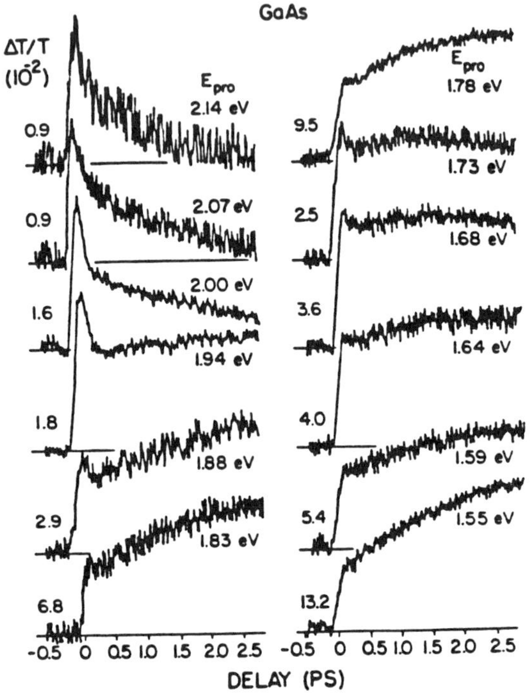

Figure 12. Experimental absorption saturation of Schoenlein et al.[3] for a pump/continuum probe experiment in GaAs using 2 eV pump pulses. The relative scale of the traces is indicated on the left.

Figure 13 shows the electron density for the Γ–valley as a function of time after photoexcitation by a 75–fs, 2–eV pump pulse. The center of the pulse occurs at t=0. While the scale shown is arbitrary, all plots have the same relative scale. We see that at –40 fs, a three–peak distribution begins to appear. This is the result of electrons generated by transitions from the heavy, light and split-off hole bands. At t=0, they become more pronounced. The electrons generated by the light and heavy hole transitions are energetically able to scatter to the satellite L (0.3 eV) and X (0.49 eV) valleys. Therefore, 40 fs after the center of the pulse, the heights of the two higher-energy peaks are suppressed compared to the split-off peak because of rapid intervalley scattering.

From 40 to 160 fs after the center of the pulse, one sees the peak heights diminish and the presence of secondary peaks that result from polar-optic phonon scattering. Also on this time scale, one sees the distribution broaden owing to carrier-carrier scattering. By 200 fs, the distribution function has just two broad peaks. At later times, the distribution is flat up to 0.3 eV and decays rapidly with increasing energy. Electrons from the L valley, however, begin to scatter back into the Γ valley in large numbers and raise the average energy of the

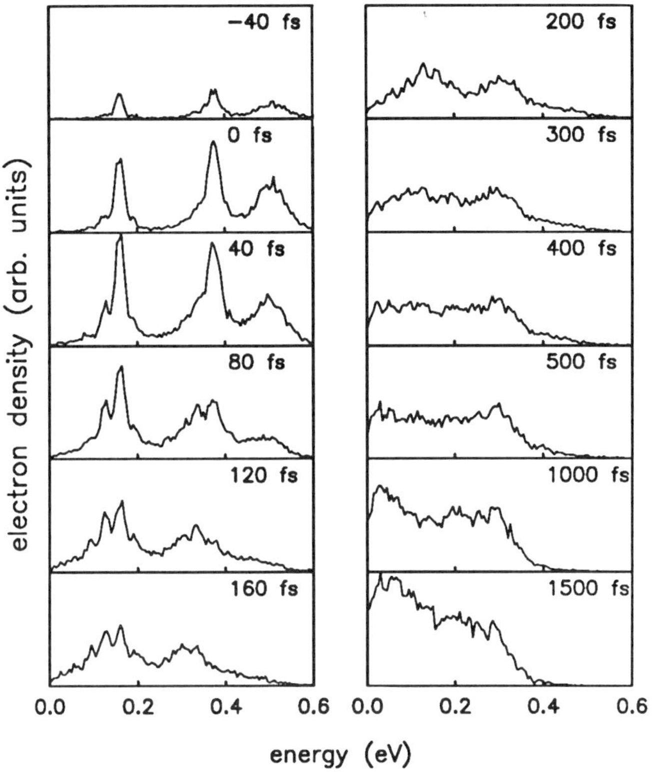

Figure 13. Time dependence of the electron density distribution from 0 to 1.5 ps. The center of the pulse occurs at 0 and the pulse width is 75 fs. All densities are on the same scale. We see three initial peaks produced by transitions from the heavy, light and split–off hole bands. From 40–160 fs, one sees a decay of the two highest energy peaks from the light and heavy hole transitions as they scatter into the satellite X and L valleys. At later times, electrons returning from the satellite valleys act as a "heat source" and considerably slow equilibration to the lattice temperature. The knee at 0.3 eV is from electrons returning to the Γ valley from the bottom of the L valleys. At long times, the distribution is characterized by a two temperature distribution.

Γ–valley electrons. The distribution is not well characterized by a Maxwellian. At longer times, 1000 fs and 1500 fs, one notes the increase in total number of electrons as they return from the L valley. At 1500 fs, the distribution function is characterized by a two temperature distribution: one temperature above the L valley threshold and one below. The temperature below the threshold is significantly hotter than the lattice temperature since the returning L valley electrons heat up the Γ–valley distribution.

Figure 14 shows the heavy hole carrier density from –40 to 1500 fs. At 0 fs, one sees that the initial photoexcited distribution is quite broad arising from both the anisotropy of the heavy hole band and the fast hole scattering rates. By

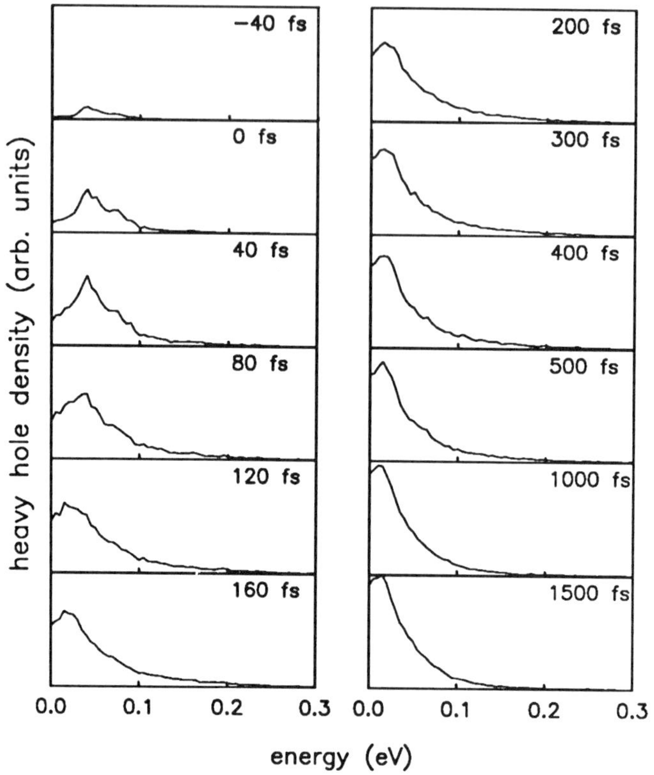

Figure 14. Time dependence of the heavy hole density from –40 to 1500 fs. At the center of the pulse, a small peaked density of holes appears that quickly scatters through hole-hole and hole-phonon scattering to reach an effective temperature which is higher than the lattice temperature. At longer times, the tail of the distribution becomes less pronounced as the temperature decreases.

120 fs, a hot Maxwellian distribution forms for the heavy holes because of high scattering rates and because the initial excitation is near the top of the band. From 160 fs to 1500 fs, the temperature of the distribution is seen to decrease as the tail becomes less pronounced.

The light and split-off hole density plots in figs. 15 and 16 show some phonon structure because the photoexcitation peaks occur away from the band edges. For the light hole band, the initial peak forms 0.2 eV away from the band edge. The light holes begin to thermalize through the emission of optic phonons and carrier-carrier scattering. By 300 fs, the light hole distribution closely resembles a Maxwellian at an elevated temperature. The distribution cools from 300 to 1500 fs. At long times, some of the light holes scatter into the heavy hole band.

For the split-off holes, the initial peak forms closer to the band edge than

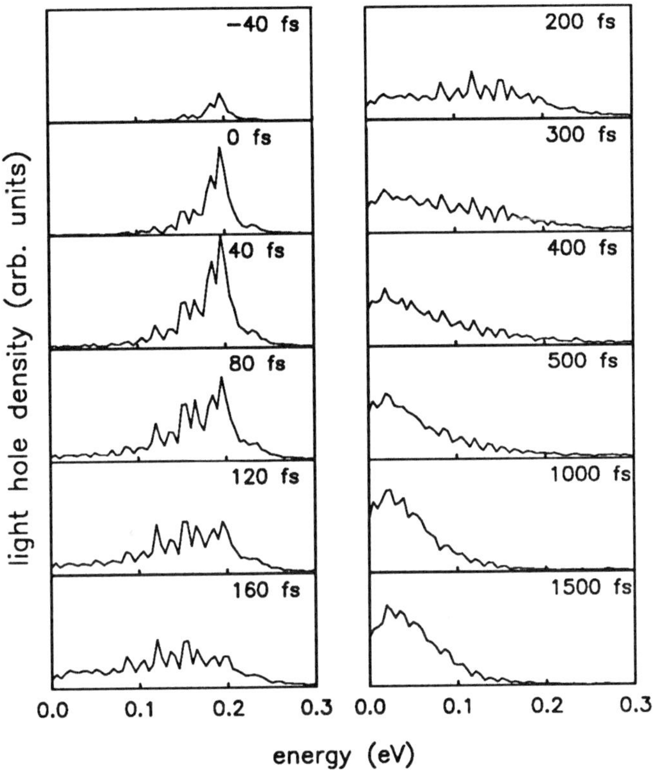

Figure 15. Time dependence of the light hole density from –40 to 1500 fs. At 0 fs, a significant peak in the light hole population develops. As time increases, the peak diminishes through hole-phonon and hole-hole scattering. At 300 fs, the light hole distribution function becomes thermal-like and can be characterized by an effective temperature, though substantially higher than the lattice. At later times, the temperature of the distribution decreases as can be seen from the increase in the slope of the tail of the distribution.

the light holes. During the first 100 fs, phonon replica peaks are clearly present in the distribution. By 200 fs, the distribution already looks Maxwellian. The split-off holes cool in energy as time increases. At longer times (> 400 fs) a depletion in the number of split-off holes is evident, much more so than for the light holes. This occurs because split-off holes scatter primarily into the heavy hole band.

In fig. 17, we reshow the evolution of all four carrier densities on single plots with density on the z-axis, energy on the y-axis and time on the x-axis. It is easier to compare magnitudes at the different energies when plotted in this way.

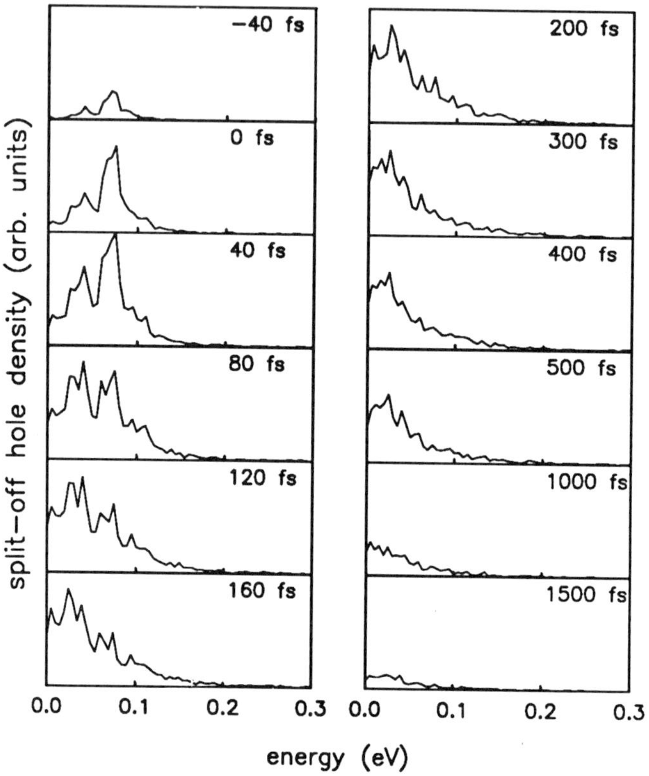

Figure 16. Time dependence of the split-off hole density from –40 to 1500 fs. An initial peak occurs at the center of the pulse at 0 fs. As time evolves, one can see the density relax through the emission of phonons and hole-hole scattering until it finally begins to become thermal-like at 200 fs. At longer times, the effective temperature cools and at very long times, the number of split–off holes decreases as they scatter to the heavy hole band.

Pump/Continuum Probe Absorption

As can be seen by looking at the carrier densities, they do not directly compare to the plots of the experimental nonlinear absorption (cf. fig. 12). To obtain the absorption or differential transmission from the distribution functions, one must use eq. 9.

Results for the calculation of differential transmission as a function of probe delay and energy are shown in figs. 18. The experimental results of Schoenlein et al.[3] are shown in fig. 18a. This is the same as fig. 12 except now all the different probe energies are shown on one plot weighted by the relative magnitude of the individual components. Figure 18b shows the results of our Monte Carlo calculations including electrons and heavy, light and split-off holes using the densities shown in figs. 13–16. Parameters in the Monte Carlo simulation have

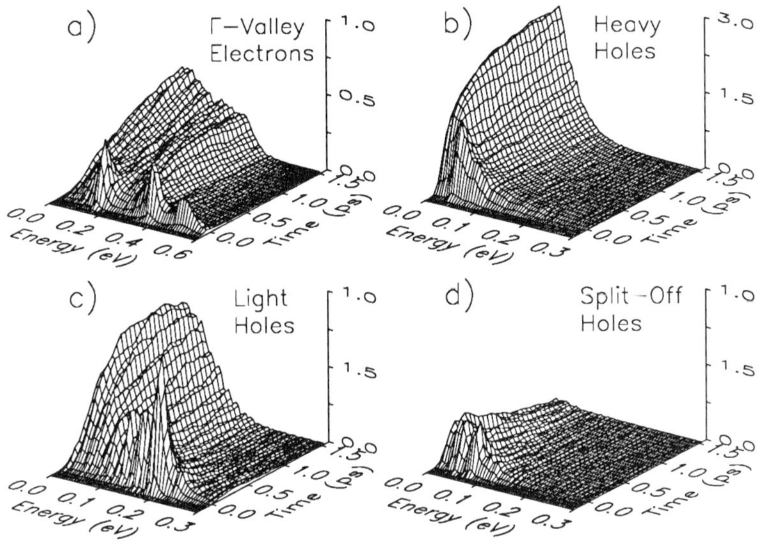

Figure 17. Carrier density as a function of energy and time for a) Γ-valley electrons, b) heavy holes, c) light holes, and d) split-off holes for photoexcitation by a 75 fs pulse centered at t=0. a) Three initial optical transitions are shown from (left–to–right) split-off, light, and heavy hole bands. Intervalley scattering occurs quickly for electrons excited from the light and heavy hole bands, suppressing the heights of the upper two peaks. b) A hot Maxwellian distribution forms quickly because of high scattering rates and because the transition occurs near the top of the band. c) Peaks resulting from phonon emission are shown. d) Intersubband scattering, mostly to the heavy hole band, causes the total density of holes in the split–off band to decrease.

not been optimized to achieve the best fit. Therefore, the agreement between theory and experiment, though not perfect, is excellent.

Figures 18c and 18d show the contributions to the differential transmission of only the holes and of only the electrons, respectively. The hole differential transmission is the sum of the contributions from heavy, light, and split–off hole bands. The hole differential transmission converges more quickly than the electron differential transmission, which shows a slowly increasing rate of rise in the contribution at low energy. The differential transmission for the electrons is lower at 1.73 eV than at 1.78 eV because the split-off transition is not allowed at the lower energy.

In fig. 19, we show the individual contributions to the differential transmission at a probe of 1.68 eV. As can be seen, the holes make a major contribution to the total differential transmission. In the key, c or v denote whether the contribution comes from carriers in the conduction or valence bands, and hh or lh denotes the optical transition as from the heavy or light hole. We see that the heavy holes contribute a substantial part to the initial change in transmission.

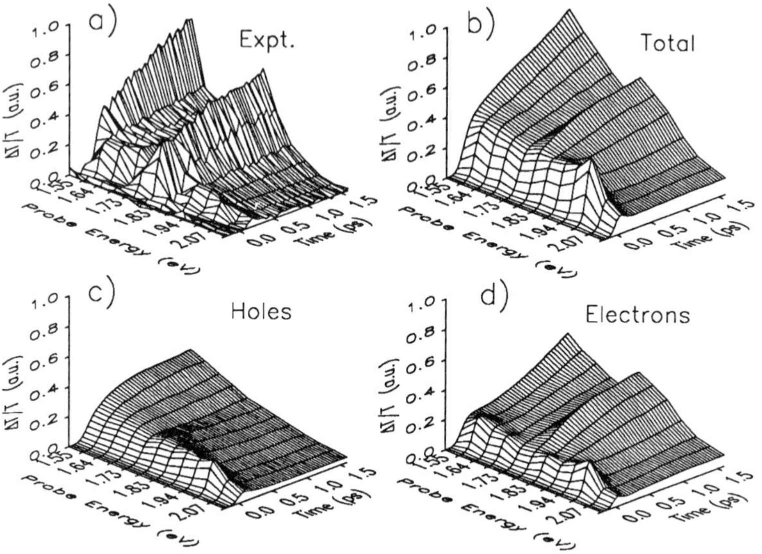

Figure 18. a) Experimental transient differential transmission spectra from Schoenlein et al.[3](cf. fig. 12) compared to b) calculated differential transmission spectra including both electrons and holes, c) calculated transmission spectra for the holes only, and d) calculated transmission for the electrons only. It is essential to include both the electrons and the holes to obtain agreement with experiment.

5. CONCLUSIONS

We have shown that the Monte Carlo technique can be an important tool for understanding and interpreting ultrafast laser spectroscopy experiments. Simple models based on fitting exponentials to the experimental response are subject to ambiguity when determining the relaxation rates. For quantitative comparison to experiment, calculating the distribution functions from Monte Carlo simulations is a good method, but by itself is insufficient. It is also important to determine the absorption (or luminescence) spectrum. This is because the optical spectra involve a sum over several allowed optical transitions weighted by the optical matrix elements and density of states.

Our results show that to accurately reproduce experimental results, the effect of both electrons and holes must be included. Also, because of warping of the hole bands and because photoexcitation can produce carriers substantially away from the band edge, a detailed calculation of the band structure is needed. In acceptor-luminescence experiments, anisotropy in the conduction band can also be seen.

To collect the most accurate information about a semiconductor system, it is helpful that more than one frequency be used in probing the system. This also provides a stringent test of any theory. While ideally one would like to isolate the optical transitions and directly measure the carrier distributions, owing to

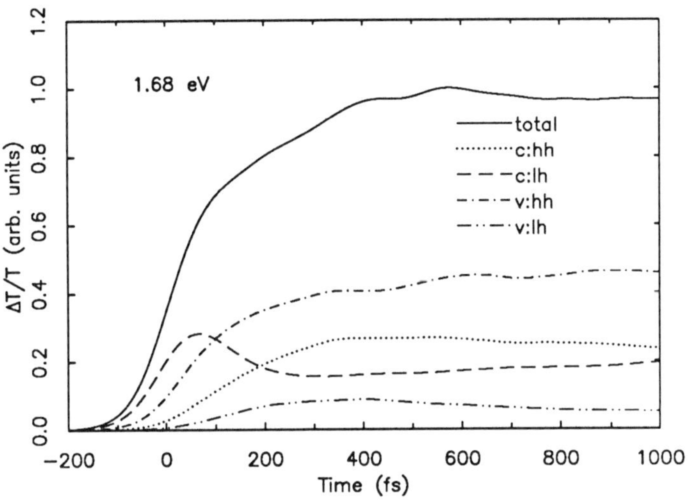

Figure 19. Relative contributions to the differential transmission in electron and hole bands for each transition for 2.0–eV pump and 1.68–eV probe. In the key, c or v indicates whether the contribution is from carriers in the conduction or valence (hole) band, and hh or lh denotes the optical transition as from the heavy or light hole. Because they are excited at the top of the band, heavy holes are responsible for a significant fraction of the initial rise in the differential transmission. On the long–time scale the electrons contribute the most to the differential transmission because they have a lower density of states than the holes.

the complicated nature of the experiments, detailed simulations using the Monte Carlo technique are needed to accurately interpret them.

6. ACKNOWLEDGMENTS

We are thankful to K. Hess, J. Fujimoto and B. Mason for useful discussions during the development of this work, and to M. Fisher for generating several of the figures. This work was supported by the National Science Foundation through Grant No. DMR8957382. Supercomputing time has been provided by the National Center for Computational Electronics at the University of Illinois.

7. REFERENCES

[1] E. M. Conwell, in *Solid State Physics, Advances in Research and Applications,*, edited by Seitz, Turnbull, and Ehrenreich (New York, 1967) Vol. 9.

[2] K. Hess, *Advanced Theory of Semiconductor Devices*, Prentice Hall, Englewood Cliffs, NJ (1988).

[3] R. W. Schoenlein, W. Z. Lin, S. D. Brorson, E. P. Ippen and J. G. Fujimoto, Appl. Phys. Lett. **51**, 1442 (1987).

[4] W. Z. Lin, J. G. Fujimoto, E. P. Ippen and R. A. Logan, Appl. Phys. Lett. **51**, 161 (1987).

[5] W. Z. Lin, R. W. Schoenlein, J. G. Fujimoto and E. P. Ippen, IEEE J. Quantum Electron. **QE-24**, 267 (1988).

[6] W. H. Knox, C. Hirlimann, D. A. B. Miller, J. Shah, D. S. Chemla and C. V. Shank, Phys. Rev. Lett. **56**, 1191 (1986).

[7] J. L. Oudar, D. Hulin, A. Migus, A. Antonetti and F. Alexandre, Phys. Rev. Lett. **55**, 2074 (1985).

[8] C. W. W. Bradley, R. A. Taylor and J. F. Ryan, Solid State Electron. **32**, 1173 (1989).

[9] M. Rosker, F. Wise and C. L. Tang, Appl. Phys. Lett. **49**, 1726 (1986).

[10] W. Z. Lin, J. G. Fujimoto, E. P. Ippen and R. A. Logan, Appl. Phys. Lett. **50**, 124 (1987).

[11] P. C. Becker, H. L. Fragnito, C. H. Brito Cruz, J. Shah, R. L. Fork, J. E. Cunningham, J. E. Henry and C. V. Shank, Appl. Phys. Lett. **53**, 2089 (1988).

[12] C. J. Stanton, D. W. Bailey and K. Hess, IEEE J. Quant. Electron. **QE-24**, 1641 (1988).

[13] P. C. Becker, H. L. Fragnito, C. H. Brito Cruz, R. L. Fork, J. E. Cunningham, J. E. Henry and C. V. Shank, Phys. Rev. Lett. **61**, 1647 (1988).

[14] M. C. Nuss, D. H. Auston and F. Capasso, Phys. Rev. Lett. **58**, 2355 (1987).

[15] B. Bebb and E. W. Williams in *Semiconductors and Semimetals*, edited by R. K. Willardson and A. C. Beers (Academic, New York, 1972), Vol. 7.

[16] B. P. Zakharchenya, D. N. Mirlin, V. I. Perel' and I. I. Reshina, Sov. Phys. Usp. **25**, 143 (1982).

[17] G. Fasol, W. Hackenberg, H. P.. Hughes, K. Ploog, E. Bauser and H. Kano, Phys. Rev. B **41**, 1461 (1990).

[18] R. G. Ulbrich, J. A. Kash and J. C. Tsang, Phys. Rev. Lett. **62**, 949 (1989).

[19] B. S. Wherrett, A. L. Smirl and T. F. Boggess, IEEE J. Quant. Electron. **QE-19**, 680 (1983).

[20] D. J. Erskine, Ph.D. dissertation (Cornell University, Ithaca, NY, 1984).

[21] C. L. Tang, I. A. Walmsley, and F. W. Wise, Appl. Phys. Lett. **52**, 850 (1988).

[22] J. Collet, T. Amand and M. Pugnet, Phys. Lett. **96A**, 368 (1983).

[23] J. Collet and T. Amand, J. Phys. Chem. Solids**47**, 153 (1986).

[24] M. Asche and O. G. Sarbei, Phys. Stat. Sol. (b) **126**, 607 (1984).

[25] M. Asche and O. G. Sarbei, Phys. Stat. Sol. (b) **141**, 487 (1987).

[26] C. J. Stanton, D. W. Bailey, K. Hess and Y.C. Chang, Phys. Rev. B **37**, 6575 (1988).

[27] M. J. Kann, A. M. Kriman and D. K. Ferry, Solid State Electron. **32**, 1831 (1989).

[28] M. A. Osman and D. K. Ferry, Phys. Rev. B **36**, 6018 (1987).

[29] F. H. Pollak, C. W. Higginbotham and M. Cardona, J. Phys. Soc. Japan **21** supplement, 20 (1966).

[30] D. W. Bailey, C. J. Stanton and K. Hess, Phys. Rev. B **42**, 3423 (1990).

[31] H.-J. Drouhin, C. Hermann and G. Lampel, Phys. Rev. B **31**, 3859 (1985).

[32] H. Shichijo and K. Hess, Phys. Rev. **B 23**, 4197 (1981).

[33] M. V. Fischetti and S. E. Laux, Phys. Rev **B 38**, 9721 (1988).

[34] K. Brennan and K. Hess, Phys. Rev. B **29**, 5581 (1984).

[35] K. Sadra, C. M. Maziar, B. G. Streetman and D. S. Tang, J. Appl. Phys. **66**, 4791 (1989).

[36] J. D. Wiley in *Semiconductors and Semimetals*, edited by R. K. Willardson and A. C. Beers (Academic, New York, 1971) Vol. 6.

[37] Even for $k = 0$, Wiley's expression is not exact since it neglects coupling to higher bands.

[38] C. Jacoboni and L. Reggiani, Rev. Mod.Phys. **55**, 645 (1983).

[39] An explanation of how the Bloch overlap factors are calculated from the eigenfunctions returned by the $\mathbf{k}\cdot\mathbf{p}$ program can be found in ref. 30.

[40] M. Costato and L. Regianni, Phys. Stat. Sol. (b) **59**, 47 (1973).

[41] M. Costato and L. Regianni, Phys. Stat. Sol. (b) **59**, 471 (1973).

[42] Y. C. Chang, D. Z.-Y. Ting, J. Y. Tang and K. Hess, Appl. Phys. Lett. **42**, 76 (1983).

[43] K. Kim, B. A. Mason and K. Hess, Phys. Rev. B **36**, 6547 (1987).

4

EXTENSIONS OF THE MONTE CARLO SIMULATION IN SEMICONDUCTORS TO FAST PROCESSES

D. K. Ferry, A. M. Kriman[a], and M.-J. Kann[b],
Center for Solid State Electronics Research
Arizona State University, Tempe, AZ 85287-6206, USA

and

R. P. Joshi
Department of Electrical and Computer Engineering,
Old Dominion University, Norfolk, VA 23529-0246, USA

The modeling of relaxation and transport in semiconductors is often performed using Monte Carlo techniques in which the scattering events are defined to include carrier-phonon interactions and Coulomb interactions among various carrier species and the ionized impurities. We consider situations in which this approach is inappropriate, and describe corresponding implementations of a more accurate technique for the Coulomb interaction in which the usual Monte Carlo technique is combined with a molecular dynamics time evolution between scattering events. In this approach, the Coulomb interaction is augmented by an exchnage Hamiltonian so that the particle dynamics incorporates quantum corrections.

INTRODUCTION

With the current interest in ultrahigh speed electronic and optoelectronic devices, it is important to understand the relaxation dynamics of nonequilibrium carriers in GaAs and other compounds. Advances in the compression techniques of laser optics have generated laser pulses as short as 6 fs, which allow us to extend the related optical techniques into the few-femtosecond regime. These techniques have been used to measure the polarization dephasing rate, and the initial exponential decay time constant for carriers in the excitation volume (in phase space) in semiconductors [1-6]. Ultrafast scattering and relaxation processes in semiconductor materials, which occur on a time scale comparable to the pulse duration, affect the behavior and contribute to the performance limits of electronic and optoelectronic devices.

While optical studies are of interest in their own right, it is important to note that as the dimensions of electronic devices reach the submicron regime, the energy and momentum losses due to carrier-carrier interactions begin to play a crucial role in device performance [7]. Carrier-carrier scattering, while conserving the total energy of the carriers, leads to evolution of the distribution function into a Fermi-Dirac-like distribution at a carrier temperature hotter than the lattice temperature. Investigations of hot photoexcited carrier relaxation in semiconductors have shown that a quasi-equilibrium energy distribution is developed at a carrier temperature far above the lattice temperature in less than 1 psec [7]. In order to understand how this thermalization is established on the femtosecond time scale through the interplay between carrier-carrier and carrier-phonon scattering, careful ensemble Monte Carlo (EMC) calculations are used. It is known that these EMC calculations are an appropriate technique for studying transient femtosecond dynamics [8,9].

The simulation of a number of systems, such as those discussed above, is beginning to require a new class of numerical models [10-12], which combine a traditional Monte Carlo treatment of phonon-charge carrier scattering with molecular dynamics simulations of the charge carrier motion between scattering events. In order to explain the situations that require this more complex modeling, we briefly review the physical assumptions that have heretofore usually permitted simpler treatments.

We consider first events that involve only non-Coulombic scattering involving single electrons (optical-phonon emission, absorption and scattering, acoustic-phonon Umklapp processes, impact ionization). In a system of charge carriers which is evolving sufficiently slowly, many such scattering events occur over a time long compared to the time scale for overall changes in the charge carriers' momentum distribution. In these quasi-equilibrium situations, an approximate ergodic behavior holds, and straight-forward Monte Carlo simulations of a single electron or hole over a short interval can be used to model the behavior of a finite number of electrons or holes (this is discussed in many chapters in this book).

On shorter time scales, which depend on the system being modeled, the electron distribution undergoes a significant perturbation in the typical time between scattering events. For example, pump-probe experiments have been conducted with laser pulses compressed to times on the order of 10 fsec, and with photon energies of 2-3 eV. During the first picosecond of relaxation, the photoexcited electrons lose roughly 10% of their energy every 100 fsec by emitting a phonon. During this initial period of ultrafast relaxation, a finite density of electrons (holes) is modeled by an "ensemble" of identical electrons (holes). This extension is distinguished as *ensemble* Monte Carlo (EMC). Because different carriers in the ensemble, with different momenta, have different rates for various scattering processes, the Monte Carlo time step would apparently depend on the element of the ensemble. Thus, time-evolving an ensemble presents a difficulty in vectorization. This difficulty is overcome by the method of "self-scattering." In this method, every carrier has a scattering time equal to that of the most-frequently-scattering carrier. At each Monte Carlo step, the same method that is used to assign randomly the particular mechanism by which scattering occurred also functions as a kind of rejection scheme: infrequent scatterers have a large probability of being assigned the fictitious "self-scattering" mechanism, which involves no actual change in momentum. With this method, EMC is vectorizable, and consequently more efficient than ordinary Monte Carlo.

If we now consider effects of the Coulomb interaction (carrier-carrier scattering, ionized-impurity scattering, and screening), it is clear that the best that can be hoped is that these introduce no further complication than the single-electron scattering mechanisms. Carrier-carrier scattering requires that multiple carrier momenta be available for sampling, so the base level of simulations in which that mechanism is significant is EMC. In general, a treatment similar to that for the non-Coulombic scattering mechanisms is possible only if the Coulombic scattering events can also be treated as essentially instantaneous. Since the Coulomb interaction is long-ranged, this is in the simplest approximation very difficult. Fortunately, the interaction between two point charges is screened not only by bound charges of the semiconductor, but also by the mobile charge of the other carriers in the system. This reduces the Coulomb interaction to one of finite range, falling off exponentially at infinity.

Various circumstances can make the simple EMC approach inappropriate. Those that we describe here fall into three partially overlapping categories: Dynamical conditions which make it impossible to compute an accurate screening function, and in which the scattering cannot be regarded as a two-particle event; large scale of the scatterer or scattering potential, relative to the screening length, which implies that there is no "time between collisions;" breakdown of the ordinary semiclassical approximation for the electronic motion. We consider the first two issues in detail in Section 2, where we describe the combined EMC/molecular dynamics (EMC/MD) approach used to overcome the deficiencies in the pure EMC approach. In Sections 3 and 4 we describe stationary and dynamic situations, respectively, which require the EMC/MD treatment for the classical-physics reasons just stated. In Section 5, we describe quantum corrections within an EMC/MD approach which make it possible to include quantum exchange effects in a Monte Carlo simulation.

DYNAMICAL EFFECTS, MULTI-PARTICLE SCATTERING, AND THE EMC/MD APPROACH

Let us assume that the Coulomb interaction leads to isolated scattering events. That is, we assume that the interaction can be treated in terms of essentially instantaneous scattering events between pairs of point charges, and that between these events the carriers have well-defined momenta. Following the discussion of the introduction, the ordinary EMC method requires only the computation of a dielectric (screening) function, $\varepsilon(\mathbf{q},\omega)$, which depends on the wave vector $\mathbf{q}$ and the frequency ω. This function is extremely complicated, and in fact depends on a knowledge of the distribution function of the charge carriers. (Cf., for example, the Lindhard-RPA formula.) As such, it must in principle be calculated concurrently in the EMC, as the carrier distribution evolves. In addition, the scattering rate for a Coulomb scattering event which occurs over a time τ must, again in principle, be computed using ε for a range of frequencies on the order of $1/\tau$. In practice, a number of approximations are made to obtain a dielectric function which is analytic, or at least tractable, and which does not require a recomputation based on the evolving carrier distribution. Using simple arguments, we can show that any approximation of this form must break down.

As a simplest approximation, we may observe that the screening length q_d^{-1} near equilibrium is given, in the limit $\mathbf{q}\to 0$, $\omega \to 0$, as

$$q_d^2 = ne^2/\varepsilon_0 E \ , \tag{1}$$

in both the Debye-Hückel (non-degenerate) and Thomas-Fermi (degenerate) limits. In (1), n is the density of carriers, ε_0 is the dielectric constant, and E is a characteristic energy: $E = k_B T$ for Debye-Hückel, and $3E_F/2$ for Thomas-Fermi screening. The dependence on energy, and on the form of the carrier energy distribution, already suggests how unlikely it is that an accurate dielectric function can be found without utilizing the distributional information in the simulation. In a semiconductor, this screening length must be taken to be the long-wavelength limit of a realistic screening function and, in fact, overestimates the screening found in most transport measurements. We discuss this below.

If we consider in particular a highly excited electron plasma of the sort occurring in a typical laser pump-probe experiment, then the density is initially nondegenerate and the electron density can be varied without appreciably changing the average energy. Holding E fixed, then, the number of electrons within a screening length varies as $n^{-1/2}$. Thus, for sufficiently high densities, the screening represents the time-averaged effect of as few as one electron. A dielectric function is therefore an extremely unrealistic way to treat the phenomenon. When the interaction between two electrons is screened, what is actually occurring is that the nearest one or two electrons are also interacting. These nearby electrons do not move any faster, on average, than the electrons whose interaction they are regarded as screening, so that the screening function is only isotropic in the average over many collisions.

What becomes clear in this analysis is that the screening of two electrons within a high-density plasma cannot be regarded as a statistical phenomenon, but rather is a problem of a few (of the many) electrons interacting simultaneously. In some respects, two-particle interactions cannot approximate this even qualitatively: Two-electron scattering does not change the (equal) magnitudes of the electron momenta, whereas multi-electron scattering does.

As an alternative to implementing screened two-particle Coulomb scattering in EMC, a different method has been developed by our group. In this method, the Coulomb interaction is not treated as a sequence of discrete collisions, but rather as a continuous-time evolution of the momenta *and positions* of the ensemble electrons, simulated by molecular dynamics (MD). For a simulation using N particles (2000-10000, typically) the particles are initially distributed randomly in a box of volume $V = N/n$; with periodic boundary conditions allowing the electron to move freely without any influence from artificial (simulation region) boundaries. The two kinds of time evolution are combined in a simultaneous EMC/MD simulation in which MD updating of the carrier coordinates is punctuated at regular intervals by EMC simulation of the scattering. In this approach, the interaction between individual charge carriers is the bare (unscreened) Coulomb interaction, and the screening enters implicitly through the cooperative motion of all the electrons. This situation is essentially that of returning from the macroscopic Maxwell equations to the microscopic ones: the polarizability is that of the vacuum, and the polarization charge is treated explicitly, rather than as a linear response of the material background.

One problem that reappears is that of the long range of the Coulomb interaction. Formally, of course, one no longer deals with cross-sections, so the infinite cross-section of the unscreened Coulomb potential is not directly a problem. However, the force exerted by an infinite gas of electrons converges only conditionally, which can lead to mathematical pathologies in a finite simulation. For example, if one computes the force on an electron at a position $x = x_0 > 0$ by considering only the (uniform) density in the region $-L < x < L$, $(L > x_0)$, then one finds a force proportional to x_0, *independent of* L.

In order to avoid these problems and obtain a physically reasonable sum of forces, we impose periodic boundary conditions on the box, and compute the force exerted not only by every charge in the simulation region, but also by all of the periodic images of each simulated charge. This sum can be performed systematically by a method of Ewald, developed for finding forces in a crystal (we do not implement the full Ewald sums, as the momentum space corrections for Fourier components introduced by the artificial periodicity are found to be negligible for the large number of particles included in the ensemble). This procedure computes the Coulomb force only through the shortest connecting vector between each pair of carriers, so that in the simulation two different primitive volume cells must be considered. The first primitive cell is the standard, lab-fixed simulation region. The second volume is defined for each carrier as being a translate of the first volume, centered on that carrier [13, 14]. This volume ensures that each of the other simulated particles interacts with the particle of interest through the shortest distance between them (so that the latter may occur in one of the replications of the first volume). Since the simulation volume may not correspond to each electron interacting through the shortest distance with the carrier of interest, replicas of the electrons that lie in the second cell must be used to compute the force through the well-known

$$\mathbf{F} = \frac{-e^2}{4\pi\varepsilon_0} \sum_{\mathrm{i}} \frac{1}{r_{\mathrm{i}}^2} \mathbf{a}_{ri} \tag{2}$$

where r_i is the distance to the interacting particle (from the reference particle) and $\mathbf{a}_{ri}$ is a unit vector pointing *to the interacting particle*, with both particles assumed to have the same charge.

It is clear that the use of a centered volume eliminates, by symmetry, the pathology described earlier. It is clear that an off-center test particle in a simulation box would experience a net force, from the other $N-1$ carriers, that was dependent upon its position within the box. The desire is to sum only over the actual N electrons used in the summation. The *total* potential arising from the interparticle potential can be written as

$$\Phi = \frac{e^2}{4\pi\varepsilon_0} \frac{1}{2} \sum_{\mathrm{i} \neq \mathrm{j}} \frac{1}{r_{\mathrm{ij}}} \tag{3}$$

where the factor of 2 arises from the double counting of each pairwise interaction in the summation over both i and j. The summation in (3) runs over all particles in the solid, both those within the basic supercell *and those that lie in any replica of the basic supercell.* The term for the particles within the basic cell and its replicas contribute to (3) the terms

$$\sum_{\text{cells}} \frac{1}{r} \rightarrow \frac{1}{r} + \sum_{L>0} \frac{1}{|\mathbf{L}+\mathbf{r}|}$$

$$\approx \frac{1}{r} + \sum_{L>0} \frac{1}{L} - r \sum_{L>0} \frac{1}{L^2} + r^2 \sum_{L>0} \frac{1}{L^3} , \tag{4}$$

where the denominator has been expanded since, in general, $L > 2^{1/2} r$ when the centered cell is used for the direct force calculation. The method of evaluating these interparticle forces and incorporating the summations over the lattice vectors to account for the longer-range replicas of the principal supercell have been studied for some time in connection with molecular dynamics calculations in other fields [15]. Note that when (4) is multiplied by L, so that the expansion is in terms of the normalized distance L/r, the resulting sums are independent of r. Rather, they are types of Ewald sums over the defining vectors of the superlattice and become constants that are well known for each choice of "crystal structure" for the boxes. With the expansion (4), the potential of (3) and the force of (2) now need only be calculated over the actual number of particles used in the basic cell, provided that the particle-centered box is used in computing these sums. What these sum terms contribute to the force (2) is correction terms that compensate for the fact that a finite simulation volume introduces certain Fourier periodicities that can upset the force calculation's accuracy, as well as a limiting of the number of particles available for screening the interaction. For example, the leading correction to the potential is a term linear in r, rather than inversely proportional to r [15]. If the number of particles is small, say less than 100, these corrections are large. However, if the number of particles used is large, the corrections are small. This is translated to the casual observation that the primitive supercell volume should have an edge that is significantly longer than the screening length, and this in turn implies that a large number of particles must be used if finite-size effects (higher-order terms in the Ewald sum) are to be avoided. Under these simulation conditions, the finite size of the simulation volume becomes unimportant in the overall EMC/MD calculations.

The EMC/MD clearly does not depend on the validity of various assumptions we questioned above: The Coulomb interaction need not occur in brief events describable as collisions, and *a fortiori* those collisions need not be approximable as involving only two charge carriers. Other assumptions of the direct EMC scheme, also remedied by the EMC/MD approach, will be discussed in subsequent sections.

IONIZED-IMPURITY SCATTERING

As a test of this approach, it is useful to consider a situation in which a proper modeling of the Coulomb interaction alone can be expected to lead to good agreement with measured data. Such a test is provided by the low temperature mobility of doped semiconductors. At low temperatures, Coulomb scattering from ionized impurities is the dominant mechanism which controls electronic transport. Low optical-phonon occupancy likewise makes phonon scattering unimportant, and so calculations do not depend on poorly-known electron-phonon matrix elements. Instead, the few parameters which need to be known accurately for a calculation — impurity concentrations, electron effective mass — are measurable and known.

When one considers, as we will do, primarily the case of n-doped semiconductors, there is a simplifying feature that might be expected to make traditional EMC approaches more accurate: Since only electrons are present, carrier-carrier scattering preserves not only momentum but average velocity, and so carrier-carrier scattering does not have a strong direct effect on the current. (It is assumed that fields are small enough that conduction band nonparabolicity can be ignored.)

The arguments above, regarding the dynamical nature of the screening by electrons, was framed for concreteness in terms of the screening between two electrons. However, the arguments still hold when the interaction is between an electron and an ionized impurity. In some respects, the situation is worse: In the case of electrons screening electrons, the average distance between electrons decreased with increasing density ($\sim n^{-1/3}$) at the same time that the static screening length decreased ($\sim n^{-1/2}$). The latter decreased more rapidly, leading to the catastrophe of less than one electron screening the interaction. In the case of ionized impurity scattering, one must deal with compensation. In a compensated semiconductor, the concentration of ionized impurities exceeds the concentration of screening electrons, so that this high-density catastrophe begins to occur at lower densities.

The Brooks-Herring approach [16] is essentially the traditional approach outlined earlier, in which scattering events occur between pairs of charges (ion and carrier), interacting via a screened potential. Refinements of this method [17,18] have improved on the screening function, or gone beyond the first Born approximation in computing the differential scattering cross section [19-21] but continue to assume two-particle scattering. Despite this simplifying assumption, the approach still has the drawback that increased accuracy comes at a cost of increasingly complicated scattering cross-sections. These approximations can also fail to satisfy fundamental constraints. For example, Blatt's treatment [19], based on a phase-shift analysis of the scattering, fails to satisfy the Friedel sum rule [22,23].

One concrete way to recognize the importance of multi-ion effects — viz., the failure of the two-particle scattering approximation — is to consider two ions very close together. If these ions have opposite charge, they are essentially a dipole, and their s-wave phase shift vanishes. If they have identical charge, their s-wave shift is increased. In a treatment that assumes two-particle scattering, however, there can be no distinction on the basis of charge sign, because the computation of scatterng cross-section assumes that the two scatterers are asymptotically far apart, and scatter incoherently.

There is, unfortunately, a superficially similar issue which obscures this failure and which we briefly address: The differential scattering cross-section (DSCS) of the bare Coulomb potential is independent of the relative sign of the scatterers. The first term of the Born series (which does not converge for this potential) also gives exactly the correct DSCS for the bare Coulomb interaction. On the other hand, the exponentially screened Coulomb potential has a DSCS which *does* depend on the relative sign of the scatterers, although its first Born approximation *does not*. The Born series converges for the exponentially screened potential, and so a Born series, as well as other methods, can be used to correct the scattering cross section and find a more accurate, sign-of-charge-dependent DSCS. Notwithstanding these improvements, it is important to bear in mind that after all of this is done, the multi-ion scattering effects described in the previous paragraph can remain completely unaccounted-for.

Previous treatments have given some consideration to multi-ion effects. For example, the classical Conwell-Weisskopf treatment [24] uses a bare Coulomb interaction; the total cross section is kept finite by imposition of a cut-off for impact parameters smaller than half the average distance between ionized impurities. Ridley [25] working in a partial wave expansion, introduced a cut-off associated with "third-body interference." In each case, the main accomodation made for multi-ion scattering was simply to assume the scattering was negligible when it was too complicated to compute. This is precisely wrong if two nearby ions of the same sign of charge are involved in scattering, and leads to an overestimate of the mobility.

We have used the EMC/MD approach to model the density-dependent ohmic mobility of n-type GaAs at 77 K [26]. The lattice temperature represents a compromise between keeping the carrier-phonon scattering rate low to emphasize impurity scattering, and avoiding complications associated with carrier freeze-out. At this temperature also, the hole concentrations are insignificant. The Monte Carlo component of the simulation takes both acoustic and optical phonon interactions into account, and k-space carrier scattering is treated through the usual energy-dependent formulations based on a Fermi Golden rule calculation [27]. In addition, treatment of non-equilibrium phonon amplification is included, as has been discussed elsewhere [28].

The initial k-space distribution of the free particles is based on an equilibrium Fermi-Dirac distribution, which is implemented using a standard rejection algorithm. In order to follow the motion of the electrons accurately, a small time step has to be used; 0.5 fsec was found to be adequate, and is short compared to the typical distance traveled between collisions (however defined). The number of electrons N used in the simulation was 2000. This is orders of magnitude larger than has been used in some previous simulations [14], and has proven adequate for other comparisons with experimental data [29,30]. A few test simulations have been run with 10000 particles, and it is found that the corresponding improvement from the $N = 2000$ results is not significant. After 3.0 psec, the system reached a steady state (under a very weak field: 200 V/cm), from which a drift velocity or mobility can be extracted.

For comparison with the present EMC/MD approach, a computation based on a modified Brooks-Herring approach of Ridley [25], and of van de Roer and Widdershoven [30], was also performed. The scattering rate used in this scheme, $\Gamma_{mod}(E)$ at an energy E, is given in terms of the usual rate $\Gamma(E)$ as

$$\Gamma_{mod}(E) = \frac{v(E)}{L_A} \times \left[1 - \exp\left(\frac{-L_A \Gamma(E)}{v(E)} \right) \right] , \tag{5}$$

where $v(E)$ is the velocity magnitude, and $L_A = N_i^{-1/3}$ is proportional to the average distance between ions (which are distributed randomly with a density N_i). The scattering rate used was that found within the static Debye approximation:

$$\Gamma_{mod}(E) = \frac{N_i Z^2 e^4}{2\pi\varepsilon_0^2 \sqrt{2m^*}} \times \frac{(1+2\alpha E)\sqrt{\gamma(E)}}{E\beta\,[\gamma(E)+E\beta]} , \tag{6}$$

where $E_\beta = \hbar^2 q_d^2 / 2m^*$, α is the (isotropic) nonparabolicity factor, and $\gamma(E) = E(1+\alpha E)$.

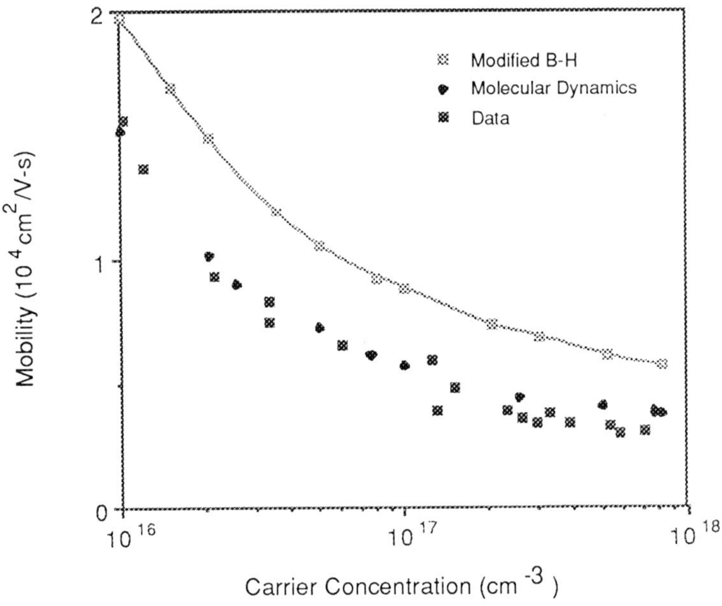

Fig. 1 Density-dependent mobility for uncompensated GaAs at 77K. Simulation results were obtained by using the EMC/MD method and by using a modified Brooks-Herring method. Experimental data of G. B. Stringfellow [29] are plotted for comparison.

Figure 1 shows a comparison of the two theoretical calculations with some experimental data for uncompensated GaAs [29]. The experimental data clearly favor the EMC/MD approach. The failure of the modified Brooks-Herring approach is also qualitatively consistent with the idea that multi-ion effects should be considered: the mobility is expected to be higher, since the scattering has been artificially cut off as a way to account for multi-ion effects.

The EMC/MD calculations were also performed for a compensated situation, for n-GaAs in which the combined concentration of acceptors and donors was three times the electron concentration. A comparison with the uncompensated semiconductor electron mobility is made in Figure 2. Not surprisingly, the mobility was reduced by compensation: For the same level of ionized impurity concentration, the free electron density is much lower with compensation, and so leads to a reduction in the screening of long-range Coulomb potentials. The role of collective contributions from the ionized impurities is therefore higher, and greater momentum randomization occurs. Consequently, mobilities at the lower impurity concentrations are reduced almost by a factor of two from their values in the uncompensated material. Based on these results, one may conclude that material compensation must have been low in the actual experimental samples, and that multi-site effects are the major cause of mobility lowering relative to the Brooks-Herring types of calculations.

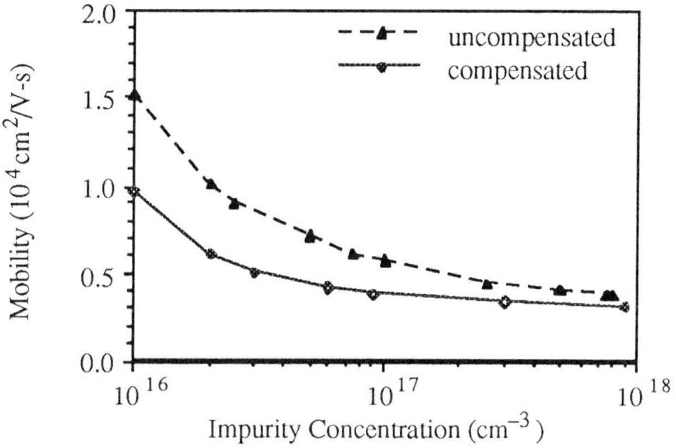

Fig. 2 Mobility of n-GaAs at 77K, as a function of the impurity concentration, for uncompensated GaAs and for a compensation ratio of 3.0.

FEMTOSECOND LASER EXCITATION

Investigations of hot photoexcited carrier relaxation in semiconductors have shown that a quasi-equilibrium energy distribution is developed at a carrier temperature far above the lattice temperature in less than 1 psec [7]. However, the initial stages of carrier relaxation occur through an interplay of both carrier-carrier and carrier-phonon scattering, so that the understanding of the initial rapid cooling observed experimentally requires the knowledge of how this thermalization is established on the femtosecond time scale [31-33]. The experiments which allow us to study this femtosecond transport regime, and to extract those parameters, are based on excitation of solid-state plasmas by laser pulses that have been compressed to times as short as 6 fsec. The relaxation of the excited plasma is subsequently studied by absorption bleaching, luminescence, and other techniques [1-6].

The experimental results [2] of probing the fast scattering of carriers out of the central valley using pump-probe techniques with two 6 fsec pulses centered at 2 eV, show that the scattering time, which is primarily determined by scattering into both the X and L valleys, was measured to be 33 fsec for densities in the range $(3\text{-}6)\times10^{18}$

cm^{-3}. The time resolution of the ultrashort 6 fsec laser pulse has made it possible for them to investigate directly the dynamics of intervalley scattering in GaAs. Shah *et al.* [34] reported a slow rise of luminescence in GaAs after excitation by a subpicosecond laser pulse, due to a slow return of electrons from the L valley to the central valley. By fitting the data with an EMC calculation, they determine the Γ-L deformation potential to be $(6.5\pm1.5)\times10^8$ eV/cm, which confirms the measured value used here (7×10^8 eV/cm) [35]. Transfer of the carriers to the satellite valleys represents a storage of energy, which is released as kinetic energy when those carriers return to the central valley. The electrons returning to the central valley act as a source of heating for the photoexcited plasma and thus slow the cooling of the electron gas.

Experiments in which faster decay rates are observed must entail another process, such as carrier-carrier scattering. With techniques of increased time resolution, Becker *et al.* [1] report the observation of femtosecond photon echoes from direct transitions in GaAs. The echo decay time constant was found to vary from 11 down to 3.5 fsec for carrier densities ranging from 1.5×10^{17} to 7×10^{18} cm^{-3}. This allowed them to determine the polarization dephasing rate (four times the echo decay time constant). The carrier density dependence of the dephasing rate indicates that carrier-carrier scattering is the dominant dephasing mechanism in the rapid initial stage of relaxation and yields previously unavailable information on Coulomb screening in the nonequilibrium plasma. Collisions involving both electrons and holes can dephase the polarization of an electron-hole pair. At high carrier densities the carrier momentum loses phase coherence primarily due to the screened Coulomb interaction between carriers. Given the time scales, it is clear that dynamical screening effects play an important role and must be incorporated in some way into Monte Carlo simulations. This problem is therefore also a sensitive test of the extent to which EMC/MD is an improvement over more traditional screening treatments.

As noted earlier, the part of the screening that is difficult to treat arises from the deformation of the charge distribution of the conduction electrons. Existing theories of screening are valid only for near-equilibrium carrier distributions. Currently a good theory does not exist to describe the process of dynamic screening in such highly nonequilibrium carrier distributions, and most analytical approaches with carrier-carrier scattering do not fully incorporate the energy scattering inherent in this process, the complex far-from-equilibrium distribution functions, and are handicapped by approximations to detailed time-dependent screening. In particular, when the assumptions of local equilibrium, linear response, or of the adiabatic motion of the screening electrons cease to be valid, the various screening models are not valid in our problem.

The present simulation incorporates only the electron-electron interaction. This is a justifiable approximation, however. Because the holes are much heavier than the electrons, and also much colder, the center-of-mass and laboratory frames for an electron-hole pair essentially coincide, and electron-hole scattering is approximately elastic. In other words, scattering from holes approximates ionized impurity scattering, so that essentially no energy exchange is involved and the event contributes primarily to momentum relaxation. This was confirmed numerically in earlier work by Osman and Ferry [7] on the role of the electron-hole interaction on the picosecond time scale, which showed that the dominant part of the distribution relaxation occurs by electron-electron scattering. Electron-hole recombination is also negligible in the first picosecond after excitation, and that process is excluded from our simulations. However, it is not at all clear that the neglect of the *screening* role of the holes is justified, and further work is necessary to assess this. Moreover, as is evident from the previous section, it would not be correct to ignore the holes if we were interested in the mobility or velocity of the electrons.

The neglect of electron-hole scattering is realized in our simulations simply by not including within the MD computation the electrostatic forces exerted upon the electrons by holes. Since we are examining the electron statistics only, the MD does not need to time-evolve hole positions. As discussed earlier, within the MD/EMC approach the screening is determined in the same way as the carrier-carrier scattering: both arise in a unified manner from the explicit trajectories of the charged-carrier ensemble. Thus, by excluding the electron-hole scattering we also exclude the holes' contribution to the screening function (which, as mentioned above, may not be valid). We justify this on the basis that the holes are heavier, so that their contribution to the high-frequency screening function is small. In general, of course, hole effects cannot be ignored. In particular, momentum relaxation effects are expected to be important in high fields, where they mediate energy exchange between carriers and field.

We simulated the time evolution of electrons in GaAs at 300K, excited by a 20 fsec pulse of 2 eV photons. We use masses, coupling constants and other physical parameters [36] that have been confirmed by several measurements using a variety of techniques. The pulse profile was rectangular, but for most analytical purposes this approximation does not have a large effect, and this process allows a more direct evaluation of scattering rates. For the photon energy chosen, the initial electron distribution is generated from all valence bands (light, heavy, and split-off). About 50% of the electrons are generated from the heavy hole band. The electron population was followed in the central Γ and satellite L and X valleys. Excitation is direct only into the Γ valley, and electrons enter the satellite valleys only by subsequent phonon scattering.

Some qualitative results of our simulations provide a purely empirical argument supporting the importance of Coulomb interactions: While at one point, more than half of the electron population is in the satellite valleys (see Fig. 3), the time for this is reached long after the excitation region is depleted to a few percent of its initial occupation. (The excitation region is the region of electron momentum space originally populated by photoexcitation.) The rate of polar-optical scattering events can be computed with greater confidence than intervalley scattering rates (if only because the electron-phonon matrix element is more accurately known), and this process also does not play an important role in the initial relaxation from the excitation region. In the pump-probe experiments, due to the large bandwidths of the ultrashort pump and probe pulses, optical phonon emission or absorption by either electrons or holes does not make a significant contribution to the initial relaxation.

Ultimately, therefore, only electron-electron interaction can explain the initial relaxation.

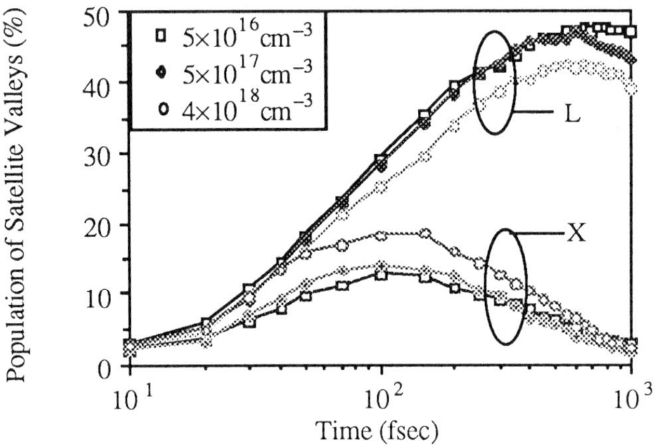

Fig. 3 Population of the satellite valleys for three densities. Upper three curves show L valley population, lower three show X valley. In this and all the subsequent graphs, symbol shapes correspond to densities according to the rule: squares – 5×10^{16} cm^{-3}, diamonds – 5×10^{17} cm^{-3}, circles – 4×10^{18} cm^{-3}.

One prediction of the model which is particularly sensitive to the electron-electron interaction, and therefore to the electron density, is the rate at which the electrons leave the Γ valley to enter the satellite valleys. From Fig. 3, we find that X valley population peaks around 160 fsec which seems to agree with results inferred by Alfano's group [37], who reported evidence of the electron scattering rate to the X valley within 170 fsec by using femtosecond pump and probe absorption measurements. The effects of a density-dependence can be seen in the different satellite-valley occupancies occurring at different excitation densities (Fig. 3).

Returning to the empirical argument, we note that in principle, electrons might relax from the excitation region by a mechanism that involved rapid transitions into and back out of the satellite valleys. This is a possibility because the short-time transition rates need not be such as to produce a distribution exponentially weighted toward lower energies. In contrast, at long times the approach to a thermal distribution makes transition rates back into the Γ valley much greater than those into the higher-energy satellite valleys, so the late occurrence of peaks in Fig. 3 immediately rules out intervalley scattering as a significant contributor to the initial relaxation.

In the highly nonequilibrium femtosecond regime, occupancy data like those shown in Fig. 3 are insufficient to determine transition rates. Our simulations allowed carriers to be exchanged between every pair of valleys. The population of the L valley, for example, includes electrons that have arrived there after passing through the X valley as well as those that arrived directly from the Γ valley. Clearly, three valley occupancies (two independent, by particle conservation) cannot yield six intervalley transition rates. It is necessary to investigate directly in the code the rate of intervalley transitions.

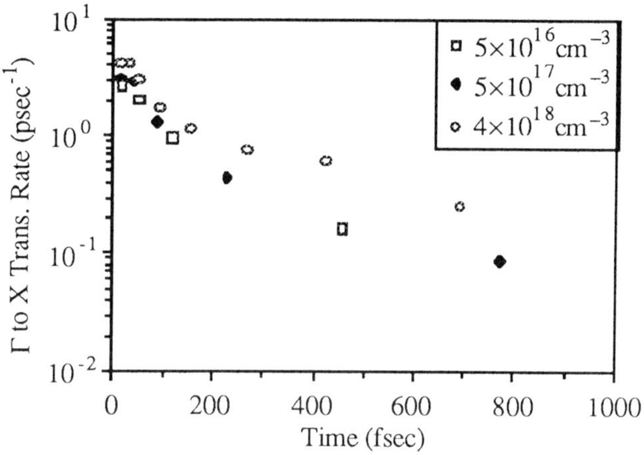

Fig. 4 Transition rates from the Γ valley to the X valley, for three electron densities.

The Γ→L rates are found to show no statistically significant density dependence. This is what one would expect in a simple picture, since the absorption of a phonon by deformation potential interaction is a single-electron process. In contrast, the Γ→X rates show a strong density dependence. Figure 4 shows the transition rate $R_{\Gamma \to X}$ for three simulations spanning almost two decades of density. The mechanism of this effect is the electron-electron scattering in the Γ valley: The threshold for transfer into the X valley lies at the upper edge of the excitation region. Even after accounting for thermal broadening, less than one tenth of the photoexcited electrons have sufficient energy to make the transition. At high density, electron-electron scattering is increased and the energy distribution is broadened. Because electron-electron scattering preserves the total energy within the electronic system, this broadening is approximately symmetrical towards greater and smaller energies. This increases the fraction of electrons above the threshold for Γ→X scattering. Transitions to the L valley do not exhibit a similar effect because, for the 2 eV laser energy used, essentially all of the photoexcited electrons are above threshold. Note that, although the Γ→L transition rates themselves are not appreciably dependent on the density, there is some density dependence in the L valley occupancies in Fig. 3 due to the effect of L↔X. At high densities, competition from Γ→X processes depletes the Γ valley, so virtually unchanged Γ→L rates yield a lower L valley population. Nevertheless, any density dependence in the $\Gamma \to L$ rates due to increased electron density in X is too small to detect at the current level of precision. The $X \to \Gamma$ and $L \to \Gamma$ rates are in the range of $(2\text{-}3) \times 10^{12}$ and $(6\text{-}10) \times 10^{11}$ sec^{-1}, respectively, for three different densities. The scattering time constants are approximately 410 fsec and 1.3 psec for $X \to \Gamma$ and $L \to \Gamma$ transitions respectively.

At later times, the decay of the transition rate represents cooling of electrons in the Γ valley as a whole. The Γ valley loses energy in two ways: polar-optical phonon (POP) scattering and preferential transfer of the hottest electrons into satellite valleys. Since electrons just excited from the heavy-hole band typically have ~10-$15\hbar\omega_0$ above the conduction band edge, and since POP scattering occurs at a rate slower than that of intervalley scattering, intervalley transfer predominates initially. It can be seen in Fig. 3 that by about 500 fsec, the population in the satellite valleys has reached a maximum; at the same time the transition rate $R_{\Gamma \to L}$ has leveled off. At these times, the $\Gamma \to L$ rate plays an important role in luminescence decay.

The general results obtained in these simulations may be summarized as follows: Scattering out of the excitation volume is dominated in the initial tens of femtoseconds by electron-electron scattering, and the scattering rate increases with increasing density. *For low excitation densities*, this rate increase agrees both in its magnitude and in its density dependence with recent experimental measurements. We leave a discussion of these particular results to the next section. The presence of electron-electron scattering modifies both the population transition rates and carrier densities in the satellite valleys, primarily by reshaping the energy distribution of carriers in the central valley. Intervalley scattering rates decrease by two orders of magnitude as the electronic system cools during the first picosecond. Thus, intervalley processes do play a role in the initial decay and the same processes play a modified role in the picosecond-scale luminescence decay. We find that rates, for particular mechanisms of carrier transfer to the satellite valleys, must be estimated carefully since the Γ-L population shift contains a significant fraction of electrons that reach the L valleys by way of the X valleys.

EXCHANGE EFFECTS

The primary motivation of our study is to understand the earliest stages of optical absorption relaxation observed by compressed-pulse optical techniques. In particular, the photon echo experiments of Becker *et al.* [1] showed a significant density dependence of the initial decay time, which apparently reflected the importance of electron-electron scattering. These experiments provide the first information on nonequilibrium screening processes on such a rapid time scale. The EMC/MD calculations described in the last section were successful in explaining the low-density decay times, suggesting that the EMC (i.e., phonon scattering) component of the simulation was accurate. The initial decrease of the decay times with increasing density was also simulated within the precision allowed by the combined uncertainties of experiment and simulation. As noted, parameters such as valley masses and deformation potentials were taken from other measurements where possible, and consistency with *ab initio* calculations has also been found.

Furthermore, our results from EMC/MD calculations also give good agreement for the various time constants. At densities of about 10^{18} cm^{-3}, however, the EMC/MD simulation began to deviate from the experimental data, and the consistency of the remainder of the picture suggested that this deviation might be significant.

In fact, at densities near 10^{19} cm^{-3}, there are strong theoretical and experimental grounds for expecting that many-electron effects will be important. In this section we describe a semiclassical approach which can be used to take account of exchange and other effects which arise from the fundamental many-body nature of the correct electron description. We incorporate the exchange interaction into an EMC/MD simulation of the general kind described earlier: all electron motion, including that involved in screening, is treated explicitly with a molecular dynamics simulation run concurrently within an ensemble Monte Carlo treatment of scattering. However, exchange and related many-body effects modify the trajectories followed in the MD part of the simulation.

The method we describe below incorporates the exchange interaction among electrons, and describes an electron distribution which satisfies both the exclusion principle and the uncertainty principle while obeying a set of equations of motion which resembles that of a classical electron gas. The need to satisfy the uncertainty principle gives rise to corrections to the usual Coulomb interaction. Finally, we test the new approach by simulating the bulk GaAs system studied experimentally by Becker *et al.* [1], finding that the new terms tend to decrease the discrepancy between the experimental results and the classical EMC/MD results that were previously found at high density.

In practice, so-called semiclassical (EMC) approaches treat the electrons as having both a well-defined plane-wave momentum and a well-defined position. To go beyond this approximation, we make an explicit choice of wavepackets, which serves as the basis for a more rigorous quantum description. To preserve as much as possible the classical picture, we choose minimum-uncertainty wave packets

$$\phi_i(\mathbf{r}) = \phi(\mathbf{r}-\mathbf{x}_i,\mathbf{p}_i) = (2\pi\sigma^2)^{-3/4} \exp\left[-\frac{(\mathbf{r}-\mathbf{x}_i)^2}{4\sigma^2} + \frac{i}{\hbar}\mathbf{p}_i\cdot(\mathbf{r}-\mathbf{x}_i)\right] . \tag{7}$$

This packet describes an electron localized about position $\mathbf{x}_i$ with uncertainty σ and about momentum $\mathbf{p}_i$ with uncertainty $\hbar/2\sigma$. When the interactions among electrons can be ignored, and the external potentials vary slowly, the value of σ does not enter explicitly into the equations of motion. One then recovers the classical equations of motion. This is roughly equivalent to the approximation $\langle\nabla\rangle V(\mathbf{r})\rangle = \langle\nabla V\rangle(\mathbf{r})$ in Ehrenfest's theorem. At the high densities we study, however, exchange gives rise to momentum-dependent forces which make a large value of σ preferable. In this situation, σ will appear explicitly, and we shall need to determine a particular value for it.

We can use packets of the form (7) as the basis for an expansion of the many-electron system:

$$a_i = \int d\mathbf{r}^3 \ \phi^*(\mathbf{r}\text{-}\mathbf{x}_i,\mathbf{p}_i)\psi(\sigma_i,\mathbf{r}), \tag{8}$$

where $\psi(\sigma_i,\mathbf{r})$ is the field operator which annihilates a particle of spin σ_i at the point $\mathbf{r}$. For any single fixed value of σ_i, if $(\mathbf{x}_i,\mathbf{p}_i)$ is allowed to vary over all phase space points, then a_i generates an overcomplete basis for describing the electron system. In the spirit of single-particle semiclassical approximations, we represent the many-electron system by a combination of wavepackets: $|\Psi\rangle = a_1^\dagger a_2^\dagger ... a_N^\dagger\, |\rangle$, where $|\rangle$ is the "vacuum" of the unexcited semiconductor conduction band, so $\langle \mathbf{x}_1,\mathbf{x}_2,...,\mathbf{x}_N|\Psi\rangle$ is a Slater determinant wavefunction. In the independent (or single-) electron approaches, $\mathbf{x}$ and $\mathbf{p}\equiv\hbar\mathbf{k}$ represent the position and momentum of a particular electron. In the present approach, on the other hand, the parameters $\mathbf{x}_i$ and $\mathbf{p}_i$ *do not* represent the position and momentum of the ith electron. Instead, they represent the ith phase-space region near which *an* electron may be expected to be found. In this way, there is no conflict between the quantum mechanical identity of the electrons and the existence of distinct positions and momenta. Classically, the non-identity of the electrons is based simply on the possibility of following their smooth trajectories exactly.

For simplicity, we will derive the equations of motion for a single parabolic band of mass m. We use a Coulomb potential $V(\mathbf{r}-\mathbf{r}') = e^2/4\pi\varepsilon|\mathbf{r}-\mathbf{r}'|$, with ε taken to be the high-frequency dielectric constant. Using this potential, the second-quantized Hamiltonian $\mathcal{H}$ yields an energy $H(\{\sigma_i,\mathbf{x}_i,\mathbf{p}_i\}_{i=1}^{N}) \equiv \langle\Psi|\mathcal{H}|\Psi\rangle$. This energy is composed of kinetic energy, direct Coulomb, and exchange interaction parts:

$$H = E_K + E_D + E_{XC} , \tag{9}$$

$$E_K = \frac{1}{2m}(\mathbf{p}_1^2 + \mathbf{p}_2^2 + ... + \mathbf{p}_N^2) + \frac{N}{8m\sigma^2} , \tag{10a}$$

$$E_D = \frac{1}{2}\sum_{i\neq j} V_\sigma(\mathbf{x}_i-\mathbf{x}_j) , \tag{10b}$$

$$E_{XC} = -\frac{1}{2}\sum_{i\neq j} \delta(\sigma_i,\sigma_j)\, \hat{V}_\sigma(|\mathbf{p}_j-\mathbf{p}_i|/\hbar)\, \Delta_\sigma(\mathbf{x}_j-\mathbf{x}_i) , \tag{10c}$$

where $\Delta_\sigma(\mathbf{x}) = (4\pi\sigma^2)^{-3/2} \exp(-\mathbf{x}^2/4\sigma^2)$, and $\delta(\sigma_i,\sigma_j)$ is a Kronecker delta. The potential

$$V_\sigma(\mathbf{x}) = \int d^3\mathbf{r}\, V(\mathbf{r}+\mathbf{x})\, \Delta_\sigma(\mathbf{r}) \tag{11}$$

is a coarse-graining over a length scale σ of the Coulomb potential, while

$$\hat{V}_\sigma(\mathbf{k}) = \int d^3\mathbf{r}\, V(\mathbf{r})\; e^{\,i\mathbf{k}\cdot\mathbf{r} - \mathbf{r}^2/4\sigma^2} . \tag{12}$$

We can incorporate exchange effects in the joint MD/EMC simulation described earlier by deriving new semiclassical equations of motion from the Hamiltonian (9):

$$\frac{d\mathbf{p}_i}{dt} = -\frac{\partial H}{\partial \mathbf{x}_i} = \sum_{\substack{j=1\\(i\neq j)}}^{N} \mathbf{F}^{D}_{ij} + \sum_{\substack{j=1\\(i\neq j)}}^{N} \delta(\sigma_i,\sigma_j)\, \mathbf{F}^{XC}_{ij} , \tag{13a}$$

$$\frac{d\mathbf{x}_i}{dt} = \frac{\partial H}{\partial \mathbf{p}_i} = \frac{\mathbf{p}_i}{m} - \sum_{\substack{j=1\\(i\neq j)}}^{N} \delta(\sigma_i,\sigma_j)\, \mathbf{G}_{ij} , \tag{13b}$$

where we have written out the form of the terms that arise from (10).

As is clear from (10), the equations of motion involve the packet size σ explicitly. For a parabolic band, the σ-dependent term in the kinetic energy clearly does not affect the equation of motion. This term is the energy cost of confining electrons in finite-size packets. It is necessary in order to satisfy the exclusion principle in a finite-density system. The gradient of the direct Coulomb interaction (10b) gives rise to the $\mathbf{F}^D$ force term in (13a), which depends only on the position parameters. For interparticle separations much greater than σ, $\mathbf{F}^D$ approaches the classical Coulomb potential. In order to maintain a picture as close as possible to the classical one, we would like the direct Coulomb force to act between essentially point-like particles, and so we will want $\sigma \ll r_s$.

The exchange interaction is more complex than either the direct interaction or the kinetic energy; it depends on every parameter in the problem and contributes terms to both (13a) and (13b):

$$\mathbf{F}^{XC}_{ij} = \hat{V}_\sigma(\mathbf{k}_j-\mathbf{k}_i)\left[\frac{\partial}{\partial \mathbf{x}_i} \Delta_\sigma(\mathbf{x}_j-\mathbf{x}_i)\right] , \tag{14a}$$

$$\mathbf{G}_{ij} = -\mathbf{G}_{ji} \equiv \frac{1}{\hbar}\left[\frac{\partial}{\partial \mathbf{k}_i} \hat{V}_\sigma(\mathbf{k}_j-\mathbf{k}_i)\right] \Delta_\sigma(\mathbf{x}_j-\mathbf{x}_i) . \tag{14b}$$

The second quantity plays a role in the position time-evolution (13b) which is essentially that of a density-dependent effective mass term. In precise contrast with the direct interaction, the appropriate limit for computing the exchange energy is $\sigma \to \infty$. This can be seen most clearly for spatially homogeneous systems (which include the one we treat here). Confining the electrons in a large box of volume V, we ignore correlation by assuming that the spatial and momentum factors in (10c) can be averaged independently. The homogeneity assumption then leads to

$$E_{XC} = \frac{1}{V} \sum_{\substack{i,j=1 \\ (i \neq j)}}^{N} \delta(\sigma_i, \sigma_j) \frac{e^2}{2\varepsilon |\mathbf{k}_i - \mathbf{k}_j|^2} , \tag{15}$$

which is the exact exchange energy (ignoring correlation). In the limit $\sigma=0$, the exchange energy given by (10c) is zero.

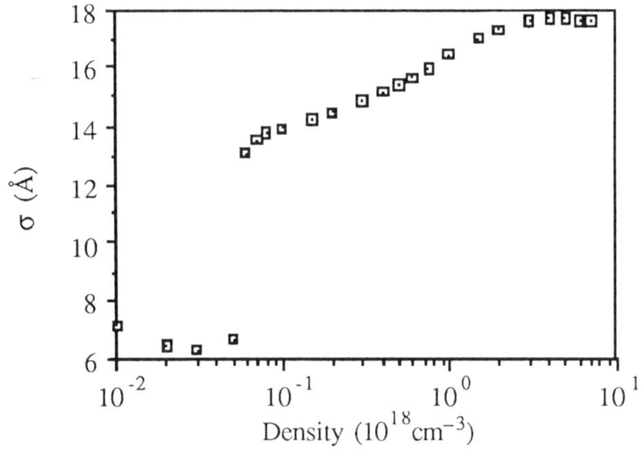

Fig. 5 The wavepacket width σ as a function of density, for initial conditions described in the text.

It is clear from the foregoing discussion that no single value of σ can yield an accurate value for both the direct Coulomb and the exchange forces. That is, the value of σ selects a class of basis states, in which either the exchange or direct energy is well defined, and thus susceptible to a classical approximation. One seeks a compromise σ value which treats both interactions with acceptable accuracy. There is no unique or best way to do this. The approach we have adopted is to compute the exact exchange energy for the initial distribution using (15), and choose σ to minimize the difference between this "exact" result and our σ-dependent approximation. Our initial distribution is uncorrelated by definition, so (9) is exact, and our approximation (10c) approaches this exact value. The differences, and justification for this approach, will be discussed elsewhere. Thus, the σ value must be computed separately for each density. The results are plotted in Fig. 5. This σ is small at low densities, rising very rapidly at a density of $\approx 5\times10^{16}$ cm^{-3}, and rising steadily in the range of densities 10^{17} to 10^{18} cm^{-3} to approach a constant value of ≈18Å; in this range of densities, thermal rather than photoexcited electrons in GaAs become strongly degenerate at room temperature. On the other hand, in a two-dimensional system it is found that σ is larger, achieving a value of approximately 4-6 nm in a quantum well of 5 nm thickness (the motion in this case lies in the plane of the quantum well) [38]. It is felt that this increase is in keeping with the general result that confinement of a wave function in one dimension *squeezes* it to expand in the other two dimensions. In our case, this increases the "uncertainty" in the plane of the quantum well.

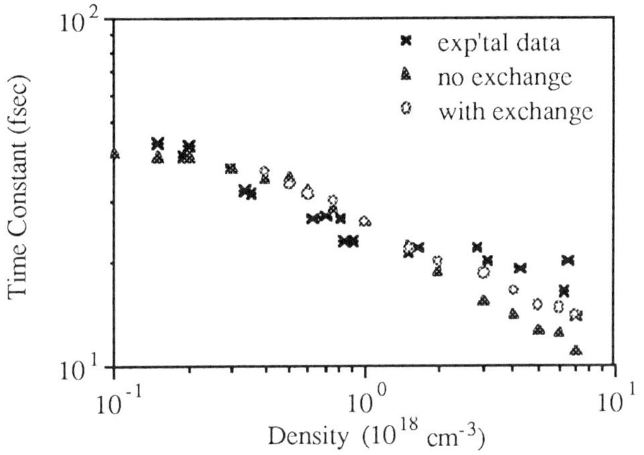

Fig. 6 Time constant for initial relaxation of photoexcited carriers as a function of density. Experimental data of Becker, *et al.* [1] are compared to simulations based on classical formalism of no exchange and on formalism with exchange.

The new MD equations were incorporated into an EMC/MD simulation that was in all other (i.e., EMC) respects identical to the simulation described above. As noted, in these earlier studies, we have found that the time constants determined from the photon echo experiments agreed well with the results of our joint MD/EMC simulations *at low densities*. With increasing density, the experiments (crosses in Fig. 6) exhibited an approximate power-law decrease of the time constant. In contrast, the simulations (triangles in Fig. 6) show a knee between 10^{17} and 10^{18} cm^{-3}, with the time constant varying slowly at the lower densities, and falling off rapidly at the higher densities. The low-density behavior simply reflects the dominance of intervalley phonon scattering processes, while strong density-dependence in the complementary regime reflects the dominance of electron-electron scattering at high density. In the range of densities below $\sim 2\times10^{18}$ cm^{-3}, the experimental and numerical simulation data were consistent. At higher densities, however, the experimental time constant was systematically larger than the simulation.

Using the formalism described above, we have included exchange corrections (open circles in Fig. 6). It is clear that inclusion of exchange significantly improves the agreement between experiment and simulation at high densities. At the low densities, where both electron-electron interaction and the exchange corrections to it are small, the previous simulations remain valid.

We can understand qualitatively why exchange increases the time constant at high density by considering the momentum time-evolution equation (13a). The dominant term on the right-hand side of (13a) is the Coulomb force. The main correction to this force is a cutoff of the force at small ($|\Delta\mathbf{r}| \lesssim \sigma$) interelectron spacings. This corresponds to the well-known "exchange hole" which arises from the Pauli exclusion principle. A second effect arises directly from the exchange energy: In equilibrium, the exchange energy of a homogeneous electron gas is an integral of the density distribution with a strictly negative kernel, so the exchange interaction is attractive. That is confirmed by the form of the exchange force (14a). The potential $\hat{V}_\sigma(\mathbf{k})$ given by (12) is strictly positive, so the exchange force acts in a direction opposite to the direct Coulomb force, partially canceling it in the short-range region where it is strongest.

Acknowledgments

The authors are indebted to S. M. Goodnick, P. Lugli, W. Pötz, J. Shah, T. Yamada, K. S. Yi and S. Zollner for useful discussions. This work was supported in part by the Office of Naval Research. In addition, one of us (R. P. J.) would like to acknowledge summer support from the Old Dominion University Research Foundation.

REFERENCES

(a) Present Address: Department of Electrical Engineering, State University of New York, Buffalo, New York.

(b) Present Address: IBM Corp., East Fishkill, N. Y.

1. P. Becker, H. Fragnito, C. Brito Cruz, R. Fork, J. Cunningham, J. Henry, and C. Shank, *Phys. Rev. Lett.* **61**, 1647 (1988).
2. P. C. Becker, H. L. Fragnito, C. H. Brito Cruz, J. Shah, R. L. Fork, J. E. Cunningham, J. E. Henry, and C. V. Shank, *Appl. Phys. Lett.* **53**, 2089 (1988).
3. M. J. Rosker, F. W. Wise, and C. L. Tang, *Appl. Phys. Lett.* **49**, 1726 (1986).
4. F. W. Wise, I. A. Walmsley, and C. L. Tang, *Appl. Phys. Lett.* **51**, 605 (1987).
5. R. W. Schoenlein, W. Z. Lin, E. P. Ippen, and J. G. Fujimoto, *Appl. Phys. Lett.* **51**, 1442 (1987).

6. W. Z. Lin, J. G. Fujimoto, E. P. Ippen, and R. A. Logan, *Appl. Phys. Lett.* **51**, 161 (1987).
7. M. A. Osman and D. K. Ferry, *Phys. Rev. B* **36**, 6018 (1987).
8. M. J. Kann and D. K. Ferry, in *Proceedings of the Picosecond Electronics and Optoelectronics Conference*, Salt Lake City, March 1989.
9. D. W. Baily, C. J. Stanton, M. A. Artaki, K. Hess, F. W. Wise, and C. L. Tang, *Solid-State Electronics* **31**, 467 (1988).
10. C. Jacoboni, in *Proc. 13th Intern. Conf. on Phys. of Semiconductors* (Marves, Rome, 1976), p. 1195.
11. P. Lugli and D. K. Ferry, *Phys. Rev. Lett.* **56**, 1295 (1986).
12. A. M. Kriman, M. J. Kann, D. K. Ferry, and R. Joshi, *Phys. Rev. Lett.* **65**, 1619 (1990).
13. See, e.g., D. K. Ferry, *Semiconductors* (Macmillan, New York, 1991) Ch. 12.
14. A. M. Kriman, M. J. Kann, D. K. Ferry, and R. Joshi, *Comp. Phys. Commun.*, in press.
15. D. J. Adams and G. S. Dubey, *J. Comp. Phys.* **72**, 156 (1987).
16. H. Brooks, *Phys. Rev.* **83**, 879 (1951).
17. N. Takimoto, *J. Phys. Soc. Jpn.* **14**, 1142 (1959).
18. G. L. Hall, *J. Phys. Chem. Solids* **23**, 1147 (1962).
19. F. J. Blatt, *J. Phys. Chem. Solids* **1**, 262 (1957).
20. A. D. Boardman and D. W. Henry, *Phys. Stat.Solidi b* **60**, 633 (1973).
21. J. R. Meyer and F. J. Bartoli, *Phys. Rev. B* **23**, 5413 (1981).
22. F. Stern, *Phys. Rev.* **158**, 697 (1967).
23. J. Friedel, *Adv. Phys.* **3**, 446 (1954).
24. E. M. Conwell and V. F. Weisskopf, *Phys. Rev.* **77**, 388 (1950).
25. B. K. Ridley, *J Phys. C: Solid State Phys.* **10**, 1589 (1977).
26. R. P. Joshi and D. K. Ferry, *Phys. Rev. B*, in press.
27. C. Jacoboni and L. Reggiani, *Rev. Mod. Phys.* **55**, 645 (1983).
28. R. P. Joshi, R. O. Grondin and D. K. Ferry, *Phys. Rev. B* **42**, 5685 (1990).
29. G. B. Stringfellow, *J. Appl. Phys.* **50**, 4178 (1979).
30. T. G. van der Roer and F. P. Widdershoven, *J. Appl. Phys.* **59**, 813 (1986).
31. M. J. Kann, A. M. Kriman and D. K. Ferry, *Phys. Rev. B* **41**, 12659 (1990).
32. M. J. Kann and D. K. Ferry, in *Proceedings of the Picosecond Electronics and Optoelectronics Conference* (Salt Lake City, March 1989).
33. D. W. Baily, C. J. Stanton, M. A. Artaki, K. Hess, F. W. Wise and C. L. Tang, *Sol.-State Electron.* **31**, 467 (1988).
34. J. Shah, B. Deveaud, T. C. Damen, W. T. Tsang, A. C. Gossard and P. Lugli, *Phys. Rev. Lett.* **59**, 2222 (1987).
35. K. Kash, P. A. Wolff and W. A. Bonner, Appl. Phys. Lett. **42**, 173 (1983).
36. S. N. Chamoun, R. P. Joshi, E. N. Arnold, R. O. Grondin, K. E. Meyer, M. Pessot and G. A. Mourou, *J. Appl. Phys.* **66**, 236 (1989).
37. A. Katz and R. R. Alfano (unpublished); W. B. Wang, N. Ockman, M. Yan and R. R. Alfano, *Sol.-State Electron.* **32**, 1337 (1989).
38. R. Joshi, A. M. Kriman, M. J. Kann, and D. K. Ferry, to be published.

5

THEORY AND CALCULATION OF THE DEFORMATION POTENTIAL ELECTRON-PHONON SCATTERING RATES IN SEMICONDUCTORS

M.V. Fischetti
IBM Research Division, Thomas J. Watson Research Center
P.O. Box 218, Yorktown Heights, New York 10598

and

J.M. Higman
Beckman Institute, University of Illinois at Urbana-Champaign
405 N. Mathews Avenue, Urbana, Illinois 61801

I. INTRODUCTION

Our ultimate goal is the simulation of carrier transport in semiconductors using the potential of the ions as the sole input. This should enable us to describe within a single framework both the carrier kinematics (band structure) and most of the dynamics (carrier-phonon collisions), and will constitute a stringent test for our understanding of semiclassical charge transport in semiconductors.

There are both fundamental and practical reasons to undertake this task. From the point of view of basic physics, Monte Carlo (MC) simulations of carrier transport in semiconductors have evolved to a very sophisticated level: the full band-structure of the crystals,[1-3] realistic density of states (DOS),[1-4] the details of the valence band,[5,6] the role of intercarrier Coulomb interaction[3,7-9] and of exchange-energy corrections,[9] and evaluations of impact ionization (inverse Auger processes) thresholds based on realistic band structure[10-12] and pair production rates[13] have been implemented. Still, in some cases, we are facing the same old problems. It is worth mentioning a few specific examples: High field transport in Si, as important as it might be for semiconductor technology, is still largely unsolved, partially because of the lack of direct experimental data. Empirical fits to low-energy transport data are still used to determine the electron-phonon coupling constants,[1-3,10-12] and these values are used for simulation well beyond their expected range of validity. In addition, selection rules and phonon dispersion are often treated poorly or completely ignored. The lack of any reliable information on the electron-phonon interaction has always affected, in turn, our ability to treat the problem of impact ionization, the two problems being obviously interdependent. Ongoing work[14] shows that the two problems can be decoupled, by describing the Coulomb interaction *from first principles.*

Moving to the III-V compound semiconductors, we seem to be able to explain the low-energy transport data or hot-carrier relaxation rates in a variety of materials and alloys,[15] but for some of these materials even the information on the band structure is, at best, fragmentary.[16] Again, empirical deformation potentials are used. As long as the transport model contains free parameters of this type, a large fraction of the predictive power of the model is lost, and along with it the possibility of assessing how well we really understand the physics. Therefore, a more fundamental approach is required. We realize that a race towards an *ab initio* description of carrier transport is doomed to be without end and subject to criticisms of incompleteness at any stage. Nevertheless, as we progress towards a more complete picture, we can hope to identify a few crucial ingredients which are necessary to describe correctly ever finer features of the problem. For instance, the work performed so far by groups in Urbana,[1,2] at IBM,[3] in Europe,[4,13] and at NTT[10-12] clearly shows that a model for the DOS as accurate as possible is really a necessary prerequisite to simulate high-energy, high-field transport.

From the practical point of view of helping the semiconductor technology through numerical modeling, as the physical models improve, the range of questions to which we can give reliable answers widens. Simple ohmic-like drift-diffusion models can generally tell us what the macroscopic terminal currents in a semiconductor device are, but fail to predict, among other quantities, switching speeds and avalanche processes. Models based on simplified band structure models may improve the predictions for switching speeds, but still fail to represent accurately the high-energy features of transport, such as, again, the avalanche process and the gate current in MOSFETs. These features could be described by models employing the full band structure[2,3] but it is yet unclear how well these models work, mainly because of their disagreement on how strong the high-energy electron-phonon interaction is. Again, a more fundamental approach is needed to give us reliable results in a range of carrier energies for which experimental data are either lacking or are too indirect.

In this chapter we describe some progress which has been made towards our goal, namely, the calculation of the electron phonon scattering rates employing a "conventional" model for the lattice dynamics (the valence shell model[17,18]) and empirical pseudopotentials.[19,20] We should stress that in computing electron-phonon matrix elements we are doing nothing new. But we believe that there is still opportunity to gain some new understanding by extending previous calculations to arbitrary electron wavevectors, analyzing intravalley deformation potentials, and explicitly pointing out the limitations of the approach.

Pseudopotentials were used to evaluate intervalley deformation potentials as early as 1971.[21] Herbert[22] treated intervalley scattering in III-V semiconductors in 1973. In 1982 Glembocki and Pollak evaluated electron-phonon matrix elements between symmetry points in Si[23] and in Ge[24] and compared these results to experimental data involving indirect (phonon assisted) light absorption in these materials. Bednarek and Rössler[25] repeated these calculations, and argued in favor of an alternate way of interpolating the potential. The interpolation question remains one of the important unresolved issues regarding these calculations, as will be discussed in Section II.3.1. Allen and Cardona[26,27] used the same formalism to compute the electron-phonon matrix elements and obtain the temperature dependence of the band gap of Si and Ge. A similar approach was employed by Gopalan and co-workers.[28] More recently Zollner, *et al.*,[29,30] have effectively revived the interest in these matters, largely in connection with the results of femtosecond spectroscopy experiments which measure intervalley relaxation times in III-V materials, and have summed up very clearly the status of such calculations. Nonlocal pseudopotentials have also been used by Grinyaev and co-workers[31,32] to evaluate intervalley deformation potentials for III-V compound semiconductors. This admittedly incomplete list of past and ongoing work reveals the unsettled nature of the matter, results from various groups being quite often in disagreement. In our opinion, despite the undeniable merit of these works, it is now clear that the goal of an *ab initio* MC simulation is, at least at present, a dream. We are facing, essentially, *the* problem of solid state physics: *to know what the potential is.* Even if we are willing (as we must) to fall back on pseudopotentials, the large freedom we have to chose the interpolation of the potential (i.e., the model for the dielectric function of the solid), the lattice dynamics model, and the choice of the local or nonlocal approximation to the pseudopotentials results in large variations for the matrix elements. We want to be frank about the inherent difficulties and resort once again to an empirical approach, but now the empiricism can be shifted to a more fundamental level than has previously been done in MC simulations.

In addition to this "fluid" state of the theory, results in the literature are almost entirely limited to the evaluation of the matrix elements for intervalley electronic transitions between symmetry points.[22-24,29-32] This is a significant limitation: On one hand, intervalley processes between symmetry points are suggestive of what the carrier-phonon scattering rates and the charge-transport properties will be, but are no more than suggestive. For example, the question of how quickly selection rules fade as we move away from symmetry points has been a matter of some concern, but has been investigated only rarely,[33,34] and not in sufficient detail, considering that in high-field transport a very large portion of the Brillouin Zone (BZ) is populated, not just the local minima at points of relatively high symmetry. On the other hand, intravalley deformation potentials, useful in nonpolar crystals to test the calculation against well established ohmic-mobility data, are consistently left unexplored, the interest often being focussed on the short-time relaxation rates in III-V materials. Therefore, only the evaluation of the total electron-phonon scattering rate over the entire BZ can really put our models to the test. We shall leave the topic of testing the model with MC simulations for future publications on the subject. Here we will focus on the analysis of the scattering rates. We believe that this analysis by itself leads to a nontrivial conclusion: *It is possible to make a unique choice for the pseudopotentials resulting in electron-phonon scattering rates which are in quantitative agreement with those previously derived empirically from MC simulations which reproduce a wide set of experimental data.* An example of this is illustrated in Section III.

To conclude this introductory section, we believe that the consistent use of a model pseudopotential for the ions in order to treat the kinematics and dynamics of carrier transport is very appealing for two basic reasons, independent of the level of physical rigor employed to arrive at the pseudopotentials: First, it is aesthetically pleasing to have a unique, albeit empirical, input to describe both the band structure and the electron-phonon interaction in a semiconductor. A second, more technical advantage is the possibility of describing charge transport accounting for the full wavevector dependence of the deformation potentials, including a rigorous adherence to the selection rules along symmetry lines and at symmetry points.

This chapter is structured as follows: In Section II.1 we review the microscopic theory of the electron-phonon interaction *without* resorting to pseudopotentials, and show that this theory reduces to the commonly used phenomenological coupling constants in the long wavelength limit. This microscopic analysis will be useful in two additional applications. First, it is helpful to note exactly what information is lost in assuming a pseudopotential for the atomic potentials, and second, we will use the more exact theory to calculate an effective potential, which we compare to the (interpolated) pseudopotentials. The latter, of course, we will ultimately use to calculate the deformation potentials and scattering rates. In section II.2, we shall briefly review the basic ideas and approximations which are at the root of the pseudopotential concept. We chose to do so, both for completeness as well as to give a feeling for how accurately the pseudo-wavefunctions, and deformation potentials calculated from them, may describe the true electronic structure. We will present in detail the parameters we have used to fit band structure data and some numerical details. We reduce the theory of Section II.1 to the case of pseudopotentials, arriving at the "rigid-ion" approximation, and in Section II.3 the form of the matrix element is given.

II. ELECTRON-PHONON INTERACTION

The theory presented below has been developed extensively over the last few decades. Since the subject here is a *computational model,* however, the details of the theory will be presented only to the extent that they are necessary to clarify the approximations involved. That is to say, the fundamentals which can be easily found in textbooks will not be derived. Having presented the formal theory, the connection to the phenomenological approach for the deformation potential interactions (the use of more-or-less adjustable coupling constants), which has been in common use in transport simulations for some time, will be made. With this connection established, we will concentrate on the "deformation potential" interaction, both the long wavelength intravalley acoustic deformation potential, and the short wavelength intervalley deformation potentials for both acoustic and optical modes. The success of the Fröhlich Hamiltonian in describing scattering by the long wavelength optical modes via the dipole interaction in polar semiconductors relieves us of that problem.

There are several early theoretical analyses of the electron-phonon interaction in semiconductors such as those by Seitz,[35] Harrison,[36] Shockley and Bardeen,[37,38] and Lawaetz.[39] Sham and Ziman reviewed the subject of the electron-phonon interaction, with most emphasis on the case of metals.[40] The work of Heine, *et al.*,[41] in defining the effective scattering potential was generalized by Vogl[42] to the case of semiconductors, and is representative of the current theoretical understanding of the problem. From the outset it is important to note that in each of these analyses at some point assume a perturbation due to a *long wavelength* lattice distortion in order to arrive at useful formulas. Unfortunately, of greater interest for modern transport simulations is the case of scattering by short wavelength modes, or in the phenomenological language, intervalley scattering. As will be shown below, this fact forces one to make assumptions regarding the scattering potential.

II.1. Microscopic Theory

We begin to develop the theory by assuming that the adiabatic approximation (Born-Oppenheimer method) holds. The idea is that we can separate the motion of the lattice from the motion of the valence electrons, i.e., the electrons adjust instantaneously to the positions of the ions. One can justify this approximation with a simple argument based on the relative masses of the ions and the electrons,[43] and more precise definitions of this approximation can be found in the literature.[42] The consequence of this assumption is that the dielectric screening of the ion core potential is represented by the frequency-independent dielectric matrix.

We divide the problem as follows. Consider the ion (the nucleus plus the core electrons, *without* the valence electrons), the potential of which will be referred to as the "bare" potential, $V_{b\alpha}$. The bare ion potential looks like a Coulomb potential of charge $Z_\alpha e$ at large distances, where Z_α is the number of valence electrons on the atom of species α, which we have stripped away from the neutral atom in defining the bare potential. In the bare potential we also have a repulsive potential at small radii due to the core electrons. The screening due to the rearrangement of the valence electrons will be described in a self-consistent way by the dielectric matrix of the material. The effects of core polarization are assumed to be small and will be neglected.

The perturbation due to the motion of the bare ions, seen by a conduction electron is written as a Taylor expansion about the equilibrium positions of the ions. We keep only the first-order term in the expansion, the second-order term giving rise to two-phonon processes, which will not be addressed here. Vogl [42] has discussed the long wavelength matrix elements for two-phonon processes. The total bare potential is

$$V_b^{\text{tot}} = -\sum_{\alpha l} \mathbf{u}_{\alpha l} \cdot \nabla_{\mathbf{r}} V_{b\alpha}(\mathbf{r}-\mathbf{R}_{\alpha l}^0) \tag{1}$$

The indices α label the atoms in the basis and l runs over all unit cells in the crystal (the subscript l will also be used later to denote the order of Bessel functions and Legendre polynomials which appear in the nonlocal pseudopotentials; this is done to conform to traditional notation and should not cause any confusion). We now use the dielectric screening theory to compute the effective potential seen by the conduction electron. Since we are interested in small displacements $\mathbf{u}_{\alpha l}$ of the ions from their equilibrium positions $\mathbf{R}_{\alpha l}^0$, the use of linear response theory is justified. We assume that the displacements $\mathbf{u}_{\alpha l}$ represent the dressed phonon variables, i.e., the screening effect of the valence electrons has been taken into account in a self-consistent way. In Section II.2.5 we describe the computational model used to arrive at these quantities. The Fourier components of the total self-consistent potential can then be written, after Fourier transforming the bare potential, as

$$\delta\phi(\mathbf{q} + \mathbf{G}) = \sum_{\mathbf{G}'} \varepsilon_{\mathbf{G},\mathbf{G}'}^{-1} V_b^{\text{tot}}(\mathbf{q} + \mathbf{G}') \tag{2}$$

The static inverse dielectric matrix $\varepsilon_{\mathbf{G},\mathbf{G}'}^{-1}(\mathbf{q}) = \varepsilon^{-1}(\mathbf{q}+\mathbf{G},\mathbf{q}+\mathbf{G}')$ represents the screening due to the valence electrons, and the off-diagonal elements of this matrix $\mathbf{G}\neq\mathbf{G}'$ contain the information about the local fields around the bare perturbing potential. The off-diagonal elements must necessarily be nonzero in an insulator or semiconductor as a consequence of the acoustic sum rule,[44] in contrast with the case of a metal where the large density of conduction electrons results in a homogeneous screening, and hence a diagonal dielectric function.

Following closely the analysis of Ref. [42] the long- and short-range parts of the total potential in (2) can be obtained. The procedure involves separating out the parts of the total potential that diverge for small $\mathbf{q}$ (the long-range Coulomb-like terms) from the part that is well behaved in the limit $q \rightarrow 0$. It must be stressed that although the division will be written formally for arbitrary $\mathbf{q}$, the distinction between the two contributions is made when $\mathbf{q}$ is far from the BZ boundaries. Unfortunately, no model exists for the different contributions to the potential which can be used for short wavelength modes. The utility of these expressions will be apparent later, however, when we use the short-range term to calculate an "effective short-range potential" which can be used to test the interpolation of the empirical pseudopotential form factors (Section II.3.2).

By introducing an inverse dielectric function for the case when the macroscopic electric field is zero, i.e., in the absence of long-range terms[42]

$$\tilde{\varepsilon}_{\mathbf{G},\mathbf{G}'}^{-1} = (1-\delta_{\mathbf{G},0}) \left[\varepsilon_{\mathbf{G}\mathbf{G}'}^{-1} - \frac{\varepsilon_{\mathbf{G}0}^{-1}\varepsilon_{0\mathbf{G}'}^{-1}}{\varepsilon_{00}^{-1}} \right] (1-\delta_{\mathbf{G}',0}) \tag{3}$$

we can rewrite the potential in (2) as

$$\delta\phi(\mathbf{q}+\mathbf{G}) = \sum_{\mathbf{G}'} \varepsilon^{-1}_{\mathbf{G},\mathbf{G}'} V_b^{tot}(\mathbf{q}+\mathbf{G}') + \frac{e}{i}\frac{\varepsilon^{-1}_{\mathbf{G},0}}{q\,\varepsilon^{-1}_{0,0}} E(\mathbf{q}) \tag{4}$$

where $E(\mathbf{q})$ is the macroscopic electric field, which can be expressed in terms of the "effective charges" of the ions. Looking first at the second term on the right hand side of (4), it has been shown[45] that in the long wavelength limit the dipole moments derived from this term correspond to the usual polar optical interaction in the polar semiconductors (the dipole terms vanish for nonpolar semiconductors). Higher moments contribute long-range terms in both polar and nonpolar semiconductors. We will not deal further with the second term on the right hand side of (4). Again, however, (4) is derived in the long wavelength limit, and so the contribution of the "long-range" term in (4), even for nonpolar semiconductors, at arbitrary $\mathbf{q}$ is not known.

We expand the ionic displacements in terms of second-quantized operators[46] according to:

$$\mathbf{u}_{\alpha l} = \sum_{\mathbf{Q}\eta} \left[\frac{\hbar}{2Nm_\alpha \omega_{\mathbf{Q}\eta}}\right]^{\frac{1}{2}} \xi^{\eta}_{\alpha}(\mathbf{Q})\,(\,a_{-\mathbf{Q}\eta} + a^{+}_{\mathbf{Q}\eta}\,)\, e^{i\mathbf{Q}\cdot\mathbf{R}^0_{\alpha l}}\, e^{i\omega_{\mathbf{Q}\eta} t} \tag{5}$$

where $\xi^{\eta}_{\alpha}(\mathbf{Q})$ is the polarization vector for the ions of type α for mode η, $a^{+}_{\mathbf{Q}\eta}$ and $a_{-\mathbf{Q}\eta}$ are the creation and annihilation operators, and N the number of cells. We note here the useful property of periodicity (in reciprocal space) of the polarization vectors,[46] namely, $\xi^{\eta}_{\alpha}(\mathbf{Q}+\mathbf{G}) = e^{-i\mathbf{G}\cdot\tau_\alpha}\xi^{\eta}_{\alpha}(\mathbf{Q})$ where $\mathbf{G}$ is a reciprocal lattice vector. The short-range matrix element between Bloch states labeled by electron wavevectors $\mathbf{k}$, electron band indices n, and phonon states in the number representation, labeled by $N_{q\eta}$, is then

$$<\mathbf{k}',n';N_{\mathbf{q}\eta}' \mid V^{sr}_{el-ph} \mid \mathbf{k},n;N_{\mathbf{q}\eta}> =$$

$$<\mathbf{k}+\mathbf{q},n' \mid \sum_{\mathbf{G}} e^{i(\mathbf{q}+\mathbf{G})\cdot\mathbf{r}}\,\delta\phi^{sr}(\mathbf{q}+\mathbf{G}) \mid \mathbf{k},n> (N_{\mathbf{q}\eta}+\tfrac{1}{2}\pm\tfrac{1}{2})^{\frac{1}{2}} \tag{6}$$

where the index η labels the phonon branch and $\delta\phi^{sr}$ represents the short-range potential, the first term on the right hand side of (4). The "+" sign is for creation of a phonon (emission), and the "−" sign is for annihilation (absorption).

In the long wavelength limit ($q\rightarrow 0$), the branches of the phonon spectrum have either purely acoustic or optical character, and furthermore, the matrix element can be expanded in powers of q. With this analysis we recover the phenomenological coupling constants.[45] Looking now at the matrix element on the right hand side of (6), which we will call $M^{sr}(\mathbf{q})$, and taking notice of the fact that the $\mathbf{G}$=0 terms in the potential $\delta\phi^{sr}$ are zero due to (3), we have

$$M^{sr}(q \to 0) = < \mathbf{k},n' \mid \sum_{\mathbf{G}} e^{i\mathbf{G}\cdot\mathbf{r}} \, \delta\phi^{sr}(\mathbf{G}) \mid \mathbf{k},n > +$$

$$i\left[\frac{\hbar}{2\omega_{\mathbf{q}\eta}}\right]^{\frac{1}{2}} \mathbf{q} \cdot \sum_{\alpha} m_{\alpha}^{-\frac{1}{2}} \, \Xi_{\alpha} \cdot \xi_{\alpha}^{\eta}(\mathbf{q}) + \quad . \; . \; . \tag{7}$$

If we specify $n = n'$, i.e., intraband scattering, the connection of the two terms on the right hand side of (7) to the phenomenological constants can be seen. For acoustic modes, the second term represents the usual deformation potential interaction, and Ξ_{α} is the acoustic deformation potential tensor.[47] In Section II.3.3 we compare some calculated results to the phenomenological acoustic deformation potentials which have been used in transport simulations.

The first term on the right hand side of (7) is the expectation value of the periodic function $e^{i\mathbf{G}\cdot\mathbf{r}} \, \delta\phi^{sr}(\mathbf{G})$, and can be shown to vanish for the acoustic modes. This term gives the intravalley optical deformation potential, which has been calculated (see Ref. [48] and references therein), but in terms of transport, they are important primarily for hole-phonon scattering, and will not be discussed further here.

II.2. The Rigid Pseudo-Ion Approximation

We now leave for the time being the "exact" theory, and present the theory of the pseudopotential description of the crystal potential. With this formalism in place we can then return to the form of the matrix element with the "exact" potential described above replaced by a pseudopotential.

II.2.1. Empirical Local and Nonlocal Ionic Pseudopotentials. The concept of pseudopotential has been introduced to reduce the complexity of the many-body problem of treating multi-electron atoms and solids.[49] Here the difficulty lies in resolving the very high-frequency components of the wave functions arising from the deep core of the bare ionic potential. To bypass this problem, at the expenses of some rigor, the real ionic potential is replaced by a potential approximating the long-wavelength features of the real potential in the "interstitial" regions away from the ion-cores. The entire information about the cores is lumped into the knowledge (to be obtained *somehow*) of the energy-dependent logarithmic derivatives of the exact wavefunctions at the spherical interface between the core and the interstitial region. Thus, we can obtain the excitation spectrum of the crystal (band-structure) and other properties depending on the outer (valence and conduction) electrons, once we are given any model potential which approximates the correct potential in the interstitial region and which yields the desired boundary conditions for the wavefunctions at the core/interstitial-region interface. In addition, not to alter the desired spectrum, one requires that no bound states in the core result from the choice of the pseudopotential (i.e., the pseudowavefunctions have no radial nodes in the core region).

The label "empirical" refers to the method chosen to determine the form of this model pseudopotential and the logarithmic derivatives at the core/interstitial-region interface. The presence of the core and outer electrons obviously affects the bare ion potential. While self-consistent approaches exist to account for the screening of the bare potential by outer electrons, empirical methods usually trade the possibility of performing total-energy calculations (and so obtain a consistent lattice dynamics picture) for a more realistic excitation spectrum. Model potentials have been found which reproduce the ionization energies of complex ions and the band structure of various solids. The seminal works by Cohen and Phillips,[50] Sham and Ziman,[40] and by Abarenkov, Animalu, and Heine[51-54] have been reviewed by Heine[49] and by Cohen and Heine[55] and we refer the reader to those excellent papers for a complete discussion.

The pseudopotential of an ion of index α is usually described as:

$$V_p^{\alpha}(\mathbf{r}) = V_L^{\alpha}(\mathbf{r}) + V_{NL}^{\alpha}(\mathbf{r},E) \tag{8}$$

the local term V_L being a simple function of position, the nonlocal term V_{NL} being expressed as:

$$V_{NL}^{\alpha}(\mathbf{r},E) = \sum_{l=0}^{\infty} A_l^{\alpha}(E)\, f_l^{\alpha}(r)\, \Pi_l \tag{9}$$

where A_l^{α} is an energy dependent well depth, $f_l^{\alpha}(r)$ accounts for the effect of the core states with l symmetry, and Π_l is the projection operator on the component of angular momentum l. The energy dependence of the well-depth reflects the attempt to improve the matching on the wavefunctions at the interface between the core and interstitial regions and to extend the energy range over which an accurate excitation spectrum can be obtained.

II.2.2. Crystal Pseudopotential and Schrödinger Equation. In a crystal, the total pseudopotential, V_p, acting on valence and conduction electrons is the sum of the potentials of each ion, modified by the different local environment. Thanks to the crystal symmetry, we can write the contribution of the local component as a Fourier series over vectors $\mathbf{G}$ of the reciprocal lattice:

$$V_L(\mathbf{r}) = \frac{1}{2}\sum_{\mathbf{G}} V(\mathbf{G})\, e^{i\mathbf{G}\cdot\mathbf{r}} \tag{10}$$

where

$$V(\mathbf{G}) = \sum_{\alpha} S_{\alpha}(\mathbf{G})\, V_{\alpha}(\mathbf{G}). \tag{11}$$

Here V_{α} is the local *atomic form factor* and

$$S_{\alpha}(\mathbf{G}) = \frac{1}{N_a} \sum_{\substack{\alpha \\ \text{in cell } l}} e^{-i\mathbf{G}\cdot\mathbf{R}_{l\alpha}^{0}} \tag{12}$$

is the *structure factor*, N_a is the number of atomic species present in the cell, $\mathbf{R}_{l\alpha}^0$ is the equilibrium position of the α-th ion in the l-th Wigner-Seitz cell ($=\mathbf{R}_l + \tau_\alpha$), $\mathbf{R}_l$ being the origin of the cell and τ_α the relative position of ion α, and

$$V_\alpha(\mathbf{G}) = \frac{1}{\Omega_a}\int d^3r \; V_L^\alpha(\mathbf{r})e^{-i\mathbf{G}\cdot\mathbf{r}} \tag{13}$$

The normalization is done over the atomic volume Ω_a. As noted by Zollner and co-workers[30] this has caused some confusion in the value of the electron-phonon deformation potentials calculated before. For semiconductors of the diamond and zinc-blende structure, with 2 atoms per unit cell (crystals we shall restrict ourselves to in the following) it is customary to write:[19,20]

$$V(\mathbf{G}) = V^S(\mathbf{G})\cos(\mathbf{G}\cdot\tau) + iV^A(\mathbf{G})\sin(\mathbf{G}\cdot\tau) \tag{14}$$

where

$$V^S(\mathbf{G}) = \frac{1}{2}[V_1(\mathbf{G}) + V_2(\mathbf{G})] \tag{15a}$$

$$V^A(\mathbf{G}) = \frac{1}{2}[V_1(\mathbf{G}) - V_2(\mathbf{G})] \tag{15b}$$

with $\tau = \tau_1 = -\tau_2 = \frac{a}{8}(1,1,1)$, a being the lattice constant. As explained by Cohen and Heine[55] a spherical symmetric approximation can be employed (i.e., $V(\mathbf{G}) = V(|\mathbf{G}|)$), and only the values of V^S at $G = \sqrt{3}, \sqrt{8}, \sqrt{11}$ (in units of $2\pi/a$) and of V^A at $G = \sqrt{3}, \sqrt{4}, \sqrt{11}$ are needed.

The nonlocal nature of the second term in (8) implies that the matrix element of the crystal potential V_{NL} over plane waves $e^{i\mathbf{k}\cdot\mathbf{r}}$ and $e^{i\mathbf{k}'\cdot\mathbf{r}}$ depends on both $\mathbf{k}$ and $\mathbf{k}'$. Choosing for $f_l^\alpha(\mathbf{r})$ the simple form:

$$f_l^\alpha(\mathbf{r}) = \begin{cases} 1 & (r < d_l^\alpha) \\ 0 & (r \geq d_l^\alpha) \end{cases} \tag{16}$$

where the radii d_l^α describe the "width" of the l-th core orbital, and restricting ourselves to the $s-$ and $d-$type core orbital, we have:

$$V_{NL}^\alpha = \frac{4\pi}{\Omega_a}\sum_{\substack{l=0,2\\ \alpha}} A_l^\alpha(E)\,(2l+1)\,P_l[\cos(\theta_{k,k'})]\;S_\alpha(\mathbf{k}-\mathbf{k}')\;F_l^s(k,k') \tag{17}$$

where $\theta_{k,k'}$ is the angle between $\mathbf{k}$ and $\mathbf{k}'$ and:

$$F_l^\alpha = \begin{cases} \dfrac{d^3}{2}[j_l^2(kd) - j_{l-1}(kd)] & (\mathbf{k}=\mathbf{k}') \\[2ex] \dfrac{d^2}{k^2 - k'^2}[kj_{l+1}(kd)j_l(k'd) - k'j_{l+1}(k'd)j_l(kd)] & (\mathbf{k}\neq\mathbf{k}') \end{cases} \tag{18}$$

where d stands for d_l^α, $j_l(x)$ is the spherical Bessel function of order l, and $P_l(x)$

is the Legendre polynomial of order l. The energy dependence of the d-wells is usually ignored, while for the s-wells it is approximated as[20]

$$A_0^\alpha = \alpha_0^\alpha + \beta_0^\alpha\{[E_0(k)E_0(k')]^{1/2} - E_0(k_F)\}, \tag{19}$$

where $E_0(k) = \hbar^2k^2/(2m)$, α_0^α, and β_0^α are empirical parameters, and k_F is the Fermi wavevector.

Expressing the wave functions as:

$$\psi(\mathbf{r}) = \sum_{\mathbf{k}} \langle\mathbf{r}|\mathbf{k}\rangle = \sum_{\mathbf{k}} \psi_{\mathbf{k}}(\mathbf{r}) = \frac{1}{\sqrt{2N\Omega_a}} \sum_{\mathbf{k}} e^{i\mathbf{k}\cdot\mathbf{r}} \sum_{|\mathbf{G}|\le G_0} u_{\mathbf{G}}(\mathbf{k})e^{i\mathbf{G}\cdot\mathbf{r}} \tag{20}$$

where G_0 is a cutoff in reciprocal space (introduced in order to handle the problem numerically) and N the number of cells, the matrix elements of the Schrödinger equation

$$\left[-\frac{\hbar^2\Delta}{2m} + \sum_{l\alpha} V_p^\alpha(\mathbf{r}-\mathbf{R}_{l\alpha}^0)\right]\psi(\mathbf{r}) = E\,\psi(\mathbf{r}) \tag{21}$$

(where $\hbar$ is the reduced Planck constant, Δ is the Laplace operator, and m is the electron mass) over the functions $\psi_{\mathbf{k}}$, are:

$$\sum_{|\mathbf{G}|,|\mathbf{G}'|\le G_0} H_{\mathbf{G},\mathbf{G}'}u_{\mathbf{G}}(\mathbf{k}) = E(\mathbf{k})u_{\mathbf{G}}(\mathbf{k}) \tag{22}$$

where:

$$H_{\mathbf{G},\mathbf{G}'} = \frac{\hbar^2}{2m}(\mathbf{k}+\mathbf{G})^2\delta_{\mathbf{G},\mathbf{G}'} + V(|\mathbf{G}-\mathbf{G}'|) +$$

$$\frac{4\pi}{\Omega_a}\sum_{\substack{l=1,2\\ \alpha}} A_l^\alpha(E)(2l+1)P_l[\cos(\theta_{KK'})]S_\alpha(\mathbf{G}-\mathbf{G}')F_l^\alpha(K,K'), \tag{23}$$

with $\mathbf{K} = \mathbf{k} + \mathbf{G}$, $\mathbf{K}' = \mathbf{k} + \mathbf{G}'$. The first two terms on the right hand side of (23) represent the local terms, while the last term represents the nonlocal corrections. The solutions, $E_n(\mathbf{k})$, of the secular equation

$$\det|H_{\mathbf{G},\mathbf{G}'}(\mathbf{k}) - E(\mathbf{k})\delta_{\mathbf{G},\mathbf{G}'}| = 0 \tag{24}$$

yield the dispersion in each band n, the eigenvectors $u_{n,\mathbf{G}}(\mathbf{k})$ yielding the pseudowavefunctions in (13). While the energy spectrum is very accurate, doubts must be raised concerning the nature of the pseudowavefunctions and how accurately they can be used to evaluate deformation potential and matrix elements. These doubts are caused by the fact that the high-frequency components of the "real" wavefunctions in the core region are washed out, the (energy-dependent) matching condition on the logarithmic derivatives at the interface between the interstitial and the core regions being the only handle on the form of the wavefunction. Actually, in the few cases of some multi-electron atoms for which one has exact solutions, the pseudowavefunctions are more accurate than expected.[55] Therefore, despite the lack of a rigorous justification, we can expect that the pseudowavefunctions provide a reasonable approximation to the "real" wavefunctions. For example, in the present context, it is interesting to recall that phonon dynamics has been predicted successfully in a few cases using empirical pseudopotentials, even for covalent semiconductors, implying that the interstitial charge density and its dynamics can be treated correctly.[55]

II.2.3. Spin-Orbit Interaction. In high–Z materials the spin-orbit interaction can be quite large. In particular, the structure and splitting of degeneracies at the top of the valence band at the zone center (Γ) is strongly affected by this interaction. Thus, if we are interested in the transport properties of holes, particularly at low energies, it is necessary to include this term in order to obtain the correct effective masses, nonparabolicity, and warping of the valence bands. Following Weisz[56] and Bloom and Bergstresser,[57] Chelikowsky and Cohen[20] have accounted for the spin-orbit contribution to the Hamiltonian by adding in Eq. (23) the term (a 2×2 spin matrix)

$$H^{so}_{\mathbf{G},\mathbf{G}'} = (\mathbf{K}\times\mathbf{K}') \cdot \sigma\{-i\lambda^S \cos[(\mathbf{G}-\mathbf{G}')\cdot\tau] + \lambda^A \sin[(\mathbf{G}-\mathbf{G}')\cdot\tau]\} , \qquad (25)$$

where σ are the Pauli matrices and:

$$\lambda^S = \frac{1}{2}(\lambda_1 + \lambda_2) ,$$

$$\lambda^A = \frac{1}{2}(\lambda_1 - \lambda_2) ,$$

$$\lambda_1 = \mu B_n^{(1)}(K) B_n^{(1)}(K') ,$$

$$\lambda_2 = \mu \frac{\mu_2}{\mu_1} B_n^{(2)}(K) B_n^{(2)}(K') ,$$

$$B_n^{\alpha}(K) = \beta \int_0^{\infty} dr\ r^2 j_1(Kr) \rho_n^{\alpha}(r) ,$$

having considered only the outermost p-type Slater orbitals for the group III, IV, and V ions we are concerned with here. λ^S and λ^A are the symmetric and antisymmetric contributions to the spin-orbit Hamiltonian, λ_α is the contribution due to the ion of index α, μ is an adjustable parameter to be fit to experimental data, μ_2/μ_1 is the ratio between the spin-orbit energies of the outermost p-type orbitals, and ρ_n^{α} are the radial functions corresponding to these orbitals. Finally, β is a normalization constant obtained from the condition[58]

$$\lim_{k\to 0} \frac{B_n^{(\alpha)}(k)}{k} = 1$$

which can be easily obtained from the asymptotic expression of the spherical Bessel function $j_1(kr)$ as $k\to 0$. The Herman-Skillman book[59] tabulates the functions ρ_n^α (shown in Fig. 1), as well as the spin-orbit energies necessary to evaluate the ratio μ_2/μ_1. Perturbation theory (degenerate and nondegenerate) is employed to evaluate the energy shift at **k** due to the spin-orbit interaction (25).

II.2.4. Band Structure Results. We have employed both the local and nonlocal approximations to determine the pseudopotential parameters necessary to fit experimental data related to the band structure of some semiconductors of the diamond (Si and Ge) and zinc-blende (GaAs, GaP, InP, and InAs) structures. We have followed essentially Cohen and Bergstresser[19] (local) and Chelikowsky and Cohen[20] (nonlocal), but we have used slightly different parameters. For one reason, as we have explained before[16] we must use a larger number of plane waves in order to obtain a very smooth dispersion as **k** scans the irreducible 1/48 wedge of the Brillouin Zone (BZ). Here, we have to fulfill an additional requirement: we want to compute the electron-phonon matrix element as rigorously as possible, with particular attention to the selection rules. We have found that a large

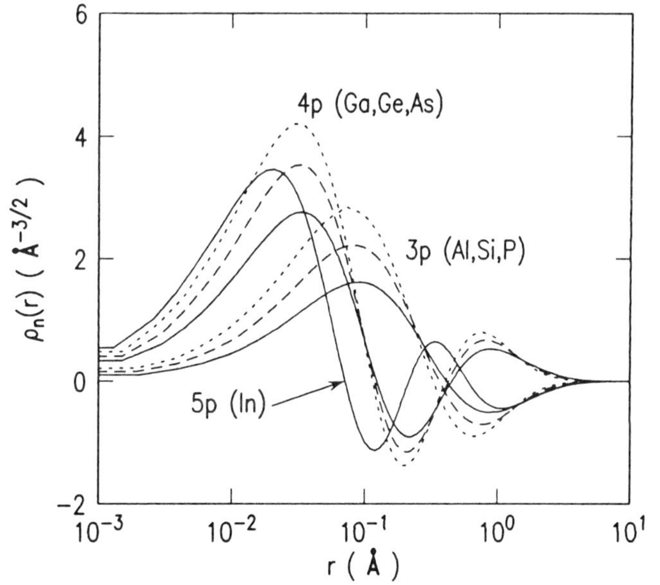

Fig. 1. Radial functions ρ_n for the outermost *p*-type Slater orbitals for the ions of interest as tabulated in Ref. [59]

number of plane waves ($\approx$ 150) is needed, in general, to obtain a numerically satisfactory evaluation of the overlap integrals to be considered below. Thus, we have used slightly larger cutoffs (G_0 in Eq. (20)). In addition, we have attempted to fit experimental data which differ from those considered in Refs. [19] and [20] In particular, we have considered the spin-orbit corrections also for Si and GaP, and we have attempted to "fine-tune" the energy separations between the minima at the Γ and L and X symmetry points in the lowest conduction band according to the most recent experimental data. Obviously, we have also attempted to reproduce the values of high-energy transitions and band-gaps. Experimental data have been obtained from Ref. [60], with some exceptions already discussed in Ref. [16]. There we have also presented our parameters and results for the local approximation. We report in Table I the nonlocal and spin-orbit parameters we have used. The number of plane waves used is indirectly given by the cutoff shown in the table: only **G** vectors such that

$$\frac{\hbar^2}{2m}|\mathbf{k}+\mathbf{G}|^2 \leq E_{cutoff}$$

have been considered. In Fig. 2 we show the band structure of 6 semiconductors obtained from the nonlocal pseudopotentials with spin-orbit interaction. In Tables II and III we list significant parameters relative to the lowest conduction and highest valence bands, respectively, compared to experimental data.[60]

II.2.5 Lattice Dynamics Models. The electron-phonon interaction can be evaluated starting from the Schrödinger equation (21) and displacing the ions from their equilibrium coordinates as indicated in Section II.1.

$$\mathbf{R}^0_{\alpha l} \rightarrow \mathbf{R}_{\alpha l} = \mathbf{R}^0_{\alpha l} + \mathbf{u}_{\alpha l} \tag{26}$$

where the displacement $\mathbf{u}_{\alpha l}$ is Fourier transformed into normal modes as indicated previously in Eq. (5). The phonon frequency, $\omega_{\mathbf{q}\eta}$, and polarization, $\xi^{\eta}_{\alpha}(\mathbf{q})$ for the 6 modes indexed by η are the eigenvalues and eigenvectors, respectively, of the the 6×6 linear system:

$$\sum_{\substack{i,i' \\ \alpha,\alpha'}} [G_{i,i',\alpha,\alpha'}(\mathbf{q}) - m_\alpha \omega^2_{\mathbf{q}} \delta_{i,i'} \delta_{\alpha,\alpha'}](\xi^{(}_{\alpha}\mathbf{q}))_{i'} = 0 \tag{27}$$

where $(\xi^{\eta}_{\alpha}(\mathbf{q}))_{i'}$ is the i' component of the polarization vector ($i,i' = x,y,z$) and

$$G_{i,i',\alpha,\alpha'}(\mathbf{q}) = \sum_l \hat{G}_{i,i',\alpha,\alpha'}(l)\, e^{i\mathbf{q}\cdot\mathbf{R}_l} \tag{28}$$

The matrix $\hat{G}$ expresses the dynamics of the lattice in the harmonic approximation. If $\Phi(\mathbf{R}_{\alpha l},\mathbf{R}_{\alpha' l'})$ is the pair potential between the ions α and α' in the cells l and l', $\hat{G}$ can be expressed as

$$\hat{G}_{i,i',\alpha,\alpha'}(l'') = \hat{G}_{i,i',\alpha,\alpha'}(\mathbf{R}_l - \mathbf{R}_{l'}) = -\frac{\partial^2 \Phi}{\partial(\mathbf{R}_{\alpha l})_i \partial(\mathbf{R}_{\alpha' l'})_{i'}} \tag{29}$$

where the subscripts i and i' indicate the components of the vectors **R**, and the first equality holds since translational symmetry ensures that that $\hat{G}$ depends only

TABLE I

NONLOCAL PSEUDOPOTENTIAL FORM FACTORS AND SPIN–ORBIT PARAMETERS
(V_S, V_A, E_{cutoff}, α_0^{α}, A_l^{α} are in Rydbergs; a, d_l^{α} are in Å)

	a	$V_S(G^2=3)$	$V_S(G^2=8)$	$V_S(G^2=11)$	$V_A(G^2=3)$	$V_A(G^2=4)$	$V_A(G^2=11)$	E_{cutoff}
Ge	5.65	−0.236	0.0019	0.056	–	–	–	7.5
Si[a]	5.43	−0.257	−0.040	0.033	–	–	–	8
GaAs	5.64	−0.235	0.018	0.050	0.075	0.060	0.005	7.5
InP	5.86	−0.235	0.0	0.051	0.090	0.075	0.025	7
InAs	6.04	−0.230	0.0	0.045	0.055	0.045	0.007	6.5
GaP	5.44	−0.240	0.020	0.057	0.100	0.070	0.025	7.5

	α_0^1	β_0^1	β_0^2	A_2^1	A_2^2	d_0^1	d_0^2	d_2^1	d_2^2	μ	$\mu^{(2)}/\mu^{(1)}$
Ge	0.0	0.0	–	0.295	–	0.0	–	1.22	–	0.00093	–
Si	0.55	0.32	–	0.0	–	1.06	–	0.0	–	0.000157	–
GaAs[b]	0.0	0.10	0.0	0.65	0.25	1.06	1.27	1.11	1.33	0.000572	0.334
InP[b]	0.30	0.05	0.25	0.35	0.55	1.06	1.27	1.15	1.38	0.0017	3.599
InAs[b]	0.0	0.25	0.35	0.50	1.00	1.06	1.27	1.19	1.43	0.00041	0.757
GaP[b]	0.32	0.05	0.30	0.45	0.40	1.06	1.27	1.07	1.29	0.0038	1.59

a. J. R. Chelikowsky and M. L. Cohen, Phys. Rev. B, vol. 14, 556 (1976).

b. Ion 1 is the anion. For the cation, α_0^2 is set to zero.

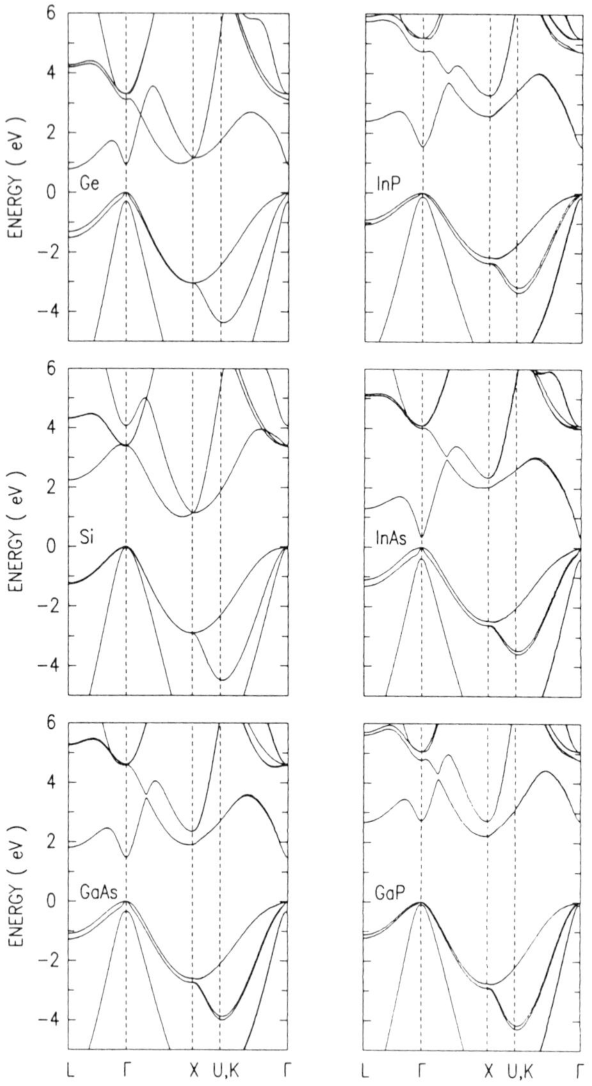

Fig. 2. Structure of the 6 highest-lying valence bands and of the 8 lowest-lying conduction bands using nonlocal empirical pseudopotentials with spin-orbit interaction.

TABLE II

CONDUCTION BAND PARAMETERS (NONLOCAL)

(E in eV, α in 1/eV, m in units of m_0)

	E_{gap}	E_{Γ}[a]	E_L[a]	E_{Δ}[a,b]	E_{X_7}[a]	m_{Γ}	α_{Γ}[c]	m_L	α_L[d]	m_X	α_X
Ge	0.79	0.138	0.0	0.180 (0.83)	–	0.049	−0.76	1.568 (l) 0.094 (t)	−0.31	1.851 (l) 0.195 (t)	0.91 (l) −0.12 (t)
Si	1.14	2.241	1.103	0.0 (0.85)	–	0.212	−0.26	1.973 (l) 0.153 (t)	−0.23	0.905 (l) 0.191 (t)	−0.44[d]
GaAs	1.48	0.0	0.330	0.418 (0.96)	0.874	0.071	−0.75	1.648 (l) 0.123 (t)	−0.28	1.460 (l) 0.243 (t)	1.82 (l)[e] −0.09 (t)
InP	1.54	0.0	0.851	1.029 (1.00)	1.719	0.104	−1.37	2.188 (l) 0.172 (t)	−0.40	0.985 (l) 0.276 (t)	−0.20 (l) −0.10 (t)
InAs	0.33	0.0	0.982	1.705 (0.90)	2.019	0.038	−1.44	1.875 (l) 0.120 (t)	−0.38	1.981 (l) 0.246 (t)	2.45 (l)[e] −0.08 (t)
GaP	2.22	0.508	0.461	0.0 (1.00)[f]	0.495	0.185	−0.72	1.821 (l) 0.159 (t)	−0.26	1.031 (l) 0.248 (t)	−0.12 (l)[e] −0.15 (t)

a. Measured from the bottom of the conduction band.

b. In parenthesis the location of the minimum along the symmetry-line Δ, in units of $2\pi/a$.

c. Averaged over (100), (110), and (111) directions.

d. Averaged over longitudinal and transverse directions.

e. Due to the camel's back structure or the strong nonparabolicity at X, $\alpha_X(l)$ has been computed only along the (100)-direction towards Γ.

f. Experiments place the Δ-minimum 0.07–0.1 $(2\pi/a)$ away from X.

TABLE III

VALENCE BAND PARAMETERS (NONLOCAL)

(Δ_{so} in eV, m in units of m_0)

	Calculated					Measured[a]			
	$\Delta_{so}(\Gamma_{7v}^+ - \Gamma_{8v}^+)$	m_h	m_l	m_{so}		$\Delta_{so}(\Gamma_{7v}^+ - \Gamma_{8v}^+)$	m_h	m_l	m_{so}
Ge	0.296	0.254	0.049	0.115	(100)	0.296	0.283	0.042	0.084
		0.477	0.056	0.102	(110)		0.377	0.044	0.095
		0.390	0.055	0.081	(111)				
Si	0.044	0.346	0.229	0.264	(100)	0.044	0.537	0.153	0.234
		0.618	0.171	0.251	(110)				
		0.732	0.163	0.251	(111)				
GaAs	0.328	0.388	0.089	0.157	(100)	0.33	0.45	0.068	0.133
		0.658	0.081	0.152	(110)	0.35	0.57	0.087	0.170
		0.920	0.079	0.144	(111)				
InP	0.130	0.564	0.173	0.194	(100)	0.108	0.45	0.12	0.121
		0.963	0.148	0.177	(110)	0.110	0.65		
		1.220	0.149	0.158	(111)				
InAs	0.380	0.341	0.042	0.121	(100)	0.37	0.33	0.024	0.14
		0.583	0.041	0.126	(110)	0.41	0.41	0.026	
		0.757	0.041	0.135	(111)				
GaP	0.083	0.472	0.201	0.208	(100)	0.080	0.54	0.16	–
		0.785	0.161	0.177	(110)	0.082	0.67	0.17	
		1.021	0.161	0.153	(111)				

a. When available, the range of the experimental data is indicated. Data from the compilation in Landolt–Boernstein, "Numerical Data and Functional Relationships in Science and Technology", vol 17, "Semiconductors", O. Madelung, M. Schultz, and H. Weiss Eds. (Springer–Verlag, Heidelberg, 1982).

on the the cell index l'', where $\mathbf{R}_l - \mathbf{R}_{l'} = \mathbf{R}_{l''}$ For a complete discussion of the harmonic formulation of lattice dynamics the reader is referred to Ref. [46].

The function Φ is the total energy and, in principle, can be calculated without any additional assumption using self-consistent[61] or even empirical pseudopotentials.[55] Although a self-consistent approach would be desirable, several complications exist in the field of lattice dynamics calculations. It is difficult to reconcile empirical pseudopotential results with empirical fits to band-structure data,[55] and the *ab initio* pseudopotential calculations rely heavily on high symmetry of the phonon wavevector **q** to make the calculations tractable. Therefore, we have adopted once more an empirical approach and obtained the dynamical matrix $\hat{G}$ from a suitable model. Two approaches of purely "fitting" nature, the *valence shell* model[17] and the *adiabatic bond-charge* model[62] are available in the public domain.[63,64] For our purpose, the fitting parameters, obtained by matching the theoretical dispersion $\omega_{\mathbf{q}\eta}$ from Eq. (27) with experimental neutron scattering data[65] are *given* constants, which we view in our computational model as completely unrelated to the electronic properties of the crystal. Of course, forcing the model to deliver reliable eigenvalues gives no guarantee as to the reliability of the eigenvectors (see comments and references given in Refs. [30] and [61]). The problem of relating the phonon spectra to the electronic properties belongs to a more fundamental class of problems which are difficult to handle even when willing to abandon the empirical approach to pseudopotentials and moving to self-consistent formulations.[66,67]

II.3. Rigid Pseudo-Ion Matrix Elements

Having obtained by independent means a model for the lattice dynamics, let us return to (1) and (2) which give for the self-consistent perturbation due to the motion of the ions

$$\delta\phi(\mathbf{q}+\mathbf{G}) = -\sum_{\mathbf{G}'} \varepsilon^{-1}_{\mathbf{G},\mathbf{G}'}(\mathbf{q}) \sum_{\alpha l} \mathbf{u}_{\alpha l} \cdot (\mathbf{q}+\mathbf{G}')\, V_{b\alpha}(\mathbf{q}+\mathbf{G}') \tag{30}$$

where $V_{b\alpha}(\mathbf{q}+\mathbf{G}')$ is the Fourier component of the bare potential which appeared in (1). In order to make a connection with the concept of a pseudopotential, we must replace the screened gradient of the bare potential, above, with the gradient of the screened potential, *i.e.*, the gradient of the pseudopotential. The total perturbation can now be written

$$\delta\phi(\mathbf{q}+\mathbf{G}) = \sum_{\alpha l} \mathbf{u}_{\alpha l} \cdot (\mathbf{q}+\mathbf{G})\, V_{\alpha}(\mathbf{q}+\mathbf{G}) \tag{31}$$

where $V_\alpha(\mathbf{q}+\mathbf{G})$ is the pseudopotential. In the case of local pseudopotentials it represents the form factor defined in (10) and (11), or if nonlocal corrections are desired, a potential of the form (17) is added to the local term as indicated in (8).

In writing (31) we have retained the assumption that the pseudopotential responds instantaneously to the motion of the ion (the "adiabatic" or "Born-Oppenheimer" approximation) discussed previously. Further, we assume that the pseudopotential does not deform as the ion moves. Clearly, by absorbing the inverse dielectric matrix into the pseudopotential in going from (30) to (31) we have neglected local field effects, represented by the off-diagonal elements of $\varepsilon^{-1}_{\mathbf{G},\mathbf{G}'}$. The latter assumption is known as the rigid pseudo-ion approximation and has been used widely, as pointed out in the introduction.

Using (8), (13), and (17) to express the perturbation in terms of its Fourier components, (5) for the ion displacements, and (20) for the electron states, the only nonvanishing matrix elements for the electron-phonon interaction between electron states in bands n and n' be written as (cf. (6))[21]

$$< \mathbf{k}',n';N_{\mathbf{q}\eta}' \mid V_{el-ph} \mid \mathbf{k},n;N_{\mathbf{q}\eta} > =$$

$$\left[\frac{\hbar}{2NM\,\omega_{\mathbf{q}\eta}}\right]^{1/2} \Delta_\eta(\mathbf{k}',n';\mathbf{k},n)\,(N_{\mathbf{q}\eta}+\tfrac{1}{2}\pm\tfrac{1}{2})^{1/2} \tag{32}$$

where $\mathbf{q} = \mathbf{k}' - \mathbf{k} + \mathbf{G}_u$, $\mathbf{G}_u$ being the reciprocal lattice vector needed to map $\mathbf{k}' - \mathbf{k}$ into the first BZ (Umklapp process), $N_{\mathbf{q}\eta}$ is the phonon average occupation, and the "+" and "−" signs refer to emission and absorption, respectively. We shall call

$$\Delta_\eta(\mathbf{k}',n';\mathbf{k},n) = -\frac{i}{2}\sum_\alpha \left[\frac{M}{m_\alpha}\right]^{1/2} \xi^\eta_\alpha(\mathbf{q})\cdot\mathbf{S}_\alpha(\mathbf{k}',n';\mathbf{k},n) \tag{33}$$

the *deformation potential,* where $M = \sum_\alpha m_\alpha$ is the mass of the primitive cell, and

$$\mathbf{S}_\alpha(\mathbf{k}',n';\mathbf{k},n) = \sum_{\mathbf{G}} (\mathbf{q}+\mathbf{G})\, e^{-i\mathbf{G}\cdot\boldsymbol{\tau}} \Bigg[V_\alpha(\mathbf{q}+\mathbf{G}) \sum_{\mathbf{G}'} u_{n,\mathbf{G}'-\mathbf{G}}(\mathbf{k})u^*_{n',\mathbf{G}'}(\mathbf{k}') +$$

$$\sum_{\mathbf{G}'} V^\alpha_{NL}(\mathbf{k}'+\mathbf{G}',\mathbf{k}+\mathbf{G}'-\mathbf{G})u_{n,\mathbf{G}'-\mathbf{G}}(\mathbf{k})u^*_{n',\mathbf{G}'}(\mathbf{k}')\Bigg]. \tag{34}$$

II.3.1. The Interpolation Problem. The evaluation of the electron-phonon matrix element (32) presents a main difficulty. It is clear from Sec. II.2.2 that the determination of the ionic pseudopotentials in the band-structure context gives us the value of the local form factors only at a finite (and small) number of **G**'s. But (34) requires the knowledge of the full **q**-dependence of the form factors $V^\alpha_L(\mathbf{q})$.

To obtain the full functional form of the pseudopotential, various approaches have been taken. The most rigorous way to proceed consists in going back to the definition of the screened form factor in terms of the bare ion potential $V_{b\alpha}$:

$$V^\alpha(\mathbf{r},\mathbf{r}') = \frac{V_{b\alpha}(\mathbf{r})}{\varepsilon(\mathbf{r},\mathbf{r}')} \tag{35}$$

which is equivalent to the definition of $\delta\phi$ given in (2). The problem is now reduced to the evaluation of the dielectric function $\varepsilon(\mathbf{r},\mathbf{r}')$ (or its Fourier transform). The constraint of simultaneously starting from the basic equation (35) and obtaining an accurate band-structure is a very strong one. Self-consistent pseudopotentials allow this, but they are often in error even by large fractions of an eV in predicting excitation spectra. Therefore, once again, empirical approaches have been taken in the past. The free electron model has been employed by Allen and Cardona[26] and by Glembocki and Pollak[23,24] to approximate the dielectric function. This implies, in the local approximation,

$$\lim_{q \to 0} V^{\alpha}(\mathbf{q}) = -\frac{2}{3} E_F \tag{36}$$

where E_F is the Fermi energy. Bednarek and Rössler [25] pointed out that, for the case of the $\Gamma \to \Delta$ transition in Si, an empirical form of the type

$$\lim_{q \to 0} V^{\alpha}(\mathbf{q}) = 0 \tag{37}$$

yields better agreement with the experimental data. Local-field corrections to the screening functions were invoked as a possible justification for (37). Allen and Cardona[27] later employed the latter limit for the form factor. Recently, Gopalan and co-workers[28] and Zollner and co-workers[29,30] have either reverted to the free-electron-model (36), or considered both cases and analyzed the different results caused by a different choice for the interpolation of the form factors. Grinyaev and co-workers[31,32] have employed a different set of nonlocal pseudopotentials screened with the expression for $\varepsilon(\mathbf{q})$ given by Penn[68] to obtain in a consistent way both the band-structure and the intervalley deformation potentials for some III-V compound semiconductors. This seems a particularly proper approximation, since the Penn model deviates from the free-electron picture particularly as $q \to 0$. Since only the diagonal part of the dielectric function is considered though, only the longitudinal response of the potential is included. Herbert[22] has chosen a similar strategy and has analyzed also the effect of the high-q cut-off used to truncate the series in (34) as well as to limit the size of the band-structure eigenvalue problem (22). While in the latter case the truncation can be justified by "folding down" the Hamiltonian[55] in (22), it is not so in the case of (34). Usually, the form factor is set to zero for $q > \sqrt{14}$ or $q > \sqrt{16}$ (in units of $2\pi/a$). If this cutoff is removed, Herbert found significantly different results.

Although it is difficult to judge the effects of the transverse local fields in screening the electron phonon interaction, it is not unreasonable to expect that they may be important, and so a more exact treatment, as shown below, is the evaluation of the contributions of the off-diagonal components of the dielectric matrix. We will see also that it appears that a calculation including the full dielectric matrix indicates that the pseudopotential should have a nonzero "tail" beyond $q = \sqrt{16}$.

II.3.2. An Effective Potential. At this point we will use the short-range potential $\delta\phi^{sr}$ defined in Section II.1 along with the (RPA) dielectric matrix to calculate an effective short-range screened potential. In other words, we will *define* the equivalent of a pseudopotential which would appear in the rigid pseudo-ion matrix element, (38), to make it consistent with (4) and (6)

$$V_{eff}(|\mathbf{q}+\mathbf{G}|) \equiv$$

$$\frac{\mathbf{q}+\mathbf{G}}{|\mathbf{q}+\mathbf{G}|^2} \cdot \sum_{\mathbf{G}'\neq 0} (1-\delta_{\mathbf{G},0}) \left[\varepsilon^{-1}_{\mathbf{G}\mathbf{G}'} - \frac{\varepsilon^{-1}_{\mathbf{G}0}\varepsilon^{-1}_{0\mathbf{G}'}}{\varepsilon^{-1}_{00}} \right] (\mathbf{q}+\mathbf{G}')\, V_b(\mathbf{q}) \tag{38}$$

We will use Si as an example, and calculate the effective potential starting with a spherical bare ion (Si^{4+}) model potential used by Schlüter, *et al.*,[67]

$$V_{b\alpha}(q) = (a_1/q^2) \left[\cos(a_2 q) + a_3 \right] e^{a_4 q^4} \tag{39}$$

where the a's are constants, $\{a_i\} = \{-.992, .791, -.352, -.018\}$; using these constants, if the wavevector is in atomic units the potential is in units of Ry. Self-consistent bulk bandstructure calculations using this potential were verified to give "excellent agreement"[67] with empirical pseudopotential results.

The dielectric matrix was calculated using a mean-value-point scheme[69] using the ten mean-value points given by Chadi and Cohen,[70] and using the bandstructure and Bloch states taken from local empirical pseudopotential calculations[20] with a spherical cutoff E_{cutoff} = 8 Ry. We have summed over 90 conduction bands in the calculation of the 90×90 dielectric matrix, which was then inverted to obtain the inverse dielectric matrix $\varepsilon^{-1}_{\mathbf{G}\mathbf{G}'}$. This formulation necessarily ignores the effects of exchange and correlation, which contribute to the local fields,[71,72] but the RPA formulation we have adopted gives us at least some indication of the effects of the local fields on the effective potential.

It can be seen that the effective short range potential given in (38) has a minimum wavevector $|\mathbf{q}+\mathbf{G}|$ for which it is nonzero. Since $\mathbf{G}\neq 0$, we have min $|\mathbf{q}+\mathbf{G}| = \sqrt{3}/2$. This is a direct result of the fact that the short range potential in (6) is derived in the long wavelength limit, and the separation of the potential indicated in (4) is not well defined for wavevectors near the BZ boundaries. Nevertheless, we calculate V_{eff} and compare it to the interpolated pseudopotentials that have been used in rigid pseudo-ion calculations.

In Fig. 3 we show the effective potential along with the empirical local pseudopotential form factors of Chelikowsky and Cohen [20], the pseudopotential obtained by interpolating the Chelikowsky-Cohen form factors with the constraints (37) and $V^{\alpha}(\mathbf{q}) = 0$ for $q \geq \sqrt{16}$, and also the form factors from *ab initio* calculations of Appelbaum and Hamann[66] who did not truncate the pseudopotential after $G = \sqrt{11}$, but rather included components up to $G = \sqrt{24}$. The inverse dielectric matrix was calculated at $\mathbf{q}$ = (.5,.5,.5), (.4,.4,.4), and (.2,0,0), and the potential (38) was evaluated at various (small) reciprocal lattice vectors $\mathbf{G}$. Although we cannot resolve the question of how to interpolate the potential as $q \to 0$, we see that a nonzero tail beyond $q = 4$ is in evidence. This is in agreement with the results of Herbert[22] who made a similar calculation, using only the diagonal elements of the inverse dielectric matrix, and found that the large-q tail of the

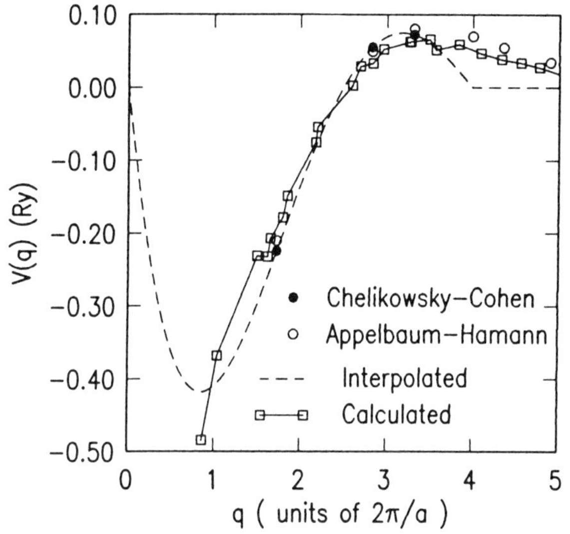

Fig. 3. The calculated effective potential V_{eff}, the Chelikowsky-Cohen local empirical pseudopotential form factors and the interpolated pseudopotential $V^{\alpha}(q)$, and the form factors of Appelbaum and Hamann.

potential can have a significant effect on the matrix elements. The form factors of Appelbaum and Hamann[66] for $G \geq 4$ also show a similar tail (Fig. 3). Clearly, there is work yet to be done regarding the form of the potential to be used, and at this point we leave the dielectric response formulation of the problem, return to the rigid pseudo-ion approach, and present our results.

II.3.3. Deformation Potential Results. In Fig. 4 we show the phonon spectra obtained from the valence shell model with the parameters from Refs. [63,65,73]. Fig. 5 illustrates the actual local ionic interpolated pseudopotentials employed in the evaluation of Eq. (34). The set of Tables IV through VII gives details of the deformation potentials we have obtained for the lowest conduction bands. In Table IV we list the intravalley acoustic deformation potentials. The acoustic phonon matrix elements have the following asymptotic forms for $q \to 0$ when $\mathbf{k}$ is at a valley minimum and $\mathbf{k}' = \mathbf{k} + \mathbf{q}$:[47]

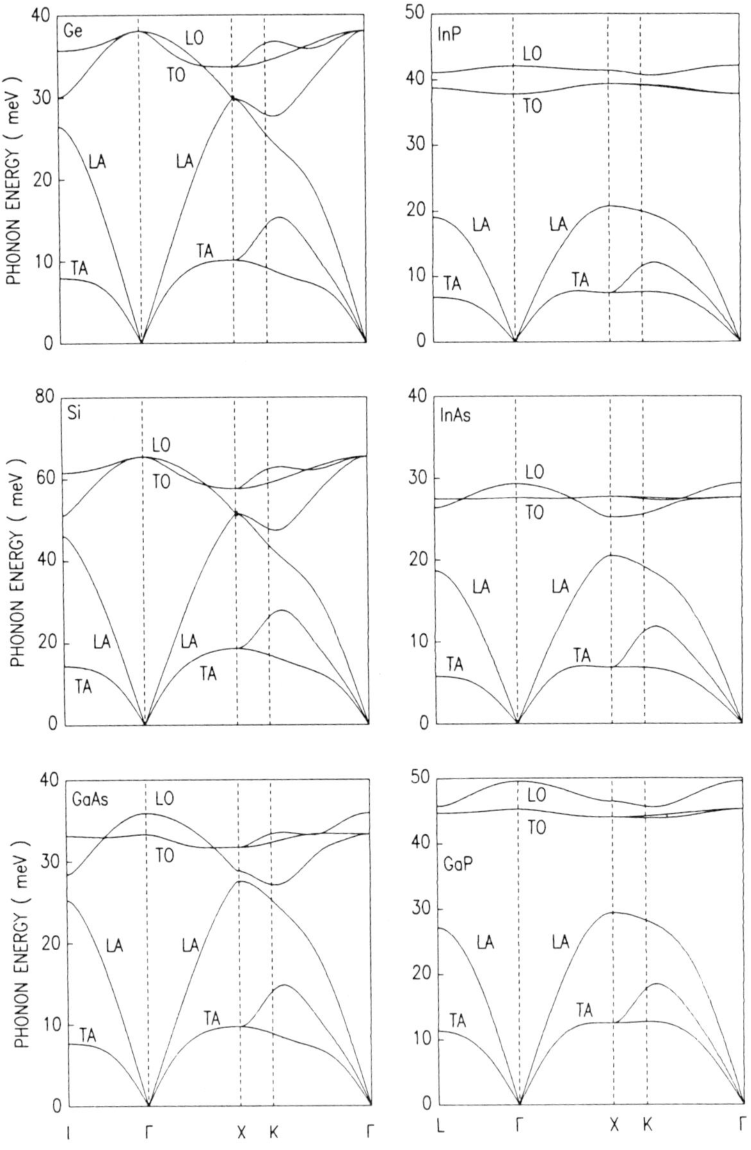

Fig. 4. Phonon dispersion from the valence shell model.

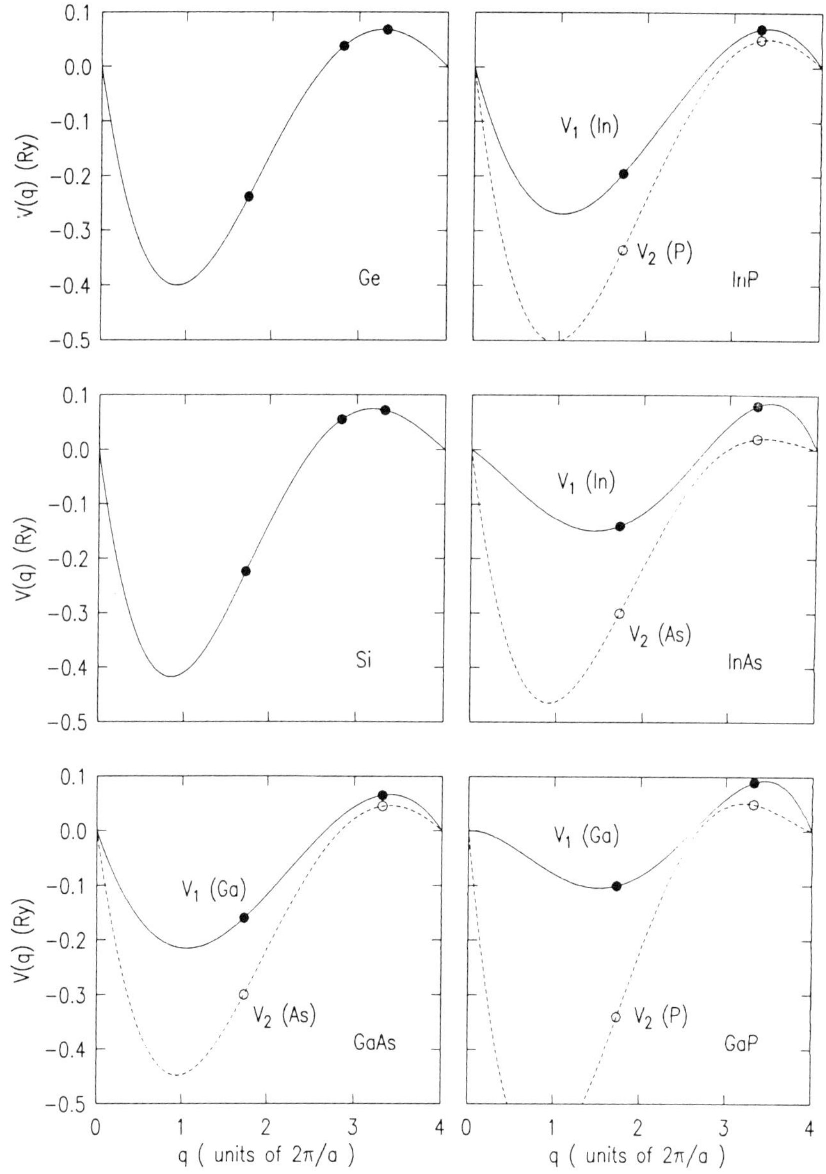

Fig. 5. Interpolated local pseudopotentials employed to compute the electron-phonon matrix elements.

TABLE IV

INTRAVALLEY DEFORMATION POTENTIALS

(Ξ and Δ in eV, $(D_tK)_{op}$ in 10^8 eV/cm)

	$\Delta_{LA}^{(\Gamma)}$	$\Xi_d^{(L)}$	$\Xi_u^{(L)}$	$\Xi_d^{(\Delta)}$	$\Xi_u^{(\Delta)}$	$(D_tK)_{op}^{(L)}$	
Ge	13.1	11.9	−6.9	8.5	−3.9	3.6	local
	20.4	12.6	−7.6	8.4	−3.6	3.8	nonlocal
	5.0	11[a]		9[a]		5.5	b
		15.9 to 19.3	−10.5 to −12.3				c
Si	5.7[e]	8.6	−2.5	9.4	−3.7	2.2	local
	12.3[e]	13.6	−11.3	10.3	−11.3	0.1	nonlocal
				9[a]			d
				8.1 to 9.2	5(?)		c
GaAs	3.2	5.8	1.5	6.7	−0.6	3.7	local
	3.1	5.3	3.7	5.5	1.6	4.0	nonlocal
	7.0	9.2[a]		9.27[a]		3.0	f
	8.6			6.3			c
InP	3.0	7.0	3.2	7.6	−0.6	6.1	local
	4.2	4.1	5.9	6.2	0.4	2.9	nonlocal
	7.0	12.0[a]				6.7	g
	8.0						h
InAs	5.8	3.7	12.4	6.3	0.3	3.7	local
	3.5	2.5	7.4	4.0	8.2	4.9	nonlocal
	8.0						h
	5.8						c
GaP	4.0	8.3	−1.9	7.5	0.1	1.9	local
	6.2	8.1	−2.0	7.4	−1.2	1.0	nonlocal
				7.7			c

a. Effective deformation potential in the isotropic approximation.
b. C. Jacoboni et al., Phys. Rev. B vol. 24, 1014 (1981).
c. From the compilation in Landolt–Boernstein.
d. C. Jacoboni and L. Reggiani, Rev. Mod. Phys. vol. 55, 645 (1983).
e. At $\Gamma_{2'}$.
f. M. A. Littlejohn, J. R. Hauser, and T. H. Glisson, J. Appl. Phys. vol. 48, 4587 (1977).
g. W. Fawcett and D. C. Herbert, J. Phys. C., Vol. 7, 1641 (1974).
h. K. Brennan and K. Hess, Solid–State Electron, vol. 27, 347 (1984).

TABLE V

CALCULATED INTERVALLEY DEFORMATION POTENTIALS (10^8 eV/cm)

	ΓL	ΓΔ	LL	LΔ	ΔΔ	Ref.
Ge	2.9 (2.9)	1.0 (1.0)	4.3 (4.0)	2.7 (2.6)	2.9 (2.7)	a
	2.8	0.6 (2.4)	0.7 (0.3)	–	2.0 (3.7)	b
	3.3 (2.1)	–	1.5 (0.2)	3.6 (3.7)	–	c
Si	2.2 (2.5)	2.4 (2.0)	5.0 (4.5)	2.5 (2.5)	2.9 (2.2)	a
	–	–	–	–	3.0 (4.3)	c
GaAs	3.4 (3.5)	3.4 (4.1)	3.8 (3.9)	2.8 (2.9)	2.9 (3.1)	a
	3.0	3.0	1.2	–	4.9	b
	3.6	3.8	0.6	4.3	6.3	d
	–	3.6 (3.1)	–	–	4.4 (5.5)	c
InP	5.6 (4.7)	4.7 (6.1)	5.9 (5.1)	5.0 (4.3)	4.0 (3.7)	a
	2.3	1.6	0.9	–	3.1	b
	2.9	2.9	0.9	2.6	4.8	d
	3.1 (4.2)	–	– (2.6)	–	–	c
	4.8	4.4	2.5	3.8	4.6	e
InAs	4.1 (3.1)	5.4 (4.4)	3.4 (3.6)	2.7 (2.5)	3.1 (2.8)	a
	2.0	2.2	1.1	–	2.5	b
	2.4	3.0	1.0	3.8	4.8	d
GaP	6.2 (5.4)	7.2 (6.2)	5.4 (5.3)	3.4 (3.2)	4.7 (4.4)	a
	1.1	1.1	0.6	–	2.6	b
	3.0	3.8	0.8	3.5	6.0	d

a. This work. In parenthesis, the values obtained using nonlocal pseudopotentials.
b. S. Zollner, M. Cardona, and S. Gopalan, J. Appl. Phys. (1990). In parenthesis are the values obtained with a different interpolation for the Fourier components of the ionic pseudopotentials.
c. D. C. Herbert, J. Phys. C, vol. 6, 2788 (1973). In parenthesis are the values obtained without any cutoff on the Fourier components of the ionic pseudopotentials.
d. S. N. Grinyaev, G. F. Karavaev, and V. G. Tyuterev, Sov. Phys. Semicond., vol. 23, 905 (1989).
e. D. C. Herbert, W. Fawcett, and C. Hilsum, J. Phys. C, vol. 9, 3969 (1976).

TABLE VI

CALCULATED INTERVALLEY SCATTERING PARAMETERS

	Nonlocal pseudopotentials $(Dq)_{ij}$ (10^8 eV/cm) and $\hbar\omega_{ij}$ (meV)[a]					Local pseudopotentials $(Dq)_{ij}$ (10^8 eV/cm) and $\hbar\omega_{ij}$ (meV)[a]				
ij =	ΓL	ΓΔ	LL[b]	LΔ[c]	ΔΔ[d]	ΓL	ΓΔ	LL[b]	LΔ[c]	ΔΔ[d]
Ge	2.90 (25.96)	0.98 (31.97)	4.03 (14.51)	2.63 (18.65)	2.71 (28.75)	2.90 (26.36)	0.97 (31.83)	4.31 (33.87)	2.66 (28.07)	2.93 (29.16)
Si	2.49 (40.07)	1.99 (45.44)	4.47 (26.69)	2.48 (27.91)	2.19 (45.03)	2.22 (40.90)	2.40 (46.81)	5.00 (59.93)	2.46 (47.05)	2.86 (50.08)
GaAs	3.51 (26.44)	4.06 (27.76)	3.89 (15.43)	2.92 (16.86)	3.14 (24.76)	3.37 (26.79)	3.42 (28.10)	3.75 (27.43)	2.77 (18.75)	2.86 (25.39)
InP	4.72 (39.41)	6.12 (39.45)	5.08 (30.21)	4.33 (30.62)	3.66 (35.14)	5.63 (39.65)	4.66 (40.38)	5.93 (39.11)	5.01 (34.67)	3.99 (37.33)
InAs	3.08 (23.53)	4.37 (24.68)	3.59 (14.91)	2.49 (16.97)	2.82 (25.62)	4.12 (25.20)	5.42 (25.09)	3.43 (25.97)	2.72 (17.13)	3.08 (26.50)
GaP	5.36 (44.52)	6.24 (45.85)	5.33 (43.12)	3.24 (35.71)	4.39 (44.57)	6.23 (45.45)	7.20 (46.48)	5.36 (45.74)	3.35 (35.10)	4.65 (44.42)

a. In parenthesis.

b. Averaged over one (1,1,1)–to–(−1,−1,−1) process, three (1,1,1)–to–(−1,1,1) processes, and three (1,1,1)–to–(−1,−1,1) processes.

c. Averaged over three (1,1,1)–to–(1,0,0) processes and three (1,1,1)–to–(−1,0,0) processes.

d. Averaged over one (1,0,0)–to–(−1,0,0) g–process and four (1,0,0)–to–(0,1,0) f–processes.

TABLE VII

INTERVALLEY SCATTERING PARAMETERS

	From Monte Carlo fits[a] $(Dq)_{ij}$ (10^8 eV/cm) and $\hbar\omega_{ij}$ (meV)[b]					Calculated $(Dq)_{ij}$ (10^8 eV/cm) and $\hbar\omega_{ij}$ (meV)[b]				
ij =	ΓL	ΓΔ	LL[c]	LΔ[d]	ΔΔ[e]	ΓL	ΓΔ	LL[c]	LΔ[d]	ΔΔ[e]
Ge	4.88 (23.21)	4.78 (23.04)	5.26 (23.95)	4.65 (22.83)	3.78 (30.71)	2.90 (26.36)	0.97 (31.83)	4.31 (33.87)	2.66 (28.07)	2.93 (29.16)
Si	2.44 (37.71)	2.42 (37.58)	2.63 (38.87)	2.34 (37.16)	2.26 (50.90)	2.22 (40.90)	2.40 (46.81)	5.00 (59.93)	2.46 (47.05)	2.86 (50.08)
GaAs	5.25 (22.69)	5.48 (23.45)	5.94 (24.97)	5.01 (21.85)	2.99 (24.31)	3.37 (26.79)	3.42 (28.10)	3.75 (27.43)	2.77 (18.75)	2.86 (25.39)
InP	5.06 (22.15)	4.98 (21.89)	5.75 (24.27)	4.68 (20.90)	2.80 (25.68)	5.63 (39.65)	4.66 (40.38)	5.93 (39.11)	5.01 (34.67)	3.99 (37.33)
AlAs	7.02 (25.07)	7.30 (25.07)	8.02 (26.97)	6.63 (23.24)	3.57 (26.34)	–	–	–	–	–
InAs	5.59 (17.45)	6.35 (19.23)	6.35 (19.23)	5.59 (17.45)	3.36 (19.26)	4.12 (25.20)	5.42 (25.09)	3.43 (25.97)	2.72 (17.13)	3.08 (26.50)
GaP	5.10 (25.56)	5.86 (28.56)	5.86 (28.56)	5.10 (25.56)	2.77 (26.61)	6.23 (45.44)	7.20 (46.48)	5.36 (45.74)	3.35 (35.10)	4.65 (44.42)

a. M. V. Fischetti, IEEE Trans. Electron Devices, vol. ED–38, 634 (1991).

b. In parenthesis.

c. Averaged over one (1,1,1)–to–(–1,–1,–1) process, three (1,1,1)–to–(–1,1,1) processes, and three (1,1,1)–to–(–1,–1,1) processes.

d. Averaged over three (1,1,1)–to–(1,0,0) processes and three (1,1,1)–to–(–1,0,0) processes.

e. Averaged over one (1,0,0)–to–(–1,0,0) g–process and four (1,0,0)–to–(0,1,0) f–processes.

$$\Delta_{TA,LA}(\mathbf{k}',\mathbf{k}) \quad \to \Delta_{LA}\, q \qquad (\mathbf{k}' \to \Gamma;\ LA)$$

$$\to O(q^2) \qquad (\mathbf{k}' \to \Gamma;\ TA) \tag{40}$$

$$\Delta_{TA,LA}(\mathbf{k}',\mathbf{k}) \quad \to (\Xi_u + \Xi_d \cos^2\theta) q \qquad (\mathbf{k}' \to L, X;\ LA)$$

$$\to \Xi_d \cos\theta \sin\theta\ q \qquad (\mathbf{k}' \to L, X;\ TA), \tag{41}$$

where θ is the angle between $\mathbf{q}$ and the major (longitudinal) axis in the ellipsoidal L and X valleys, and Δ_{LA}, Ξ_d, and Ξ_u are the deformation potentials. (We will refer to "X valleys" generally, as minima which lie near the X point, along the Δ symmetry line, to be consistent with the common language in the literature.) For the optical-phonon matrix elements we have:

$$\Delta_{TO,LO}(\mathbf{k}',\mathbf{k}) \quad \to (D_t K)_{op} \qquad (\mathbf{k}' \to L;\ TO, LO)$$

$$\to 0 \qquad (\mathbf{k}' \to \Gamma, X;\ TO, LO), \tag{42}$$

where the TO and LO contributions in the L-valley vary in different directions, but, symbolically, $LO^2 + TO^2 = (D_t K)^2_{op}$ is invariant. Notice in Table IV how the nonlocal corrections have a large effect, particularly in the high-lying valleys. This is due to the modifications of the low-q behavior of the form factors due to the nonlocal potential (17). In all fairness, it must be said that in some materials local and nonlocal empirical pseudopotential fits to the band structure yield different results. For some interband optical transitions we lack experimental data, or the experimental data are widely scattered, making the empirical fit uncertain, and causing the discrepancy between the local and nonlocal calculations of the bandstructure. This may well account for part of the difference we see going from a local to a nonlocal approximation in the calculation of the deformation potentials. Notice also that while the analytical limit (41) represents remarkably well our numerical results for the X valleys, significant deviations are found even at relatively small q ($\approx 0.01(2\pi/a)$) in the case of the L valley. Data taken either from direct experiments or used in previous MC simulations[74-78] are also shown in Table IV.

In order to simplify the Tables V through VII for the intervalley deformation potentials in the lowest-lying conduction band, we have lumped together the matrix elements for an intervalley process from valley i to valley j for the various phonon modes and defined an effective intervalley deformation potential, $(Dq)_{ij}$, and an effective phonon frequency, ω_{ij} as follows:

$$(Dq)_{ij} = \left[\sum_{\eta} \Delta^2_{\eta}(\mathbf{k}_i, \mathbf{k}_j) \right]^{1/2} \tag{43}$$

$$(\hbar\omega_{ij})^{-1} = \frac{1}{(Dq)^2_{ij}} \sum_{\eta} \frac{\Delta^2_{\eta}(\mathbf{k}_i, \mathbf{k}_j)}{\hbar\omega_{\eta}(\mathbf{q}_{ij})} \tag{44}$$

where $\mathbf{q}_{ij} = \mathbf{k}_i - \mathbf{k}_j + \mathbf{G}_u$, and $\mathbf{k}_i$, $\mathbf{k}_j$ are the electron wavevectors at the minima, $i,j = \Gamma, L, X$, (again, we use X to denote minima along the Δ symmetry line). Thus, the global quantity $(Dq)^2_{ij}/\hbar\omega_{ij}$ is related to the strength of the intervalley phonon (emission) processes at zero temperature. We have collected data from the literature[29,31,32,79] and processed them according to (41) and (42) for a direct comparison. Average intervalley deformation potentials computed within the local and nonlocal approximations and the relative effective phonon-energies are shown in Table V. A comparison of our local and nonlocal results with previously computed values are reported in Table VI, while a comparison between our local results and values derived from previous MC work[16] is presented in Table VII.

Despite the uneasiness we have expressed above with the freedom in constructing the pseudopotential by interpolating on the empirical form factors, Table VI shows that, on average, different approaches result in similar results in important cases (such as the now popular Γ-to-L deformation potentials for GaAs). Major deviations are seen for those compound semiconductors, such as GaP, InP, and InAs, for which even the band structures used by various authors may differ. Therefore, at least part of the difference seen in Table VI may be attributed not only to the (physically meaningless) interpolation chosen, but also to the use of different pseudopotential form factors.

More pleasing, at least to us, is the message we extract from Table VII. The values we have derived from MC fits show the same qualitative trend obtained here. Si appears to have the lowest deformation potentials (also in excellent quantitative agreement with our empirical results), with values increasing as we move to the heavier III-V compounds, InAs and GaP showing in both cases the largest coupling constants. Excellent quantitative agreement is also seen for GaAs and InP. At first, one may think that this is a "forced" result, because of the way we have chosen the interpolation scheme for the ionic pseudopotentials. However, the fact that a single choice would yield a satisfactory agreement with our loose target for 5 intervalley processes for each of 6 semiconductors is not an obvious result. Perhaps this is some indication that we are on the right track.

III. Scattering Rates

The final and most important step we shall consider here is the calculation of the electron phonon scattering rates accounting for the band-structure (24), the phonon spectrum (27), and the matrix element (32). The scattering rate, $1/\tau^{\pm}_{\eta}(\mathbf{k},n)$ for emission (+) or absorption (−) of phonon of mode η for an electron with wavevector $\mathbf{k}$ in band n is computed from the Fermi Golden Rule (we now use $N_{\mathbf{q}\eta}$ to represent the thermal equilibrium Bose-average for the phonon occupation number)

$$\frac{1}{\tau_\eta^{\pm}(\mathbf{k},n)} =$$

$$\frac{2\pi}{\hbar} \sum_{n',\mathbf{k}'} |\langle \mathbf{k}',n' | V_{el-ph} | \mathbf{k},n \rangle|^2 \, \delta[\, E_n(\mathbf{k}) - E_{n'}(\mathbf{k}') \pm \hbar\omega_{\mathbf{q}\eta} \,](N_{\mathbf{q}\eta} + \tfrac{1}{2} \pm \tfrac{1}{2}) \quad (45)$$

where, as before, $\mathbf{q} = \mathbf{k} - \mathbf{k}' + \mathbf{G}_u$ and only the matrix element corresponding to emission/absorption of the phonon branch η, as in Eq. (32), is selected. The numerical evaluation of Eq. (45) has been described in detail in a previous publication[16] In order to limit the central processing unit (CPU) time required to evaluate Eq. (45) for 916 **k** points in the 1/48 irreducible wedge of the BZ, we had to make full use of the symmetry of the crystal. The eigenvectors $u_{n,\mathbf{G}}(\mathbf{k})$ and $\xi_\alpha^\eta(\mathbf{q})$ have been computed over a shifted mesh of spacing 0.05 $2\pi/a$ in the wedge and stored together with their eigenvalues $E_n(\mathbf{k})$ and $\omega_\eta(\mathbf{q})$ The sum over final states $\mathbf{k}'$ was performed by including these mesh points, but extended over the entire BZ. Therefore, whenever a **k** or **q** outside the wedge was encountered, the electronic and phonon eigenvectors had to be transformed outside the wedge. Only the local approximation to the band structure, ignoring spin-orbit corrections has been employed for this computation (so that each band in the sum (45) is doubly degenerate): because of the large CPU requirements, we cannot experiment freely with the effect of various approximations on the total scattering rates.

We show in Fig. 6 the electron-phonon scattering rate for Si at 77 K and 300 K averaged over all directions as a function of the electron energy E

$$\frac{1}{\tau(E)} = \frac{1}{D(E)} \sum_{\mathbf{k},n,\eta,\pm} \frac{1}{\tau_\eta^{\pm}(\mathbf{k},n)} \delta[\, E_n(\mathbf{k}) - E \,], \quad (46)$$

where

$$D(E) = \sum_{\mathbf{k},n} \delta[\, E_n(\mathbf{k}) - E \,] \quad (47)$$

is the density of states at energy E. The total scattering rates are in excellent agreement with the 'old' values obtained empirically from MC fits in the range 0.5 to 2 eV. Below about 0.1 eV, the scattering rate appears larger because of the high deformation potential for TA phonons in the first conduction band, approaching the old values as the second conduction starts at about 0.13 eV. Nonlocal corrections, at least judging from Table IV, would enhance this difference. We still have to assess its effect on the low-field mobility and diffusion constants with MC simulations.

IV. CONCLUSIONS

The ability to treat electron transport starting from one of the most fundamental quantities of a semiconductor, namely the ionic potential, is very appealing for aesthetical and practical reasons. We have shown here that the evaluation of

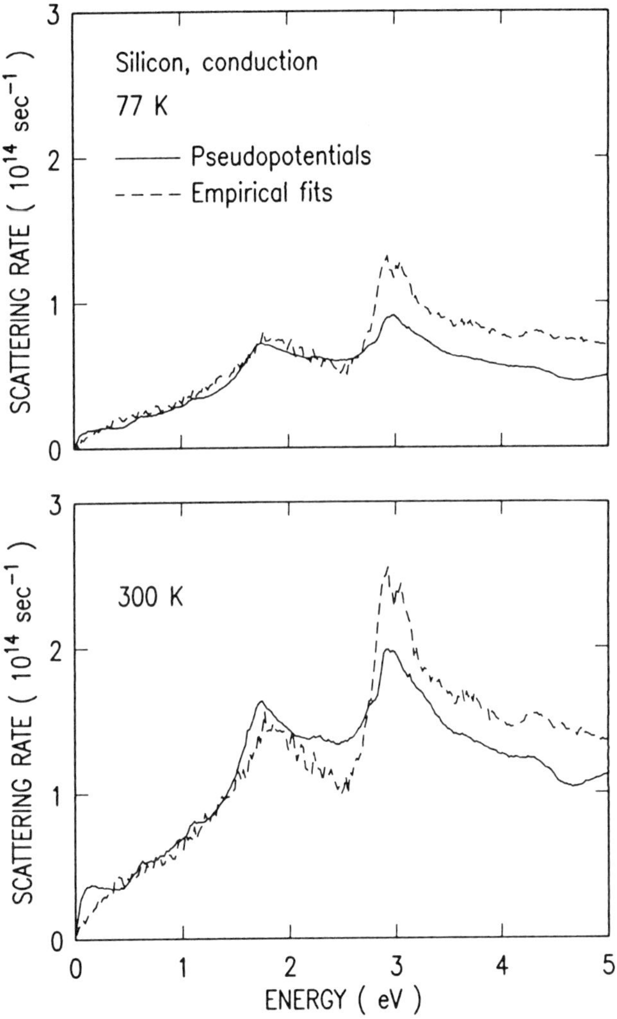

Fig. 6. Total electron-phonon scattering rate in Si as a function of electron energy. The scattering rates have been averaged over all directions, as in (46).

the electron-phonon scattering rates in the entire BZ of many semiconductors using a consistent scheme for interpolating the pseudopotentials results in a good agreement with the rates used in previous MC simulations. Thus, electron transport in semiconductors need not be treated with fitting parameters, as previously done. Unfortunately, a few uncertainties remain: The interpolation scheme has been chosen after "*having shamelessly abandoned rigor,*" to quote Ref. [22]. This choice affects significantly the scattering rates and we cannot provide any rigorous justification for it. A self-consistent approach would re-introduce rigor in the model, and would make it possible to treat also the dynamics of lattice (phonons) in the same unified framework of pseudopotentials, thus removing a second important source of uncertainty in our results. Unfortunately, this remains a problem too formidable to be solved, at least at present.

We are currently testing the fully anisotropic scattering rates in a specially constructed Monte Carlo simulation program, which will truly be the test of *any* model we choose to calculate the matrix elements.

Acknowledgment: The authors would like to acknowledge D.J. Arnold for useful discussions. One of us (J.M.H.) would like to thank R.M. Martin for sharing his valuable insights on the electron-phonon interaction, and J. Bude and Y. Tanimura for several helpful comments. Work done at Urbana was supported by the National Science Foundation through the National Center for Computational Electronics.

References

1. H. Shichijo and K. Hess, *Phys. Rev. B* **23** p. 4179 (1981).
2. J.Y. Tang and K. Hess, *J. Appl. Phys.* **54** p. 5139 (1983).
3. M.V. Fischetti and S.E. Laux, *Phys. Rev. B* **38** p. 9721 (1988).
4. R. Brunetti, C. Jacoboni, F. Venturi, E. Sangiorgi, and B. Ricco, *Solid-State Electron.* **32** p. 1663 (1989).
5. C.J. Stanton, D.W. Bailey, and K. Hess, *Phys. Rev. Lett.* **65** p. 231 (1990).
6. K. Sadra, C.M. Maziar, B.G. Streetman, and D.S. Tang, *J. Appl. Phys.* **66** p. 4791 (1989).
7. M.A. Osman, M. Cahay, and H.L. Grubin, *Solid-State Electron.* **32** p. 1911 (1989).
8. T. Wang and K. Hess, *J. Appl. Phys.* **57** p. 5336 (1985).
9. A.M. Kriman, M.J. Kann, D.K. Ferry, and R. Joshi, *Phys. Rev. Lett.* **65** p. 1619 (1990).
10. N. Sano, M. Tomizawa, and A. Yoshi, *Appl. Phys. Lett.* **55** p. 1418 (1989).
11. N. Sano, M. Tomizawa, and A. Yoshi, *Appl. Phys. Lett.* **56** p. 653 (1990).
12. N. Sano, T. Aoki, M. Tomizawa, and A. Yoshi, *Phys. Rev. B* **41** p. 12122 (1990).
13. R.Thoma, H. Peifer, W.L. Engl, W. Quade, R. Brunetti, and C. Jacoboni, *Proceedings, Workshop on Numerical Modeling of Processes and Devices for Integrated Circuits-NUPAD III*, (June, 1990).
14. J. Bude, K. Hess, and G.J. Iafrate, submitted to *Phys. Rev. B*. See also the chapter by J. Bude in this book.
15. J. Shah, *Solid-State Electron.* **32** p. 1051 (1989).
16. M.V. Fischetti, *IEEE Trans. Electron Devices* **ED-38** p. 634 (1991).
17. L.A. Feldkamp, D.K. Steinman, N. Vagelatos, J.S. Sing, and G. Venkataraman, *J. Phys. Chem. Solids* **32** p. 1573 (1971).
18. A.A. Maradudin, E.W. Montroll, G.H. Weiss, and I.P. Ipatova, "Theory of Lattice Dynamics in the Harmonic Approximation," in *Solid State Physics, Suppl. 3, 2nd edition*, ed. H. Ehrenreich,Academic Press, New York (1971).
19. M.L. Cohen and T.K. Bergstresser, *Phys. Rev.* **141** p. 789 (1966).
20. J.R. Chelikowsky amd M.L. Cohen, *Phys. Rev. B* **14** p. 556 (1976).
21. M.L. Cohen and Y.W. Tsang, "Theory of the Electronic Structure of Some IV-VI Semiconductors," pp. 303-317 in *Proceedings of the Conference on Physics of Semimetals and Narrow-Gap Semiconductors*, ed. D.L. Carter and R.T. Bate,Pergamon Press (1971).
22. D.C. Herbert, *J. Phys. C* **6** p. 2788 (1973).

23. O.J. Glembocki and F.H. Pollak, *Phys. Rev. Lett.* **48** p. 413 (1982).

24. O.J. Glembocki and F.H. Pollak, *Phys. Rev. B* **25** p. 7863 (1982).

25. S. Bednarek and U. Rössler, *Phys. Rev. Lett.* **48** p. 1296 (1982).

26. P.B. Allen and M. Cardona, *Phys. Rev. B* **23** p. 1495 (1981).

27. P.B. Allen and M. Cardona, *Phys. Rev. B* **27** p. 4760 (1982).

28. S. Gopalan, P. Lautenschlager, and M. Cardona, *Phys. Rev. B* **35** p. 5577 (1987).

29. S. Zollner, S. Gopalan, and M. Cardona, *Appl. Phys. Lett.* **54** p. 614 (1989).

30. S. Zollner, S. Gopalan, and M. Cardona, *J. Appl. Phys.* **68** p. 1682 (1990).

31. S.N. Grinyaev, G.F. Karevaev, V.G. Tyuterev, and V.A. Chaldyshev, *Sov. Phys. Solid State* **30** p. 1586 (1988).

32. S.N. Grinyaev, G.F. Karevaev, and V.G. Tyuterev, *Sov. Phys. Semicond.* **23** p. 905 (1989).

33. S. Zollner, J. Kircher, M. Cardona, and S. Gopalan, *Solid-State Electronics* **32** p. 1585 (1989).

34. J.M. Higman, J. Bude, and K. Hess, *Computer Physics Commun.*, (1991). to appear.

35. F. Seitz, *Phys. Rev.* **73** p. 549 (1948).

36. W.A. Harrison, *Phys. Rev.* **104** p. 1281 (1956).

37. W. Shockley and J. Bardeen, *Phys. Rev.* **77** pp. 407-408 (1950).

38. J. Bardeen and W. Shockley, *Phys. Rev.* **80** p. 72 (1950).

39. P. Lawaetz, *Phys. Rev.* **183** p. 730 (1969).

40. L.J. Sham and J.M. Ziman, "The Electron-Phonon Interaction," pp. 223-298 in *Solid State Physics, vol. 15*, ed. F. Seitz and D. Turnbull,Academic Press, New York (1963).

41. V. Heine, P. Nozieres, and J.W. Wilkins, *Philos. Mag.* **13** p. 741 (1966).

42. P. Vogl, *Phys. Rev. B* **13** p. 694 (1976).

43. O. Madelung, *Introduction to Solid-State Theory*, Springer-Verlag, New York (1978).

44. R.M. Pick, M.H. Cohen, and R.M. Martin, *Phys. Rev. B* **1** p. 910 (1970).

45. P. Vogl, "The electron-phonon interaction in semiconductors," pp. 75-115 in *Physics of Nonlinear Transport in Semiconductors*, ed. D.K. Ferry, J.R. Barker, and C. Jacoboni,Plenum Press, New York (1980).

46. P. Brüesch, *Phonons: Theory and Experiments I, Lattice Dynamics and Models of Interatomic Forces*, Springer-Verlag, New York (1982).

47. C. Herring and E. Vogt , *Phys. Rev.* **101** p. 944 (1956).

48. W. Pötz and P. Vogl, *Phys. Rev. B* **24** p. 2025 (1981).

49. V. Heine, "The Pseudopotential Concept," pp. 1-36 in *Solid State Physics, vol. 24*, ed. F. Seitz and D. Turnbull,Academic Press, New York (1970).

50. M.L. Cohen and J.C. Phillips, *Phys. Rev.* **124** p. 1818 (1961).

51. V. Heine and I. Abarenkov, *Phil. Mag.* **9** p. 451 (1964).

52. A.O.E. Animalu, *Phil. Mag.* **11** p. 379 (1964).

53. I. Abarenkov and V. Heine, *Phil. Mag.* **12** p. 529 (1965).

54. A.O.E Animalu and V. Heine, *Phil. Mag.* **12** p. 1249 (1965).

55. M.L. Cohen and V. Heine, "The Fitting of Pseudopotentials to Experimental Data and Their Subsequent Application," pp. 37-248 in *Solid State Physics, vol. 24*, ed. F. Seitz and D. Turnbull,Academic Press, New York (1970).

56. G. Weisz, *Phys. Rev.* **149** p. 504 (1966).

57. S. Bloom and T.K. Bergstresser, *Solid State Commun.* **6** p. 465 (1970).

58. J.P. Walter, M.L. Cohen, Y. Petroff, and M. Balkansky, *Phys. Rev. B* **1** p. 2661 (1970).

59. F. Herman and S. Skillman, *Atomic Structure Calculations,* Prentice Hall, Englewood Cliffs, NJ (1963).

60. Landolt-Börnstein, *Numerical Data and Functional Relationships in Science and Technology* New Series, Ed. in chief K.-H. Hellwege, Group III: "Crystal and Solid State Physics," vol. 17: "Semiconductors," O. Madelung, M. Schultz, and H. Weiss eds. subvolume a: "Physics of Group IV Elements and III-V Compounds," O. Madelung, ed., Springer-Verlag, Berlin (1982).

61. K. Kunc and R.M. Martin, "Ab Definitio Calculation of Structure and Lattice Dynamics of GaAs," pp. 65-99 in *Ab Initio Calculation of Phonon Spectra*, ed. J.T. Devreese, V.E. Van Doren, and P.E. Van Camp,Plenum Press, New York (1983).

62. W. Weber, *Phys. Rev. B* **15** p. 4789 (1977).

63. K. Kunc and O.H. Nielsen, *Computer Physics Commun.* **17** p. 413 (1979).

64. O.H. Nielsen and W. Weber, *Computer Physics Commun.* **18** p. 101 (1979).

65. D.L. Price, J.M. Rowe, and R.M. Nicklow, *Phys. Rev. B* **3** p. 1268 (1971).

66. J.A. Appelbaum and D.R. Hamann, *Phys. Rev. B* **8** p. 1777 (1973).

67. M. Schlüter, J.R. Chelikowsky, S.G. Louie, and M.L. Cohen, *Phys. Rev. B*, p. 4200 (1975).

68. D. R. Penn, *Phys. Rev.* **128** p. 2093 (1962).

69. A. Baldereschi and E. Tosatti, *Phys. Rev. B,* **17** p. 4710 (1978).

70. D.J. Chadi and M.L. Cohen, *Phys. Rev. B* **8** p. 5747 (1973).

71. M.S. Hybertsen and S.G. Louie, *Phys. Rev. B* **35** p. 5585 (1987).

72. M.S. Hybertsen and S.G. Louie, *Phys. Rev. B* **35** p. 5602 (1987).

73. P.H. Borcherds and K. Kunc, *J. Phys. C* **11** p. 4145 (1978).

74. W. Fawcett and D.C. Herbert, *J. Phys. C* **7** p. 1641 (1974).

75. M.A. Littlejohn, J.R. Hauser, and T.H. Glisson, *J. Appl. Phys.* **48** p. 4587 (1977).

76. C. Jacoboni, F. Nava, C. Canali, and G. Ottaviani, *Phys. Rev. B* **24** p. 1014 (1981).

77. C. Jacoboni and L. Reggiani, *Rev. Mod. Phys.* **55** p. 654 (1983).

78. K. Brennan and K. Hess, *Solid-State Electron.* **27** p. 347 (1984).

79. D.C Herbert, W. Fawcett, and C. Hilsum, *J. Phys. C* **9** p. 3969 (1976).

6

Ensemble Monte Carlo Investigation of Nonlinear Transport Effects in Semiconductor Heterostructure Devices

Kevin F. Brennan
School of Electrical Engineering
Georgia Tech, Atlanta, Georgia 30332

INTRODUCTION

The advent of precise epitaxial crystal growth techniques, such as molecular beam epitaxy, MBE, and metal-organic chemical vapor deposition, MOCVD, have fomented the development of new semiconductor devices which contain two or more dissimilar materials grown in contact with one another. Depending upon the type of alignment of the conduction and valence bands as well as the doping differential, various effective potential discontinuities can be formed across the interface. The heterostructure discontinuities are often used to create devices in which the position of the carriers can be controlled in real space. Devices such as heterostructure and quantum well lasers [1] exploit heterostructure band-edge discontinuities to enhance carrier confinement in order to affect higher device efficiency. Spatial control of the free carrier concentration can be further exploited through modulation doped heterostructures in which the free carriers are spatially separated from their parent donor atoms [2]. As a result, high mobility at high carrier concentration can be attained in these structures.

Though heterostructures have become of great importance in modern semiconductor devices, by virtue of the additional control they provide over charge localization, they can be alternatively used to effect carrier transport dynamics. The introduction of a heterostructure at the emitter-base junction of a bipolar junction transistor was first suggested by Shockley [3] as a means of suppressing reverse hole injection into the emitter. As a consequence, a high doping concentration in the base can be retained providing high frequency operation of the transistor.

Though the above mentioned advantages of a heterostructure in a bipolar transistor were recognized early, a heterostructure can also effect the carrier drift velocity in a more subtle manner. If the heterostructure at the emitter-base junction is sufficiently abrupt, the electrons are injected across the interface with a considerable kinetic energy which can result in a further increase in the cutoff frequency of the device. This further enhancement of the high frequency performance arises from an increased electron velocity well above the saturation value called velocity overshoot [4]. The injected high energy electrons can transit the narrow base region well in excess of the steady-state velocity due to the different energy and momentum relaxation times present.

High energy injection across a heterointerface not only alters the carrier drift velocity but can effect the impact ionization rate as well. Chin, et al., [5] first recognized that the electron ionization rate can be enhanced above both the bulk ionization rate as well as the corresponding hole ionization rate through use of a multi-quantum well structure. The electron ionization rate, α, in a device with a spatially periodic electric field superimposed on a dc bias field is enhanced above its respective value in the absence of the periodic modulation as a result of the strong nonlinear dependence of α on the electric field and the existence of a threshold energy in the impact ionization process [6]. Recent experimental measurements have confirmed that the electron ionization rate can be increased over the hole ionization rate in some cases [7-10].

As we have seen from the above examples, the introduction of a heterostructure can alter carrier transport in semiconductors in a nonlinear manner. In order to understand the resulting nonlinear dynamics, typically numerical methods must be employed. Analytical models, though highly useful in many applications, often require some degree of linearization in order to affect a solution. The nonlinear physics one seeks to solve is then quite often removed. Preservation of nonlinear effects in their entirety in a device model is generally only possible through use of numerical techniques. This need coupled with the tremendous technical advances in computers has lead to extensive numerical approaches to semiconductor device analysis.

The most flexible and inherently complete numerical computational scheme for studying semi-classical transport in semiconductors is the ensemble Monte Carlo method. The Monte Carlo method consists of a simulation of the trajectory of one or more carriers in a crystal or device structure subject to the action of applied electric fields, boundary conditions, scattering mechanisms and the crystalline band structure. The carrier free flight time, phonon scattering agents and the final electron states are determined stochastically in accordance with some given probabilities describing the microscopic process. The Monte Carlo method, which draws its name from the stochastic aspect of the analysis, provides a rigorous solution of the Boltzmann equation limited only by the extent to which the physics of the problem is included. Therefore, the Monte Carlo method naturally lends itself to the study of nonlinear transport phenomena in semiconductors, such as that affected by heterostructures.

In this chapter, we analyze through use of the ensemble Monte Carlo method, three different nonlinear transport phenomena which arise in devices containing heterostructures. These are: impact ionization across a potential step [11], velocity enhancement through extended gamma valley confinement via heterostructures [12], and carrier heating during lateral transport in heterostructure devices [13]. The following section presents a brief discussion of the effect a heterostructure has upon the impact ionization rate in a semiconductor. Here we describe a simplified analytical theory as well as more detailed numerical Monte Carlo calculations. Section 3 presents a discussion of repeated velocity overshoot through extended gamma valley confinement focusing on the use of potential steps formed by heterostructures. In Section 4 a discussion of carrier heating via lateral transport in a heterostructure layer is presented. Finally, a summary is presented in Section 5.

IMPACT IONIZATION ACROSS A POTENTIAL STEP

Low noise, high bandwidth operation of an avalanching device requires that the electron and hole ionization rate coefficients be vastly different. Ideally, optimal performance of an avalanching device is obtained for single carrier initiated, single carrier ionization conditions. The presence of only a few percent of secondary carrier ionization can result in a substantial nonlinear increase in the excess noise factor of the device [14-16]. Unfortunately, the electron and hole impact ionization rates are roughly comparable in most compound semiconductors. As a result, avalanching devices made from these materials exhibit poor noise and bandwidth operation.

It was first suggested by Chin, et al., [5] that the electron to hole ionization rates ratio could be altered by introducing a periodic potential structure via a series of quantum wells into the avalanching region of a device (Figure 1). They recognized that owing to the difference in the conduction and valence band edge discontinuities in a GaAs/AlGaAs multi-quantum well structure, as well as the difference in the carrier relaxation rates, that the electron ionization rate could, in principle, be increased above both its bulk value and the corresponding hole ionization rate.

It at first appears that the enhanced ionization rate in a multi-quantum well structure is unphysical on the basis of energy conservation. The net energy gain from the multi-quantum wells is zero when averaged over a full period. Therefore, it appears surprising that the ionization rate, which depends upon the carrier energy, would be enhanced from the bulk rate since a carrier, on average, gains no more energy within a multilayered structure than it would in bulk material at the same applied electric field. Nevertheless, both theoretical predictions [6,11,17,18], and experimental measurements [7-10], indicate that the electron ionization rate is indeed higher in a multi-quantum well device than in bulk material. The question is then, what is the physical origin of the enhancement?

The enhancement can be understood qualitatively in the following way. The impact ionization process has a threshold energy; carriers whose energy lie

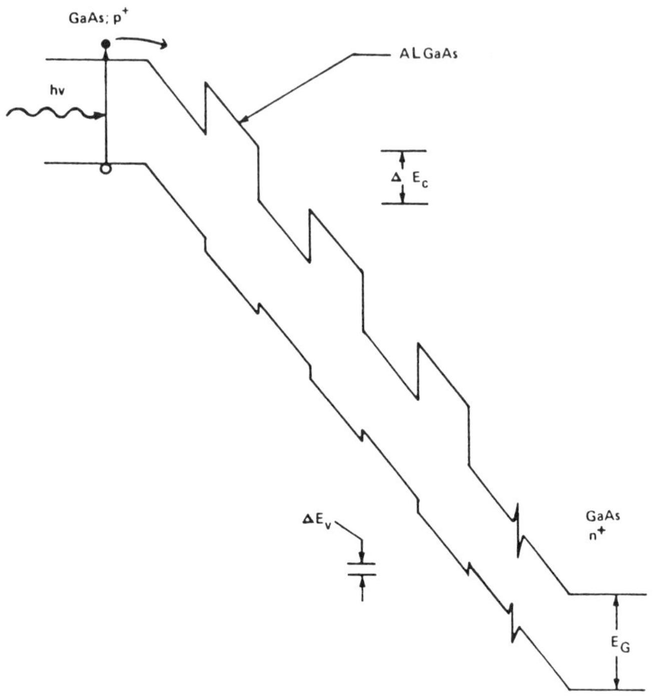

Figure 1. Energy band diagram of a multi-quantum well APD.

below this threshold cannot initiate any impact ionization events. If a carrier is heated, even locally to an energy greater than the threshold, it is possible that this carrier can impact ionize. Superposition of a periodic potential, or equivalently an electric field, onto a uniform applied electric field can then result in localized carrier heating such that impact ionization can occur where it otherwise would not. In other words, ionization can occur in a multi-quantum well device even when the applied field in bulk material leads to no ionization events since the electrons get an energy boost from the potential discontinuity in the immediate vicinity of the step supplying sufficient energy for the electron to attain or exceed the threshold. In this way, the potential discontinuities act to periodically heat the electrons locally causing impact ionization events.

The enhancement can be understood quantitatively from use of Shockley's lucky electron theory. The impact ionization rate, measured as the inverse mean distance a carrier travels prior to ionizing, can be expressed in terms of the mean free path, l, and the probability that the carrier drifts without collision as [18],

$$\alpha_z = \frac{eF}{E_{th}} \exp\left[-\frac{\hbar}{eF}\int_{E}^{E_{th}} \left(\frac{dE}{dk}\right)^{-1} \frac{1}{\tau_{tot}^{ph}(E)}\, dE\right] \tag{1}$$

where E is the average electron energy, F is the electric field, and $1/\tau^{ph}{}_{tot}(E)$ is the energy-dependent phonon scattering rate. The above expression for the rate is obtained from the product of the probability that the carrier drifts without collision from the average energy to the threshold energy (the exponential term in equation 1) and the reciprocal of the minimum ionization mean free path (the prefactor in eq. 1).

Here we are interested in the effect on the impact ionization rate of the superposition of a periodic square well potential, arising from the multi-quantum well structure, onto a linear potential due to a constant applied overall field. Being that the multi-quantum well potential is periodic it can be expanded in terms of a Fourier sine series [6]. The periodic multi-quantum well potential gives rise then to a periodic field which is superimposed onto the constant applied field. The periodic field can be determined from the derivative of the Fourier sine expansion of the multi-quantum well potential.

Combining the periodic and constant components of the field gives an expression for the net field acting on the carriers. Notice that the field appears in both the prefactor and the exponential term in equation 1. The contribution of the periodic component of the field to the prefactor can be neglected since it averages out in many cases [6]. Retaining the spatial variation of the field in the exponent however, has an interesting effect on the ionization rate. Keeping only the first two Fourier components in the expansion of the field, the average ionization rate (defined as the integral of eq. 1 with respect to z) over the full superlattice period becomes,

$$\alpha = \frac{1}{2\ell}\frac{e}{E_{th}}\int_{-\ell}^{\ell} F_0 \exp\left[-\frac{C}{F_0 + \dfrac{2V_0}{\ell}\cos\left(\dfrac{\pi z}{\ell}\right)}\right] dz \tag{2}$$

where C is the z-independent part of the exponent,

$$C = \frac{\hbar}{e}\int_{\bar{E}}^{E_{th}}\left(\frac{dE}{dk}\right)^{-1}\frac{1}{\tau_{tot}^{ph}(E)}\,dE \tag{3}$$

After a Taylor Series expansion of the denominator in the exponent, the ionization rate reduces to,

$$\alpha = \frac{eF_0}{E_{th}}\,e^{-\frac{C}{F_0}}\,I_0(\beta) \tag{4}$$

where $Io(\beta)$ and β are, the modified Bessel function of zero order and $2V_0C/\ell F_0^2$, respectively. It is important to realize that the average rate is always enhanced over a full period since the modified Bessel function of zero order is always greater than 1.

Though the analytical formulation of the ionization rate derived above aids the understanding of the physical origin of the impact ionization rate enhancement in multi-quantum well devices, it is of limited usefulness in calculating the ionization rate in realistic systems due to the complicated dependence of the parameter C on the band structure and phonon scattering rate. A more complete theory needs address the details of the carrier flight to the threshold energy and beyond taking into account the detailed action of the band structure and phonon scatterings.

The ensemble Monte Carlo method provides a natural means of including much of the details of the scattering rates and band structure with little approximation in calculating carrier transport [19] within a large energy range. The Monte Carlo model makes few assumptions and retains most of the important physics needed to describe high energy, high electric field transport. In the Monte Carlo calculations we review here, the full details of the conduction band structure are included based on an empirical pseudopotential calculation [20]. Initially, numerous mesh points for the E(k) structure are calculated throughout the reduced Brillouin zone (the reduced zone can completely reproduce the entire first Brillouin zone through the allowed symmetry operations of the crystal [19]). During the simulation, the energy corresponding to a particular k-point is determined from the numerically generated E(k) mesh by interpolation. In this way, proper account of the nonparabolicity and warping of the band structure at high energies is assured.

The electron-phonon scattering rates are calculated at low energy using Fermi's golden rule. A reasonably accurate formulation of the rates is obtained if the integration is performed numerically over the final density of states derived from the pseudopotential calculation as discussed by Fischetti and Laux [21]. This is particularly important in the case of the valence bands since the warping of these bands is great even at low energy. Nevertheless, phonon scattering calculations which include the numerically generated density of states still suffer from the inevitable lack of information and imprecision of the coupling constants, phonon energies and other parameters needed. The only viable way in which these parameters can be assessed and the model be calibrated is to compare the steady-state calculations of the various macroscopic quantities, such as the drift velocity, threshold field, and peak steady-state drift velocity to experimental measurements [22]. We have performed a series of "computer experiments" in order to assess a set of material parameters which, when used with our ensemble Monte Carlo simulator, reliably reproduces steady-state experimental data [23].

In the calculations presented here, the low energy phonon scattering rates are calculated assuming a nonparabolic analytical model for the band structure [19]. Though this approach is perhaps not quite as accurate it is still reasonably useful at low energies within the first conduction band. In our model, the high energy phonon scattering rates are corrected for quantum mechanical collision broadening effects by numerically solving the full order self-energy equation subject to the density of states generated from the pseudopotential calculation [24]. In this way, much of the essential physics of high energy transport in semiconductors is included.

In order to appreciate the difficulty of studying high energy electron transport it is useful to examine a typical electron trajectory in a crystal under the action of a high applied electric field (Figure 2). Examination of Figure 2 shows that the electron history is extremely complicated between impact ionization events. The energy of the carrier fluctuates around some average value with no apparent order. On average, the energy gain from the field is balanced by energy loss to the lattice via phonon scatterings. However, the electron occasionally undergoes local fluctuations from the average in which its energy can be vastly different from the average steady-state value. It is during these excursions from the mean behavior that an impact ionization event occurs. In order to account for this complicated behavior it is necessary then to employ numerical simulation.

Brennan, et al., [11] first demonstrated theoretically that the electron ionization rate can be enhanced over its bulk and the corresponding hole ionization rate in a multi-quantum well device similar to that sketched in Figure 1. In general, the net ionization rate is found under constant electric field from [18,25],

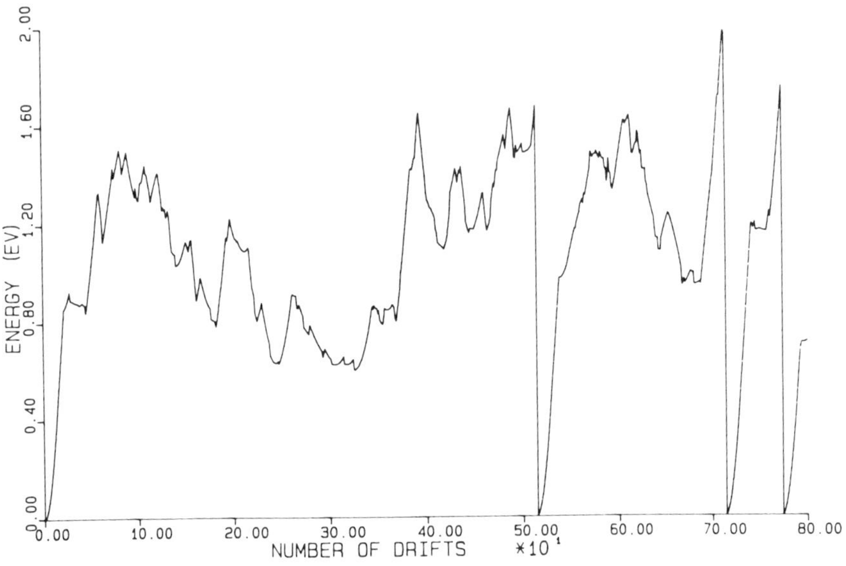

Figure 2. Electron energy as a function of intercollisional drifts in bulk GaAs at an applied field of 500 kV/cm.

$$\alpha = \frac{\alpha_{GaAs} L_W + \alpha_{AlGaAs} L_B}{L_W + L_B} \tag{5}$$

where L_W and L_B are the well and barrier widths respectively. In this case these layers are comprised of GaAs and AlGaAs. Eq. 5 asserts that the net ionization rate is determined from the weighted average of the GaAs and AlGaAs layer rates. It is important to realize that the rates used in eq. 5 are not necessarily the bulk ionization rates. As discussed above, the ionization rate within the GaAs layer is always enhanced above its bulk value. Depending upon the extent of the enhancement, the net ionization rate in the entire multi-quantum well structure can be larger than that in a corresponding bulk GaAs sample. The enhancement of the ionization rate within the narrow gap layer, GaAs in Figure 1, must be quite large in order that the net ionization rate be larger than the bulk rate. This is obvious since within the multi-quantum well structure, the wide gap layers, AlGaAs in Figure 1, do not contribute significantly to the ionization rate at most electric field strengths, and hence act as a dead layer. Therefore, since the net ionization rate is found by averaging over the AlGaAs layers as well as the GaAs layers, in order that the total net rate in the multi-quantum well structure exceed that within bulk GaAs, the rate within the GaAs layers must be very much enhanced over its corresponding bulk value.

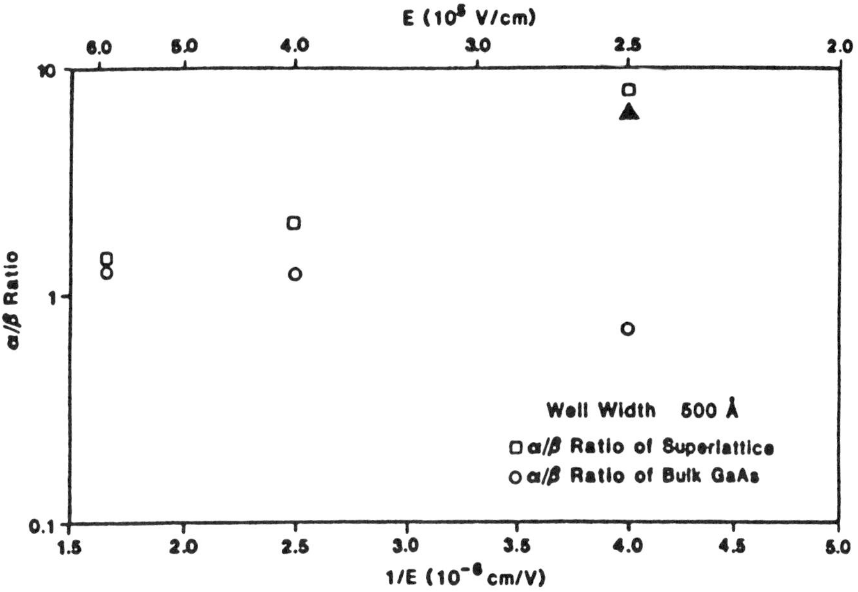

Figure 3. Calculated electron/hole ionization rates ratio in bulk GaAs and a 500/500 Å well/barrier GaAs/AlGaAs (45% Al) multi-quantum well APD. Triangle represents experimental data [26].

The calculated electron to hole impact ionization rates ratio as a function of inverse applied electric field for the multi-quantum well structure shown in Figure 1 is presented in Figure 3 for a structure with 500 Å wide well and barrier layers respectively. For comparison purposes, the calculated bulk ionization rates ratio is also included in the Figure. The largest value of the ionization rates ratio occurs at low applied electric field strength, ~ 250 kV/cm, for this example. The increased ratio is due predominantly to the enhanced electron ionization rate. Comparison to recent experimental measurements of the electron to hole ionization rates ratio [26] made on a similar device structure show good agreement with the calculations. Inspection of Figure 3 shows that the electron ionization rate is enhanced by roughly a factor of 8 from its bulk value.

Further inspection of Figure 3 shows that at low applied electric fields the net electron ionization coefficient is significantly enhanced over both its bulk value as well as the corresponding hole ionization coefficient. As the electric field strength increases however, the multi-quantum well rates ratio shows no significant difference from that for the bulk. This is not surprising since at high applied electric fields, the effect of the potential step on the carrier temperature is less important. At high applied fields, the electrons gain the largest fraction of the energy needed to reach threshold from the field and not from the potential step. The electrons are already heated to energies near or at the threshold by the field. Hence, in this case the electrons are at the threshold anyway and the step makes little difference. At low fields, however, the potential step can have a much greater effect on the electron ionization rate since it can provide a larger fraction of the energy needed to reach threshold.

The calculated hole impact ionization rate in the representative multi-quantum well devices considered here does not show any appreciable difference from the weighted average of the corresponding bulk rates. As discussed above, the electron ionization rate within the multi-quantum well structure is given by the weighted average of the layer rates which are typically very different from the corresponding bulk rates. The hole ionization rate enhancement within the GaAs layer in these structures is too small to offset the reduced ionization rate within the AlGaAs layer. Hence, the net hole ionization rate is not observed to be much different from the weighted average of the bulk rates. The hole ionization rate is less enhanced than the electron ionization rate apparently as a result of the smaller valence band edge discontinuity as well as the larger hole relaxation rate in these materials.

In order to fully appreciate the effect on the ionization rate of the introduction of the periodic potential it is instructive to plot the spatial ionization rate profile in a multi-quantum well device [27]. The number of ionization events at each point along the device is determined for both the electrons and holes in bulk and multi-quantum well devices. We choose a sufficiently large number of initiating carriers such as to give a statistically reliable picture of the spatial ionization profile. After an ionization event occurs, a secondary carrier of the same species as the initiating carrier is added to the simulation at the position corresponding to the ionization event. The complete ensemble, parents and all subsequent daughters born of them, are then simulated until each carrier is collected.

The number of electron ionization events are plotted as a function of distance traveled in bulk GaAs at various applied electric field strengths, 350, 400, and 500 kV/cm in Figure 4. As can be seen from this figure, the ionization process in bulk material is completely random spatially. The number of

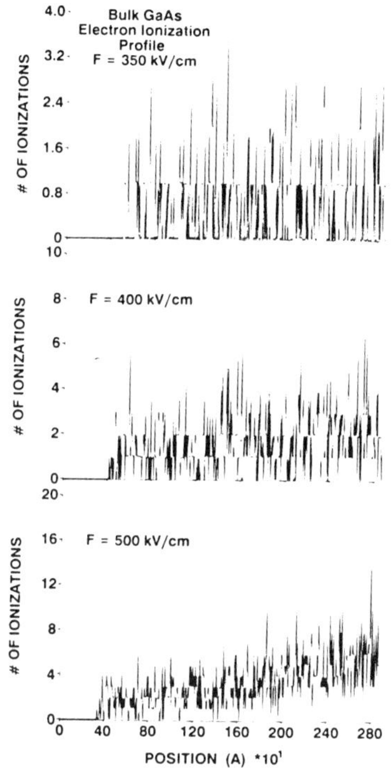

Figure 4. Number of electron ionization events plotted versus distance in bulk GaAs.

ionizations increases as expected with increasing electric field. No correlation in the number of events with spatial location is apparent. The electron ionizations occur strictly at random owing to the inherently stochastic nature of the ionization process in bulk material.

Upon the introduction of a spatially periodic potential, the electron impact ionization rate becomes strikingly localized as is evident from inspection of Figure 5. Notice that the vast majority of electron impact ionization events occur in the GaAs layer near the heterojunction interface as expected. This is particularly apparent at low applied fields where it is clear that the potential step provides the main impetus for impact ionization. As the electric field increases

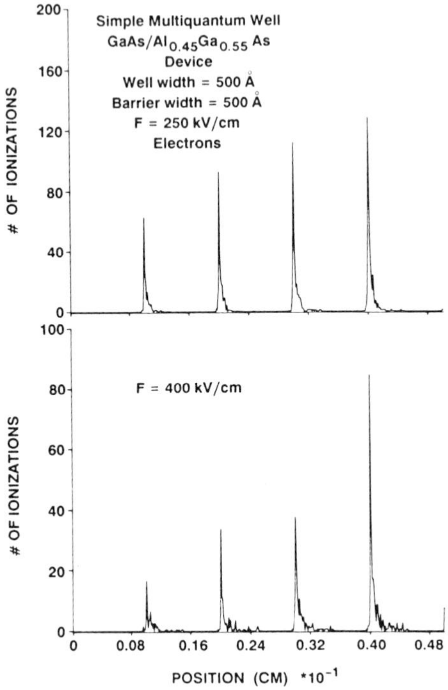

Figure 5. Number of electron ionization events plotted versus distance in a GaAs/AlGaAs MQW-APD.

more ionizations occur away from the interface as is evident from the appearance of smaller peaks in Figure 5b.

The hole spatial ionization profile in bulk GaAs and in a multi-quantum well structure are plotted in Figures 6 and 7, respectively. The hole ionization profile in bulk GaAs is similar to the electron ionization profile; the events occur spatially at random. The total number of events depicted are fewer since fewer parent holes are launched in the simulation. It is interesting to note that hole ionization occurs within the multi-quantum well GaAs layers at random in stark contrast to the highly localized electron ionization in these structures. Notice the absence of the sharp peaks in the number of events near the interface in the hole ionization case. Again, the explanation appears to be due to the different potential discontinuities as well as the different transient behaviors of the carriers.

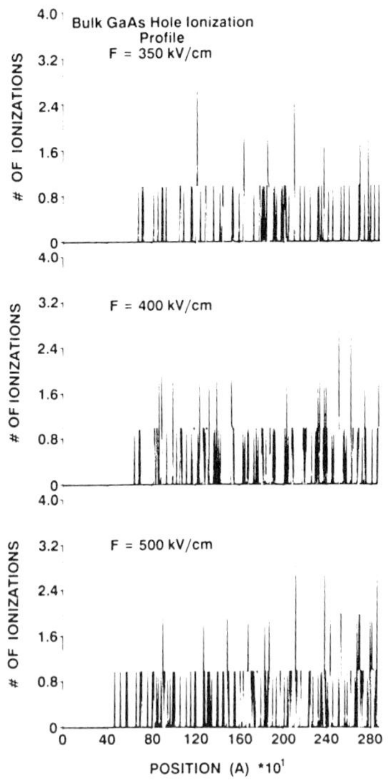

Figure 6. Number of hole ionization events plotted versus distance in bulk GaAs.

Hence the potential steps in the multi-quantum well device plays less of role in the hole ionization process than that for electrons in GaAs/AlGaAs APDs.

As mentioned above, the presence of secondary hole ionization need be avoided in most device applications as it introduces a feedback mechanism into the multiplication process which in turn amplifies noise fluctuations in a nonlinear manner. Interestingly, within a unipolar ionizing device, the excess noise can vanish altogether. Capasso [28] first recognized that by partitioning the multiplication region into a series of stages via the use of a superlattice or by a multi-quantum well structure that the excess noise could be reduced below that possible in a conventional APD even if both devices exhibit single carrier ionization. As seen from Figures 4-7, the multi-quantum wells impart a spatially deterministic aspect to the ionization process. If properly designed, such a device

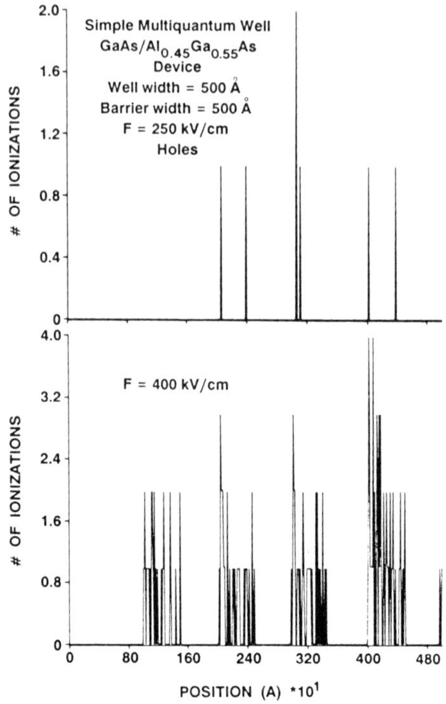

Figure 7. Number of hole ionization events plotted versus distance in a GaAs/AlGaAs MQW-APD.

could exhibit fully deterministic ionization resulting in completely noise free multiplication.

Though the simple multi-quantum well structure has been demonstrated to show some enhancement of the electron ionization rate, the overall improvement of the noise properties of these APDs will probably not be great. Further improvement requires structures in which the electron distribution is selectively heated over the corresponding hole distribution. The basic unit cell of such a device along with its corresponding electric field profile is sketched in Figure 8 [29-32]. The basic unit cell of the device sketched in Figure 8 consists of a multi-quantum well formed of five separate layers made from two different bandgap materials, in this case GaAs and AlGaAs. By doping the layers as shown, p+ - i - n+ (AlGaAs) followed by near intrinsic GaAs and AlGaAs layers, the net electric field is made asymmetric within the unit cell. The device is

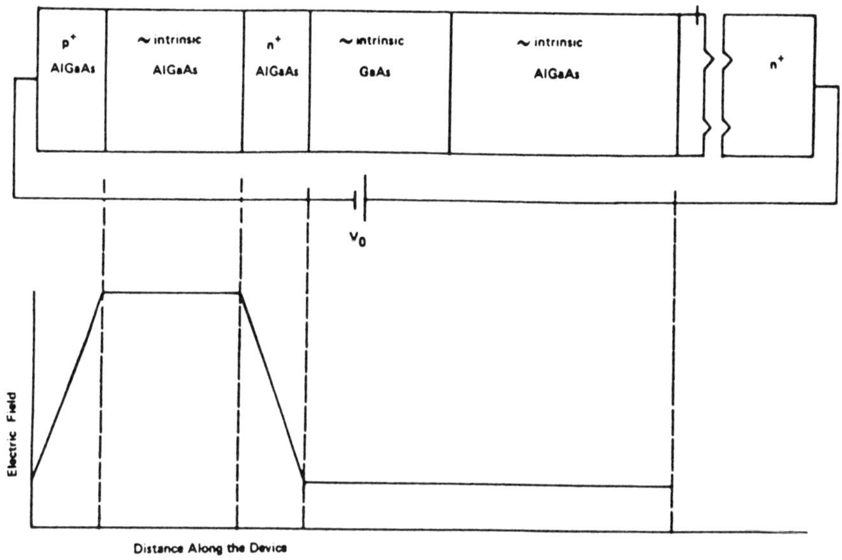

Figure 8. Schematic representation of a p-i-n heterojunction APD and the corresponding electric field profile.

operated at sufficient reverse bias to fully deplete the p-i-n layer and any residual bias produces an overall constant electric field throughout the structure. Electrons enter the unit cell from the high field side (left side in Figure 8) and holes from the low field side. The electrons are subsequently heated by the combined action of the built-in field, overall bias field and potential discontinuity prior to being injected into the narrow gap GaAs layer. The combined action of these mechanisms acts to "superheat" the electron distribution to energies at which impact ionization occurs. Conversely, the holes undergo heating only from the constant overall bias field within the intrinsic AlGaAs layer prior to injection into the GaAs layer, and far fewer, if any, ionization events result. The holes of course, also drift through the high field p-i-n region but then enter the low field AlGaAs layer where the energy bandgap is much wider than in the GaAs layer. Consequently, the hole ionization rate is low and the holes can cool prior to entering the narrow gap GaAs layer. Judicial choices of layer thicknesses, doping concentrations, and overall bias can provide substantial electron ionization at negligible hole ionization [30-32].

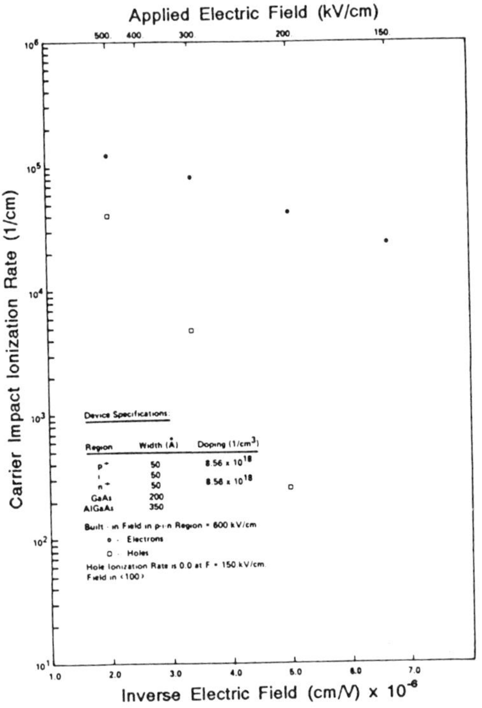

Figure 9. Calculated electron and hole ionization rates as a function of inverse electric field for the device geometry listed in the figure.

Calculations of the electron and hole ionization rate in a sample structure are presented in Figure 9. As can be seen from this figure, the hole ionization rate can be completely suppressed while retaining a significant electron ionization rate. Based on these calculations, it appears that a low noise interband APD can be made.

EXTENDED VELOCITY OVERSHOOT VIA POTENTIAL STEPS

Electronic transport in central valley dominated semiconductors, such as GaAs and InP, has the unusual property that as the applied bias increases beyond a fixed threshold the steady state electron drift velocity decreases leading to a substantial decrease in the current. This effect, called negative differential resistance, arises from electron transfer at high energies from the low mass central valley to the much higher effective mass satellite valleys in k-space.

Sustained high electron drift velocity in these materials is only possible if intervalley transfer can be frustrated throughout the entire flight of the carriers.

Through velocity overshoot the electron drift velocity can be substantially increased over the corresponding steady state velocity. Overshoot arises from the difference in the energy and momentum relaxation times [4] and arises whenever the electron distribution is heated rapidly to energies much greater than the steady state average energy. Being that the energy relaxation of the carrier distribution is not instantaneous, the electrons maintain a high average energy for a period of time during which their velocities can well exceed their steady-state value. Rapid carrier heating can occur either from the injection of cold electrons into a relatively high field region or through high energy injection into a low field region. In either case, the electron distribution is initially much hotter than its corresponding steady-state value.

The Monte Carlo method provides a useful tool for investigating velocity overshoot in central valley semiconductors since it contains all of the essential physics of the process. Tang and Hess [33] and later Brennan, et al., [34] investigated transient electronic transport in GaAs and InP respectively using the Monte Carlo method. They plotted the electron drift velocity as a function of distance after launch to clearly illustrate the effects of velocity overshoot which are reproduced here in Figures 10 and 11. As can be seen from Figure 10 velocity overshoot of injected cold electrons in bulk GaAs is substantial at high applied fields (30, 50 and 100 kV/cm) but rapidly decays back to the steady state soon after injection owing to significant intervalley transfer [33]. At very low applied fields, 1 kV/cm, the velocity overshoot is insignificant. Similarly, if the electrons are injected, as over a heterointerface, at energies above their corresponding steady-state values, overshoot is observed, as shown in Figure 11. However, if the carriers are injected near the intervalley separation energy then substantial intervalley transfer soon occurs leading to a sharp decrease in the drift velocity as shown by Figure 11. Conversely, if the electrons are injected at very low energies, no substantial velocity overshoot is observed. There exists then only a narrow range of launching energies and applied electric fields for which substantial velocity overshoot can be attained.

The primary factor limiting extended velocity overshoot in central valley semiconductors is the onset of intervalley transfer into the high mass satellite valleys. As shown in Figures 10 and 11 high speeds are obtainable from velocity overshoot only over very small distances, < 1500 Å, and only under very limited conditions in applied electric field strength and injection energy. Extended velocity overshoot in central valley semiconductors can be made possible if a mechanism exists which limits the carrier energies below the intervalley threshold energy throughout the entire device. Brennan and Hess [35] have shown, based

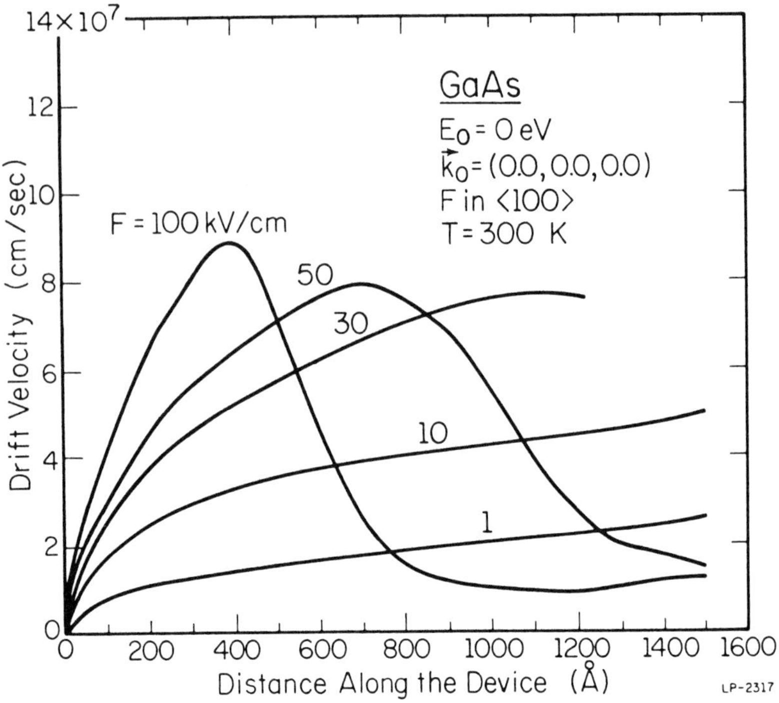

Figure 10. Calculated electron drift velocity as a function of distance at various applied electric fields at zero launching energy. [Copyright 1982, IEEE, ref. [33]]

on Monte Carlo calculations, that impact ionization in narrow gap semiconductors with large intervalley threshold energies, such as InAs, does just this. Impact ionization typically occurs prior to intervalley transfer in InAs thereby preventing transfer out of the central valley. The resulting average electron velocity can be very high [35]. However, this approach is of little practical use since it is undesirable to have carrier multiplication in most high speed device applications.

A similar effect can be achieved if the electrons "climb" a step-like potential structure, introduced through a series of heterostructures, under the influence of an overlaid applied electric field as shown in Figure 12. After being launched from a potential barrier the electrons are accelerated by an applied electric field. They gain kinetic energy from the field until they reach the first potential step. Provided the electrons have sufficient kinetic energy to cross into

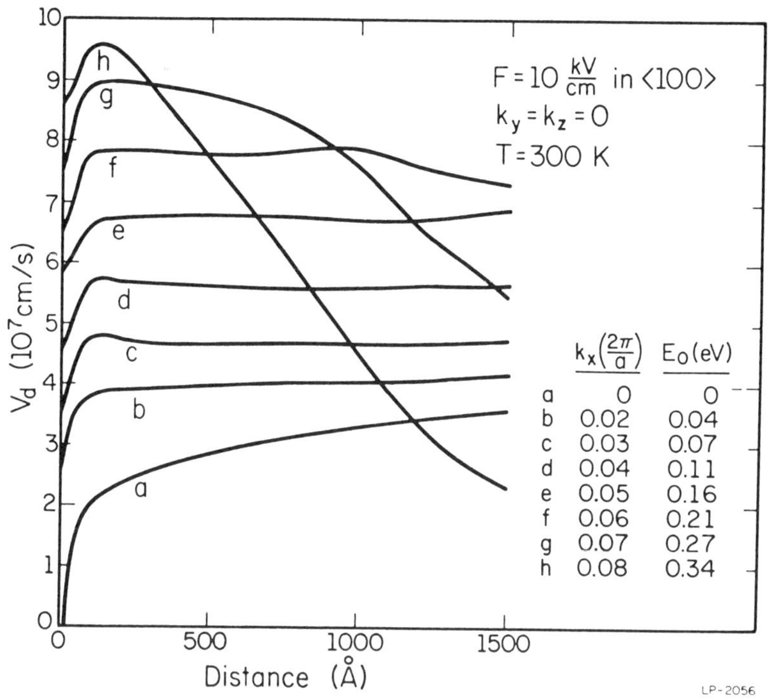

Figure 11. Calculated electron drift velocity as a function of distance at fixed electric field (10 kV/cm) for various launching energies in GaAs.

the potential barrier, their kinetic energy is then lowered by an amount equal to the potential barrier. The electrons continue to drift under the action of the applied field until they reach the next step where their kinetic energy is lowered again. In this way, the electrons' excess kinetic energy is periodically removed by the potential steps thus ensuring continued gamma valley confinement.

The effect of continued gamma valley confinement on the electron drift velocity can be shown using the Monte Carlo method [36]. Calculated results of the electron drift velocity in a staircase GaAs potential structure are shown in Figure 12. As can be seen from Figure 12, the average electron drift velocity for electrons which climb the steps is substantially higher throughout the entire device than for either the case without the steps or for electrons which descend the steps. In the case where the electrons descend the steps the calculated drift velocity for the given conditions here is much lower. The reason for this is that

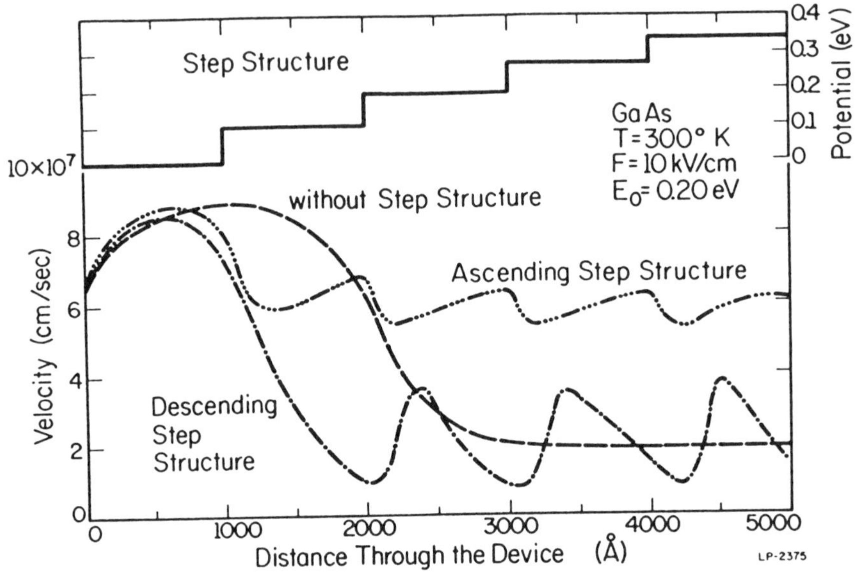

Figure 12. Average drift velocity as a function of distance for three cases, "climbing" up, "climbing" down, and no stairs. The top part of the illustration shows the potential structure to scale.

the electrons gain kinetic energy as they descend the steps but their velocity decreases due to intervalley transfer. This is clearly shown in Figure 13 where the relative gamma valley population of the electrons is plotted as a function of device length for the two cases, ascending and descending. The electrons are confined completely within the gamma valley throughout the entire device in the ascending step case, while substantial intervalley transfer is observed when the electrons propagate down the stairs. Therefore, we find a surprising result; extended velocity overshoot can be attained in structures in which the electrons lose excess kinetic energy such that they are continuously confined within the gamma valley. The underlying reason for this paradoxical result is that by maintaining the electron energies below the intervalley threshold, the electrons are forced to remain within the gamma valley in which they have a low effective mass and a high velocity.

Alternative strategies for extended gamma valley confinement have recently been proposed in which the electrons lose energy continuously by moving against an inverted collector field [37,38] in a similar manner as described above

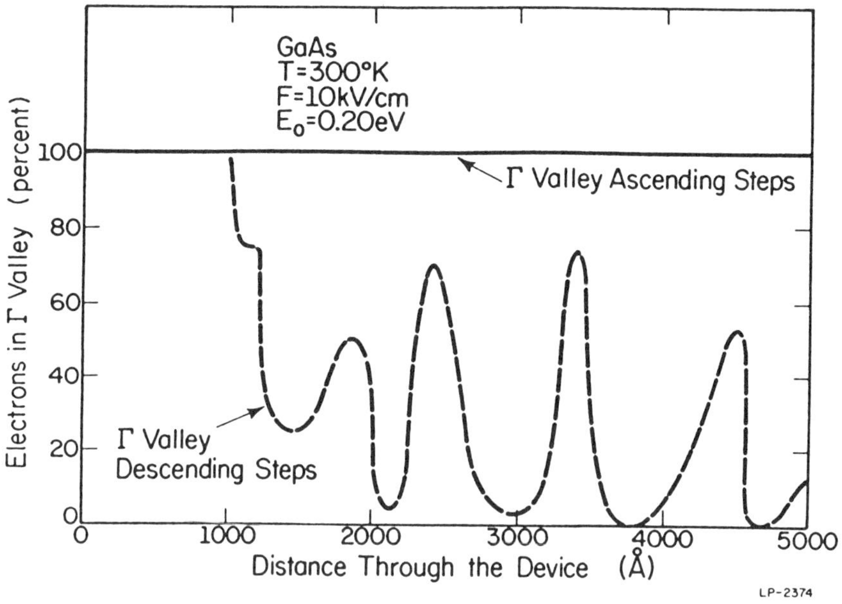

Figure 13. Percent time the electrons spend in the gamma valley for the cases "climbing" up and down the stairs.

for the potential steps. In these structures the electron's excess kinetic energy is continuously removed by the action of an inverted collector field as opposed to the potential steps in which the excess kinetic energy is removed in a discontinuous manner. Nevertheless, the physical features are identical; the electrons are confined within the gamma valley during most of their flight resulting in substantial overshoot. Recent experiments on inverted collector HBTs have verified this effect [39]. These experiments demonstrated a substantial improvement, ~ 30 %, in the cutoff frequency of the inverted collector HBT compared to an identical conventional HBT [39]. Therefore, owing to the nonlinear properties of electron transport in central valley semiconductors, highest carrier speeds are obtained paradoxically by limiting the energy gain such that the electrons are always confined within the gamma valley.

CARRIER HEATING THROUGH LATERAL MOTION IN A HETEROSTRUCTURE

Negative differential resistance can be induced in an alternative manner from k-space transfer through the application of an electric field parallel to a heterostructure layer. This mechanism, called real space transfer [40], is again a field induced hot electron transfer process but involves the transfer of electrons from a high mobility semiconductor layer into a low mobility semiconductor such as from GaAs to AlGaAs [13]. In a real space transfer device, a field is applied parallel to the heterostructure direction. The applied field parallel to the interface heats the electrons such that they are able to redistribute themselves within the potential well formed by the heterostructure through real space transfer. Real space transfer and the subsequent redistribution of the electrons are therefore strongly influenced by the transverse electric field. An analogous situation occurs in a system which is spatially quantized in energy levels. Subband population depends upon the electron temperature through the parallel electric field as well as upon the transverse field through which the energy of the subbands is determined [41]. In this section we discuss the influence of transverse electric fields or potential barriers on nonlinear transport parallel to heterointerfaces.

It has been well established that the action of the gate field in MOSFETs degrades the electron mobility within the conducting channel formed between the underlying oxide and semiconductor layers. The mobility degradation, which has been found to be well in excess of that expected from interface roughness [42], can be explained by carrier heating via the gate field. However, the gate field, by itself, does no work on the electron gas since it is applied perpendicular to the carrier motion. Nevertheless, experimental noise measurements [43] indicate the presence of carrier heating which apparently is caused by the gate field.

Brennan and Hess [13] first proposed an explanation for the origin of carrier heating by the combined action of longitudinal and transverse electric fields or potentials by recognizing that the parallel field heating enables the redistribution of the electron gas within the potential established by the transverse field or heterostructure. In order to demonstrate this effect, the ensemble Monte Carlo method can be used since it properly treats nonlinear semi-classical effects. The carrier heating can be demonstrated by studying either the impact ionization rate in a system with a transverse electric field or potential barrier or the real space transfer rate in a system with band bending. In either case the presence of hot electrons is easily demonstrated since both processes have a threshold below which no events are observed. Therefore, if a significantly higher ionization or transfer rate is observed in the presence of a transverse field as opposed to in its absence, carrier heating must have occured in some manner by the transverse field.

In an attempt to optimize the design of new APD structures, Brennan [44] applied the ensemble Monte Carlo method to the study of a device first proposed by Capasso [45], called the channeling avalanche photodiode. The channeling APD consists of interdigitated p-n heterojunctions comprised of alternating layers of GaAs and AlGaAs. Upon the application of a reverse bias, the interdigitated p-n junctions become depleted until punch-through is attained. Thereafter, further reverse bias results in a constant electric field applied parallel to the layers. Brennan [44] observed, using Monte Carlo calculations of the impact ionization rate, that the electron ionization rate was greatest in a multilayered GaAs/AlGaAs undoped channeling APD in which the electrons could freely transfer between layers. This is contrary to what was expected since the device is designed to confine the electrons within the narrow gap GaAs layers while sweeping the holes into the wider gap AlGaAs layers by the action of the built-in field arising from the fully depleted interdigitated p-n heterojunctions. Hence it is expected that the electron ionization rate would be greatest within a structure for which the confinement in the GaAs layers is highest. Nevertheless, the Monte Carlo calculations showed that the electron ionization rate is greatest when the electrons could freely transfer between the GaAs and AlGaAs layers via real space transfer induced by the action of the parallel applied field.

In an attempt to explain the enhanced ionization rate calculated within the channeling APD, Brennan and Hess [13] applied a similar Monte Carlo analysis to study the impact ionization rate in silicon in the presence of combined transverse and longitudinal electric fields. The ionization rate was computed in two different potential configurations as shown in Figures 14 (a) and (b). A control on the calculations is provided by first calculating the ionization rate in bulk silicon. These calculations along with recent experimental measurements [46] (the range of measured values lies between the solid lines) are presented in Figure 15. In both configurations shown in Figure 14, a longitudinal electric field is applied. The electrons drift along the device subject to the applied longitudinal field and can move within the well or triangular barrier region subject to the constraints applied by energy conservation. Inspection of Figure 15 shows that the electron ionization rate is significantly larger than the bulk silicon rate in either the square well or triangular barrier designs.

The electron heating can be explained as follows. Under the action of an applied electric field, the electrons move through a bulk semiconductor and ultimately steady-state is established. In steady-state the energy gain from the field is balanced by the energy lost to the lattice via inelastic phonon scatterings. If a transverse potential is introduced, as it is in the devices shown in Figure 14, as well as in the undoped channeling APD, the electron motion continues as before until scatterings redirect their momentum into the direction of the transverse potential. Provided the electrons have sufficient kinetic energy, gained from the longitudinal electric field, they can move into the higher potential region where they lose kinetic energy to the potential differences. Upon transfering the electrons lose kinetic energy to the transverse potential. All other factors, being

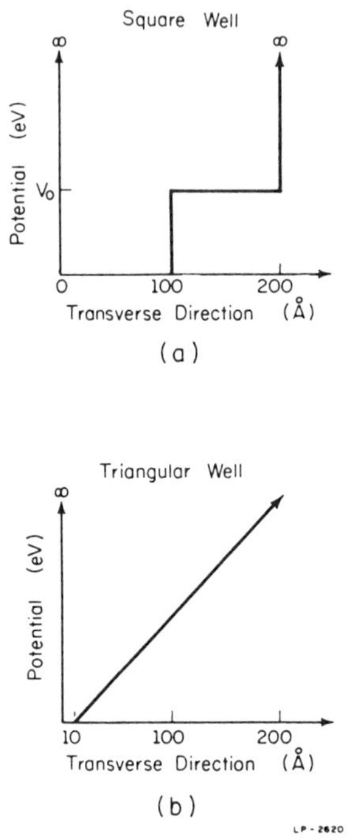

Figure 14. Potential energy diagrams: a) square well, b) triangular well used in the calculations.

equal, the longitudinal electric field will reheat the electrons within the high potential region to an average energy consistent with that expected from the longitudinal field. Hence, the electrons within the high potential regions are heated to an average energy which is roughly equal to the average energy in the low potential region but is now measured from the minimum potential in the high potential region. Therefore, the electrons in the high potential regions are heated to the usual average energy plus the potential.

Of course, the heating to this energy serves as an upper bound to what is possible since electrons will transfer back to the low potential region and lose their excess kinetic energy by phonon scatterings. The average electron energy as measured from the minimum of the conduction band edge will nevertheless be larger than the case in which the transverse potential is absent. It is important to recognize that the transverse potential, by itself, does not add any net energy

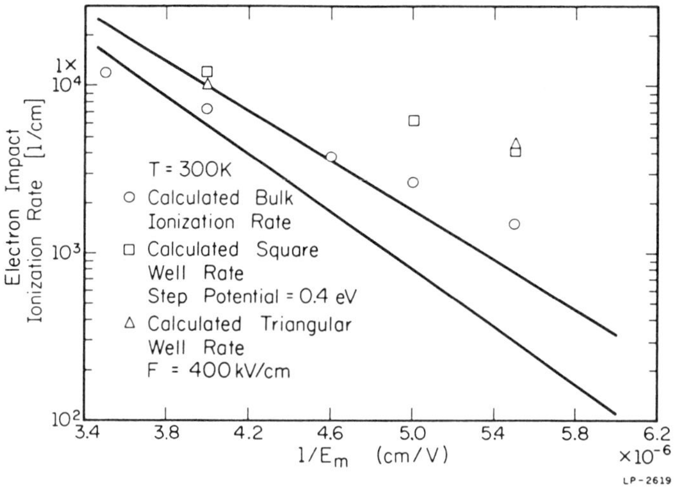

Figure 15. Calculated electron impact ionization rate in bulk silicon and in the square and triangular well potentials of Figure 14. The solid lines represent the range of experimental data from [46].

to the carriers. The action of the longitudinal field on the carriers within the high potential region is responsible for supplying the excess carrier energy. Consequently, upon reentering the low potential region, the electrons are hotter than they would have been without having transfered to the high potential region. In this way, a transverse field or potential can significantly alter the carrier temperature and hence the ionization rate.

CONCLUSIONS

In this chapter, we have analyzed three different nonlinear transport effects in semiconductor heterostructures, impact ionization enhancement and velocity overshoot through perpendicular injection and carrier heating through real space transfer. Owing to the inherent nonlinearity of each mechanism, the most complete formulation of these effects is provided by the Monte Carlo method. Based on Monte Carlo calculations, several surprising results are obtained. It is found that the electron ionization rate can be enhanced in a multi-quantum well structure when the carriers are injected perpendicular to the heterointerface. In addition, by confining the electron distribution within the gamma valley through repeated flight against a series of potential steps, the average electron drift velocity can be made exceedingly high. Finally, it is found that carrier heating can occur through lateral transfer of the carriers via real space transfer.

REFERENCES

[1] N. Holonyak, Jr., R. M. Kolbas, R. D. Dupuis, and P. D. Dapkus, IEEE J. Quantum Electron., QE-16, 170 (1980).

[2] R. Dingle, H. L. Stormer, H. L. Gossard, and W. Wiegmann, Appl. Phys. Lett., 33, 665 (1978).

[3] W. Shockley, U. S. Patent 2,569,347 (1951).

[4] J. G. Ruch, IEEE Trans. Electron Dev., ED-19, 652 (1972).

[5] R. Chin, N. Holonyak, Jr., G. E. Stillman, J. Y. Tang, and K. Hess, Electron. Lett., 16, 467 (1980).

[6] K. Brennan, K. Hess, and F. Capasso, Appl. Phys. Lett., 50, 1897 (1987).

[7] F. Capasso, W. T. Tsang, A. L. Hutchinson, and G. F. Williams, Appl. Phys. Lett., 40, 38 (1982).

[8] T. Kagawa, H. Iwamura, and O. Mikami, Appl. Phys. Lett., 54, 33 (1989).

[9] F. Osaka, T. Mikawa, and O. Wada, IEEE J. Quantum Electron., QE-22, 1986 (1986).

[10] F.-Y. Juang, U. Das, Y. Nashimoto, and G. K. Bhattacharya, Appl. Phys. Lett., 47, 972 (1985).

[11] K. Brennan, T. Wang, and K. Hess, IEEE Electron Dev. Lett., EDL-6, 199, (1985).

[12] K. Brennan and K. Hess, IEEE Electron Dev. Lett., EDL-4, 419, (1983).

[13] K. Brennan and K. Hess, IEEE Electron Dev. Lett., EDL-7, 86 (1986).

[14] A. S. Tager, Sov. Phys.-Solid State, 6, 1919 (1965).

[15] R. J. McIntyre, IEEE Trans. Electron Dev., ED-13, 164 (1966).

[16] K. Matsuo, M. C. Teich, and B. E. A. Saleh, IEEE Trans. Electron Dev., ED-32, 2615 (1985).

[17] B. K. Ridley, IEEE Proc. J. Optoelectron., 132, 177 (1985).

[18] F. Capasso, in Semiconductors and Semimetals, edited by R. K. Willardson and A. C. Beer (Academic, New York, 1985) vol. 22D, pp. 1-172.

[19] H. Shichijo and K. Hess, Phys. Rev. B, 23, 4197 (1981).

[20] M. L. Cohen and T. K. Bergstresser, Phys. Rev., 141, 789 (1966).

[21] M. V. Fischetti and S. E. Laux, Phys. Rev. B, 38, 9721 (1988).

[22] T. Wang, Ph.D. Thesis, Univ. of Ill. Urbana, IL. 1986.

[23] K. F. Brennan, D. H. Park, and Y. Wang, IEEE Trans. Electron Dev., 37, 536 (1990).

[24] K. F. Brennan, N. Mansour, and Y. Wang, presented at the Minnesota Supercomputer Institute Symposium on Supercomputer Simulation of Semiconductor Devices, Minneapolis, MN, November, 1990.

[25] K. F. Brennan and Y. Wang, IEEE Trans. Electron Dev., 35, 634 (1988).

[26] T. Kagawa, H. Iwamura, and O. Mikami, Appl. Phys. Lett., 54, 33 (1989).

[27] K. Brennan, IEEE J. Quantum Electron., 24, 2001 (1988).

[28] F. Capasso, W. T. Tsang, and G. F. Williams, IEEE Trans. Electron Dev., ED-30, 381 (1983).

[29] H. Blauvelt, S. Margalit, and A. Yariv, Electron. Lett., 18, 375 (1982).

[30] K. Brennan, IEEE J. Quantum Electron., QE-22, 1999 (1986).

[31] K. Brennan, IEEE Trans. Electron Dev., ED-34, 782, (1987).

[32] K. Brennan, IEEE Trans. Electron Dev., ED-34, 793 (1987).

[33] J. Y-F. Tang, and K. Hess, IEEE Trans. Electron Dev., ED-29, 1906 (1982).

[34] K. Brennan, K. Hess, J. Y. Tang, and G. J. Iafrate, IEEE Trans. Electron Dev., ED-30, 1750 (1983).

[35] K. Brennan and K. Hess, Solid-State Electron., 27, 347 (1984).

[36] K. Brennan and K. Hess, IEEE Electron Dev. Lett., EDL-4, 419, (1983).

[37] C. M. Maziar, M. E. Klausmeier-Brown, and M. S. Lundstrom, IEEE Electron Dev. Lett., EDL-7, 483 (1986).

[38] J. Hu, K. Tomizawa, and D. Pavlidis, IEEE Electron Dev. Lett., 10, 55 (1989).

[39] K. Morizuka, R. Katoh, M. Asaka, N. IIzuka, K. Tsuda, and M. Obara, IEEE Electron Dev. Lett., 9, 585, (1988).

[40] K. Hess, H. Morkoc, H. Shichijo, and B. G. Streetman, Appl. Phys. Lett., 35, 469 (1979).

[41] K. Hess, Advanced Theory of Semiconductor Devices, Prentice Hall, Englewood Cliffs, 1988.

[42] S. M. Goodnick, R. G. Gann, J. R. Sites, D. K. Ferry, C. W. Wilmsen, D. Fathy, and O. L. Krivanek, J. Vac. Sci. Technol. B, 1, 803 (1983).

[43] M. Tanimoto and D. K. Ferry, IEEE Electron Dev. Lett., EDL-4, 246 (1983).

[44] K. Brennan, IEEE Trans. Electron Dev., ED-32, 2467 (1985).

[45] F. Capasso, IEEE Trans. Electron Dev., ED-29, 1388 (1982).

[46] V. M. Robbins, T. Wang, K.F. Brennan, K. Hess, and G. E. Stillman, J. Appl. Phys., 58, 4614 (1985).

7

MONTE CARLO SIMULATION OF QUASI-ONE-DIMENSIONAL SYSTEMS

D. Jovanovic and J.P. Leburton

The Beckman Institute for Advanced Science and Technology
and
The Department of Electrical and Computer Engineering,
University of Illinois, Urbana-Champaign, IL

1 INTRODUCTION

Advances in epitaxial technologies and fine-line lithography have made possible the fabrication of low-dimensional semiconductor structures with controlled degrees of vertical and lateral confinement[1-7]. When it was suggested that the reduced degrees of freedom in quasi-one dimensional (1D) systems may lead to a suppression of elastic scattering and a subsequent enhancement in the carrier mobility[8, 9], intense research activity was stimulated to investigate the prospect for high performance device applications[10]. Mesoscopic structures such as quantum rings and wires have been purposely designed to evidence a wide variety of new phenomena ranging from universal conductance fluctuations[11, 12] and quantum interferences[13-16] to the Aharanov-Bohm effect[17, 18]. In addition, quantum wire-based lasers have recently been fabricated using novel techniques for achieving transverse confinement by direct growth on vicinal GaAs substrates[7, 19, 20]. Meanwhile, the observation of Shubnikov-de-Haas type oscillations in the transconductance of an array of electrostatically confined quantum wires has confirmed the existence of 1D quantized subbands above the liquid nitrogen temperature[21]. Normal device operation, however, requires a room temperature environment and voltage biases significantly higher than those needed to observe quantum interference phenomena. While transverse quantization can be maintained over a wide temperature range, it is doubtful that the long-range phase-coherence causing quantum interference effects can be preserved in the presence of inherent dissipative mechanisms; confined particle transport appears more realistic. The latter phenomenon can be described in terms of semi-classical scattering theory in analogy with longitudinal magnetotransport[22].

Among the various theoretical approaches developed for investigating particle transport in confined systems[23, 24, 25], the Monte Carlo method is a powerful technique for analyzing particle dynamics in general confinement configurations and for various field conditions and temperatures. In this chapter we focus on the simulation of quantum wires and quasi-1D systems because of their importance in nanostructure technologies. Although the implementation of a Monte Carlo code suitable to quasi-1D systems may seem rather straightforward, the apparent simplicity of 1D carrier motion is offset by the complex multi-subband structure arising from the high level of quantization attainable in double confinement geometries. As a result, the computation of the various scattering rates in these complex configurations becomes the most expensive part of the simulation and makes the quasi-1D numerical model fundamentally different from bulk and two- dimensional (2D) Monte Carlo methods.

2 MODEL

2.1 Electronic Properties

There are several ways to achieve double particle confinement and realize quantum wire structures. They range from complete structural confinement between layers of wide-gap materials[19, 20] to a combinations of electrostatic and structural confinement[4, 6, 21]. From a theoretical point of view, the 1D quantized energy levels should be determined within the Hartree self-consistent potential approximation which involves solving systems of coupled two-dimensional Poisson-Schrodinger equations. In this analysis, however, we are essentially concerned with the dynamics of 1D carriers and extreme precision in generating energy levels is of secondary importance. We therefore assume simple confinement configurations arise from the elementary GaAs-AlGaAs potential wells which are decoupled along the two transverse y-and z-directions. Electrons are then free to move along the x-direction with their energies and wave functions given by

$$E_{ij}(k_x) = \frac{\hbar^2 k_x^2}{2m^*} + E_i + E_j \qquad i, j = 1, \ldots N, \tag{1}$$

and

$$\psi_{ij}(k_x, \vec{r}) = \frac{1}{\sqrt{L_x}} e^{ik_x x} \xi_i(y) \phi_j(z), \tag{2}$$

where k_x and L_x are the wave vector and the wire length in the longitudinal direction and $\xi_i(y)$ and $\phi_j(z)$ are the transverse wave functions which correspond to the quantized energy levels E_i and E_j, respectively. N is the maximum number of levels considered in a particular simulation and is determined by computational constraints. The remaining term in Eq. (1) is the kinetic energy of the particle resulting from the free-electron component of the overall electronic wavefunction (Eq. (2)). The advantage of dealing with such elementary configurations is that the decoupled electronic y- and z-wave functions are in analytical forms which facilitates the computation of the scattering rates. We are therefore free to choose structures consisting of simple potential geometries such as the infinite square well, the parabolic well and the triangular potential which suitably model the confinement potentials occurring in real devices. For instance, the combination of a square well and a triangular potential corresponds to a configuration similar to the V-groove structure field effect transistor (V-FET) proposed by Sakaki[8] (Fig.1.a.). The parabolic and triangular potential is similar to grated-gate modulation doped GaAs-AlGaAs (G-FET) wire structure fabricated by focused ion beam techniques[26] or ultra-fine mesa etching[21, 27] (Fig.1.b.). A combination of parabolic and square-well potentials has also been investigated[28].

In the former configurations the triangular well approximates the electric field, F_z, applied by the gate electrode perpendicular to the heterojunction (V-FET) or the potential gradient arising near the band-edge discontinuity of the GaAs-AlGaAs interface (G-FET). The triangular-well wavefunctions, $\phi_j(z)$, are computed using a variational approach with exponentially damped polynomials as trial functions such that

$$\phi_j(z) = e^{-\alpha_j z} \sum_{k=1}^{j} c_{kj} z^k \qquad j = 1, ...N. \tag{3}$$

Here α_j and c_{kj} are determined from the variational calculation. For E_j, the triangular-well potential energy, a useful expression is[29]

$$E_j = \left(\frac{\hbar^2}{2m^*}\right)^{1/3} \left(\frac{2}{3}\pi q F_z\right)^{2/3} (j - 1/4)^{2/3}, \qquad j = 1, ...N. \tag{4}$$

This approximation is valid if the energy levels lie deep in the well since charge screening flattens out the triangular potential at higher energies.

Electrons are confined in the y-direction by a GaAs- AlGaAs quantum well (QW) of width L_y (V-FET), or a parabolic potential with proper frequency ω (G-FET). The respective wavefunctions and energy spectra are then given by

$$\xi_i(y) = \sqrt{\frac{2}{L_y}} sin\left(\frac{i\pi y}{L_y}\right) \tag{5}$$

with

$$E_i = \frac{(i\hbar\pi)^2}{2m^* L_y^2} \qquad i = 1, ...N \tag{6}$$

for the V-FET and for the G-FET,

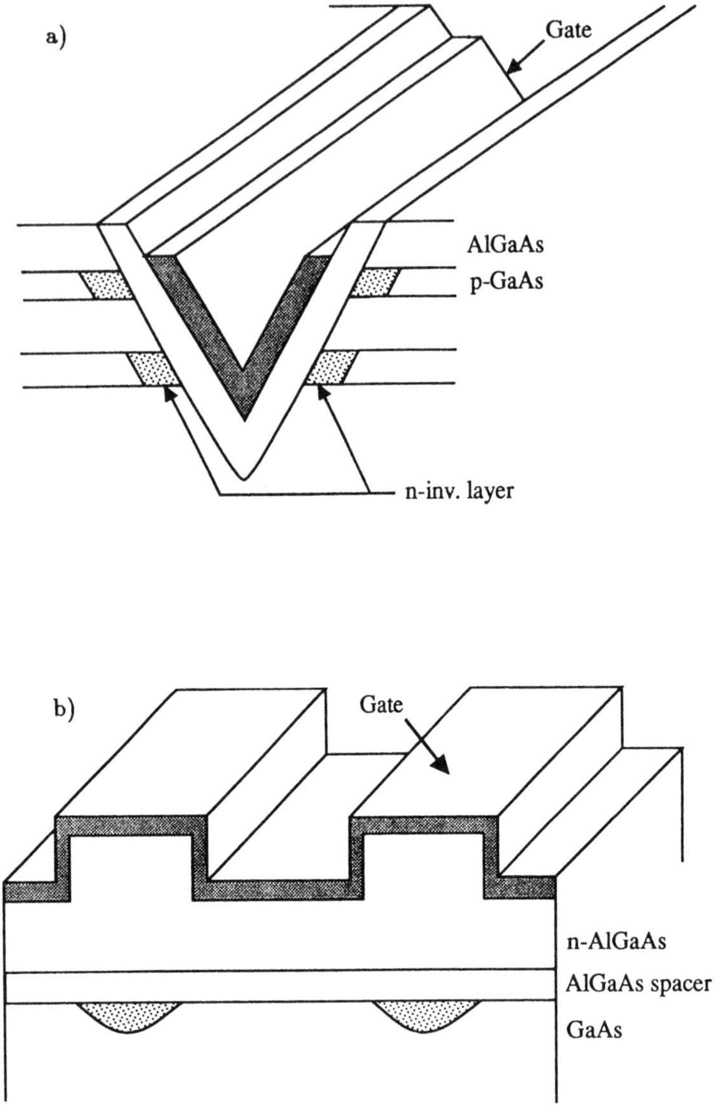

Figure 1: (a) Schematic cross-section of a quasi-1D structure fabricated by the anisotropic etching of a V-groove into a multiple-well heterostructure (V-FET) (after Sakaki, ref. 8) and (b) a MODFET based structure with a grating etched into the AlGaAs cap layer (G-FET) (after Ismail et al., ref. 21).

$$\xi_i'(y) = \left[\frac{m^*\omega}{\pi\hbar(n!)^2}\right]^{1/4} H_i(u)e^{-u^2/4} \tag{7}$$

with

$$E_i' = (i+1/2)\hbar\omega \qquad i = 1, ...N' \tag{8}$$

where

$$u = \left(\frac{2m^*\omega}{\hbar}\right)^{1/2} y \tag{9}$$

and $H_i(u)$ is the Hermite polynomial of degree i. In our model, F_z ranges from an upper limit of $200kV/cm$ down to $20kV/cm$ while the QW width varies from $50\mathring{A}$ to $250\mathring{A}$, and the energy, $\hbar\omega$ of the parabolic potential ranges from $5meV$ to $40meV$. Lower F_z fields as well as wider square-well and parabolic potentials require too many levels and make the code computationally prohibitive due to memory constraints. We would like to point out, however, that the overhead associated with the computation of the scattering rates is greatly alleviated by the choice of a harmonic-oscillator confining potential (Eqs. 6 and 7) which permits an efficient algorithm based on the Hermite polynomials recurrence relation[30] (see Sec. 2.2)

$$H_{i+1} = uH_i(u) - iH_{i-1}(u). \tag{10}$$

Although the lowest energy level of the system (Eq. 1) is clearly E_{11}, the ordering of the higher subbands is not obvious. In order to simplify the notation, the subbands will henceforth be numbered with a single subscript ν, which will range in order of increasing energy from the lowest to the highest subband. The y- and z-quantum numbers (i,j) will, in general, not be used.

When high-fields or strong confinement conditions are required, the treatment must account for non-parabolicity[31] in the Γ-valley and make an allowance for intervalley transitions to states which are effectively bulk-like due to their high effective masses. For the restricted **k**-vectors existing in quasi-1D systems, the non-parabolicity condition takes the form[32]

$$\frac{\hbar^2}{2m^*}(k_\nu^2 + k_x^2) = (E_\nu + E)(1 + \alpha(E_\nu + E)) \tag{11}$$

where we define k_ν as

$$k_\nu = \sqrt{\frac{2m^*}{\hbar^2}E_\nu(1 + \alpha E_\nu)}, \tag{12}$$

and E and E_ν are the kinetic energy and minimum energy of subband ν, respectively. It should be pointed out that k_ν is not a wavevector but merely a convenient way of expressing the nonparabolicity condition at the subband minima. The free wavevector k_x is therefore given by

$$k_x = \sqrt{\frac{2m^*}{\hbar^2}E[1+\alpha(E+2E_\nu)]}. \tag{13}$$

2.2 Scattering Rates

As in bulk simulations, the computation of the scattering rates is essential for the stochastic determination of free-flight durations and the selection of a particular scattering event from a background of phonons, impurities, and defects. Some of these mechanisms play a minor role in the determination of relaxation and response times in quasi-1D systems. For instance, impurity scattering is insignificant in contemporary modulation-doped GaAs structures, especially under high field conditions[33, 25]. In addition, electron-electron (e-e) scattering vanishes for 1D intrasubband transitions since indistinguishable particles simply exchange energy and momenta during binary collisions[34]. For intersubband transitions, however, e-e scattering is expected to contribute to the scattering background but due to the large momentum exchanges involved, this coupling will be weak and can therefore be neglected in most analyses. As a result, the large-scale transport properties of quasi-1D III-V systems are primarily determined by deformation-potential and polar phonon scattering.

In the semiclassical limit, scattering events are well separated temporally as well as spatially and Fermi's golden rule can be invoked to obtain transition probabilities for the scattering mechanisms commonly encountered in quasi-1D systems. The expression of the transition probability, $S_{\nu\nu'}(k_x, k_x', \mathbf{q})$, reads

$$S_{\nu\nu'}(k_x, k_x', \mathbf{q}) = \frac{2\pi}{\hbar} \mid M_{\nu\nu'}(k_x, k_x', \mathbf{q}) \mid^2 \delta[E_{\nu'}(k_x') - E_\nu(k_x) \pm \hbar\omega_{\mathbf{q}}]. \tag{14}$$

Here, $M_{\nu\nu'}(k_x, k_x', \mathbf{q})$ represents the matrix element for a transition from the initial k_x, ν state to the final k_x', ν' state mediated by a phonon with wavevector $\mathbf{q}$. The $\pm$ sign in the energy-conserving δ-function accounts for the emission (+) and absorption (-) of phonons between the subbands with indices ν and ν'. The total scattering rate from an initial state with energy E is therefore given by

$$\lambda_\nu(E) = \frac{A}{(2\pi)^2} \sum_{\nu' k_x'} \int d\mathbf{q}_\perp S_{\nu\nu'}(k_x, k_x', \mathbf{q}), \tag{15}$$

where A is the wire cross-section and the summation over the final k_x', ν' states takes into account all the possible intersubband transitions along with those occurring within the same band, ν. The integral over the transverse phonon wave-vectors, $\mathbf{q}_{\perp}$, sums over all the phonons that give rise to a particular final k'_x state.

The general form of the electron-phonon Hamiltonian reads[35]

$$H_{e-ph} = \sum_{\mathbf{q}} C_q \left[a_{\mathbf{q}} \, e^{i\mathbf{q}.\mathbf{r}} - a^+_{\mathbf{q}} \, e^{-i\mathbf{q}.\mathbf{r}} \right]. \tag{16}$$

The nature of each particular lattice-interaction is conveyed in the constant, C_q, which determines the electron-phonon coupling strength in the quantized active region. Although recent work[36] has demonstrated the importance of treating confined phonon modes in highly quantized systems, only bulk modes are treated here. This is a good approximation as long as well widths are not narrower than 50Å .

In modulation-doped GaAs channels, the dominant scattering agents are polar-optical phonons (POP). After taking the matrix element of the Hamiltonian (Eq. 16) between phonon states, the relevant coupling constant for scattering transitions becomes V_q which is characterized by[37]

$$V_q = i \left[\frac{2\pi}{V_{ol}} e^2 \hbar\omega_{pop} \left(\frac{1}{\epsilon_\infty} - \frac{1}{\epsilon_{dc}} \right) \right]^{1/2} \frac{1}{q} \left(N_q + 1/2 \pm 1/2 \right)^{1/2} \tag{17}$$

and a constant dispersion relation ($\hbar\omega_{pop}$=36meV) for the POP interaction. N_q is the phonon occupation number and the $\pm$ takes into account phonon emission (+) or absorption (-). The $1/q$ dependence of V_q implies that collisions involving small exchanges of momentum are the most favorable for the POP interaction.

At low temperatures, polar acoustic (piezoelectric) scattering becomes significant in GaAs, especially below the POP emission threshold. This interaction is characterized by[38]

$$V_q = i \left[\frac{\hbar e^2 h_{14}^2}{2 V_{ol} \rho v_s} \right]^{1/2} \frac{1}{q^{1/2}} \left(N_q + 1/2 \pm 1/2 \right)^{1/2}, \tag{18}$$

and an approximately linear dispersion relation; $\hbar\omega_{pzo} = \hbar v_s q$ where v_s is the sound velocity in the material system. The strength of this coupling is determined in large part by the piezoelectric constant, h_{14}, which is spherically averaged over the longitudinal and transverse modes. Although piezoelectric scattering rates are often calculated under the equipartition approximation, the full form of the occupation number is retained here to accommodate low temperature simulations where quantum effects are most easily observed. In addition, the influence of screening on the piezoelectric interaction has not been included in Eq. 18. This issue is important because it has been shown that in 2D modulation-doped structures, screening is responsible for an overall suppression of piezoelectric scattering[39]. In the present analysis, It would be undesirable to exclude any low-energy scattering mechanisms which can have an influence on carrier dynamics. Therefore, an unscreened piezoelectric interaction will be considered throughout our model with the understanding that the effects of this mechanism will be somewhat exaggerated but will not significantly affect our conclusions as most of the transport issues considered here are for relatively high longitudinal fields.

Finally, deformation potential scattering is included since it provides a strong mechanism for both intersubband (acoustic) and intervalley (optic) transitions. The acoustic branch has a coupling given by[38]

$$V_q = i\left[\frac{\hbar D_a^2}{2V_{ol}\rho v_s}\right]^{1/2} q^{1/2}\,(N_q + 1/2 \pm 1/2)^{1/2}, \tag{19}$$

and the same dispersion relation as the piezoelectric interaction while for the optic branch,

$$V_q = i\left[\frac{\hbar D_o^2}{2V_{ol}\rho\omega_{od}}\right]^{1/2} (N_q + 1/2 \pm 1/2)^{1/2}, \tag{20}$$

where D_a and D_o are the deformation potentials for the acoustic and optical branches respectively, ρ is the material mass density, and ω_{od} is the frequency of the optical phonon.

With these coupling constants, the explicit form of the matrix element becomes

$$\begin{aligned} M_{\nu\nu'}(k_x, k_x', \mathbf{q}) &= V_q\, \delta_{k_x - k_x', \pm q_x} \int dy\, \xi_{i'}^*(y)\xi_i(y)e^{\pm iq_y y} \\ &\quad \times \int dz\, \phi_{j'}^*(z)\phi_j(z)e^{\pm iq_z z}. \end{aligned} \tag{21}$$

The indices i, i', j, and j' in Eq. (21) refer to the quantum numbers of the transverse wavefunctions which together determine the subband indices ν and ν'. Scattering rates for the POP, piezoelectric, and acoustic deformation-potential interaction are displayed in Fig. 2.a. The data are taken at T=30K for which $N_q \simeq 0$ and phonon absorption is negligible compared with emission. This is especially evident in the POP scattering rate which exhibits a large increase when emission to the subband bottom becomes possible. The inelastic nature of the POP interaction implies that this is not an important scattering mechanism for energies below the POP emission threshold. In this regime, the piezoelectric and acoustic interactions are essentially quasi-elastic and therefore dominate.

The validity of Eq. (15) is restricted to small longitudinal fields for which electrons are mostly near the Γ-valley minimum and appreciable deviations from the electron effective mass are not likely. When large longitudinal fields or high confinement conditions are investigated, the non-parabolicity of the band structure needs to be accounted for (see Sec. 2.1). In the 3D case, non-parabolicity

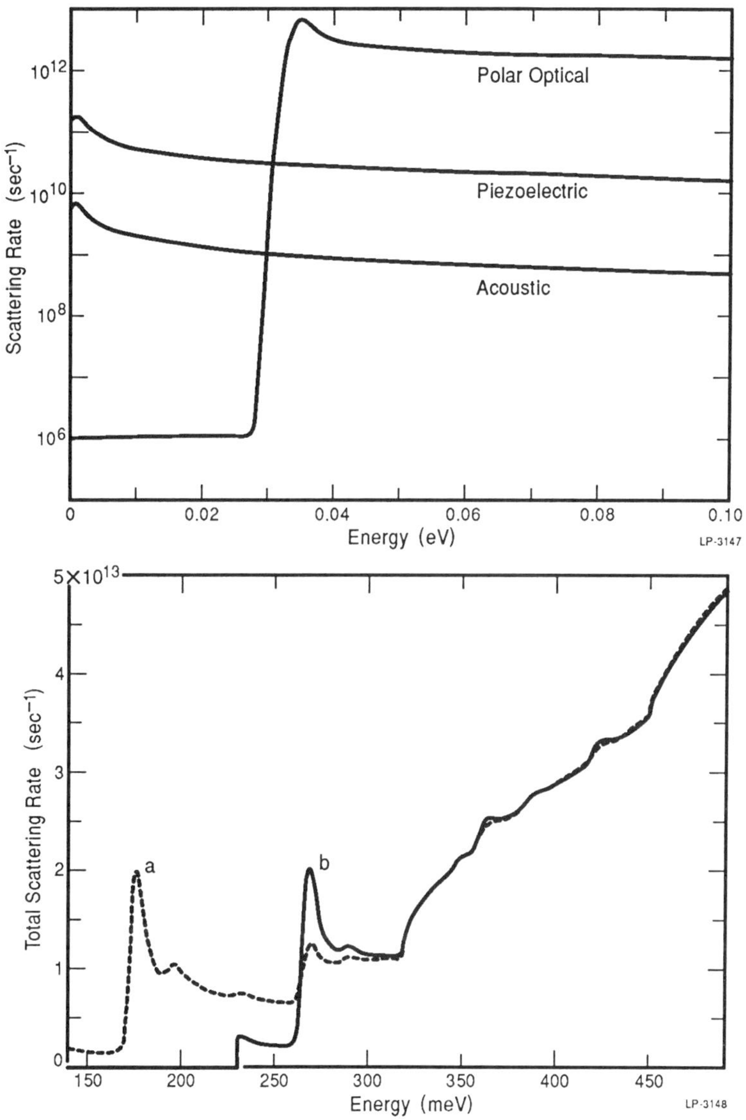

Figure 2: Scattering rates for (a) POP, acoustic, and unscreened piezoelectric phonon scattering at T=30K (F_z=120kV/cm, $\hbar\omega$=5meV) and (b) POP, acoustic, and intervalley scattering at T=300K (F_z=120kV/cm, L_y=135Å). The rates in (a) are for a single subband with the bottom of the subband taken as the energy origin. Those in (b) are for a seven subband simulation (only the first two subbands are shown) with the Γ-valley taken as the energy origin.

introduces a correction term to the scattering rates[40]

$$G(\vec{k},\vec{k'}) = (a_{\vec{k}}a_{\vec{k'}} + c_{\vec{k}}c_{\vec{k'}}cos\beta)^2, \quad (22)$$

where

$$a_{\vec{k}} = \sqrt{\frac{1+\alpha E(\vec{k})}{1+2\alpha E(\vec{k})}}, \quad (23)$$

$$c_{\vec{k}} = \sqrt{\frac{\alpha E(\vec{k})}{1+2\alpha E(\vec{k})}}, \quad (24)$$

and β denotes the angle between the 3D wavevectors $\mathbf{k}$ and $\mathbf{k'}$. For the quasi-1D case, Eq. (22) has to be spherically averaged over the transverse wavevectors since k_y, k'_y, k_z, and k'_z are indeterminate. Applying the constraint

$$k_\nu^2 = k_y^2 + k_z^2, \quad (25)$$

spherical averaging gives

$$G(\vec{k},\vec{k'}) = (a_{\vec{k}}a_{\vec{k'}})^2 + 2a_{\vec{k}}a_{\vec{k'}}c_{\vec{k}}c_{\vec{k'}}\frac{k_x k'_x}{|k||k'|} + (c_{\vec{k}}c_{\vec{k'}})^2\frac{(k_x^2 {k'_x}^2 + \frac{1}{2}k_\nu^2 {k'_\nu}^2)}{|k|^2|k'|^2}, \quad (26)$$

for intersubband transitions and

$$G(\vec{k},\vec{k'}) = (a_{\vec{k}}a_{\vec{k'}})^2 + 2a_{\vec{k}}a_{\vec{k'}}c_{\vec{k}}c_{\vec{k'}}\frac{k_x k'_x + k_\nu^2}{|k||k'|} + (c_{\vec{k}}c_{\vec{k'}})^2\frac{(k_x^2 k'_x + k_\nu^2)^2}{|k|^2|k'|^2}, \quad (27)$$

for intrasubband transitions[32].

To accurately accommodate intervalley scattering, a full band-structure treatment would be ideal, but computationally prohibitive, if one considers the complexities introduced by the quasi-1D double confinement. A three valley (Γ, L, X) model is more tractable and still realistic if one accounts for non-parabolicity. In the intervalley scattering regime the high effective masses of the satellite valleys cause weak quantization and result in bulk forms of the dispersion relation and density of final states. Scattering is therefore dominated by transitions from initial states in the quasi-1D Γ-valley to bulk states in either the X- or L-valleys. The reverse processes are not considered in the present analysis because we assume that once electrons are in the L or X valleys, they escape the wire with a negligible probability of being recaptured. The total intervalley scattering rate takes the form

$$\lambda_\nu^\Gamma(E) = \sum_{\mu'}^{X,L} \sum_{\nu' k'_x} \alpha_{\mu'} \int d\mathbf{q}_\perp \mid \langle \mathbf{k}' | V_q e^{i\mathbf{q}\cdot\mathbf{r}} | k_x, k_\nu \rangle \mid^2 G(\vec{k}, \vec{k'})$$
$$\times \delta[E_{\nu'}^{\mu'}(\mathbf{k}') - E_\nu^\Gamma(k_x) \pm \hbar\omega_{\mathbf{q}}] \qquad (28)$$

where $\mathbf{k}' = (k'_x, k'_y, k'_z)$ and V_q is the optical deformation potential coupling (Eq. 20). The factor $\alpha_{\mu'}$ accounts for the satellite valley degeneracy in the summation over final valleys μ'. Fig. 2.b shows the scattering rates for a quantum wire structure including POP, acoustic, and intervalley phonons. Clearly visible are the thresholds for phonon emission to the L-valley (318meV) and absorption to the X-valley (450meV). Phonon absorption to the L-valley also occurs but is much less obvious since the threshold for this process falls in the regime of POP emission to the minimum of subband **b**.

Although not immediately apparent in Eq. (28), the scattering rates follow roughly the same energy dependence as the final density of states. In the case of intervalley scattering to the bulk-like X and L valleys, the rates take on a roughly $\sqrt{E}$ dependence (see Fig. 2.b) while scattering within the Γ valley varies as $1/\sqrt{E}$ and poses the problem of dealing with the singular behavior of the rates near $E = 0$. Essentially, this functional dependence on the 1D density of states is a consequence of the semi-classical expression of energy conservation and ignores any quantum mechanical corrections between scattering events.

In order to circumvent the shortcomings of the golden rule, one must turn to a more accurate description of the electron-phonon interaction which consider the self energy of the polaron, $\Sigma(k_x, E)$. A suitable method is the Fock approximation[41, 42] which, after tailoring to the quasi-1D problem[43], gives:

$$\Sigma(k_x, E) = \int \frac{dq_x}{2\pi} \frac{g_+^2(k_x - q_x)}{E - \hbar\omega_q - \epsilon(q_x) - \Sigma(q_x, E - \hbar\omega_q)}$$
$$+ \int \frac{dq_x}{2\pi} \frac{g_-^2(k_x - q_x)}{E + \hbar\omega_q - \epsilon(q_x) - \Sigma(q_x, E + \hbar\omega_q)} \qquad (29)$$

with

$$g_\pm^2(k_x - q_x) \equiv \int \frac{dq_y\, dq_z}{4\pi^2} \mid \langle \mathbf{k}' | V_q e^{\pm i\mathbf{q}\cdot\mathbf{r}} | k_x, k_\nu \rangle \mid^2 . \qquad (30)$$

Even though Eq. (29) treats the issue of collision broadening in much more detail than the golden rule, it has its own shortcomings in that it yields an incorrect binding energy for the 1D polaron and does not include the overall lowering of the subband energies due to the transverse phonon component. In

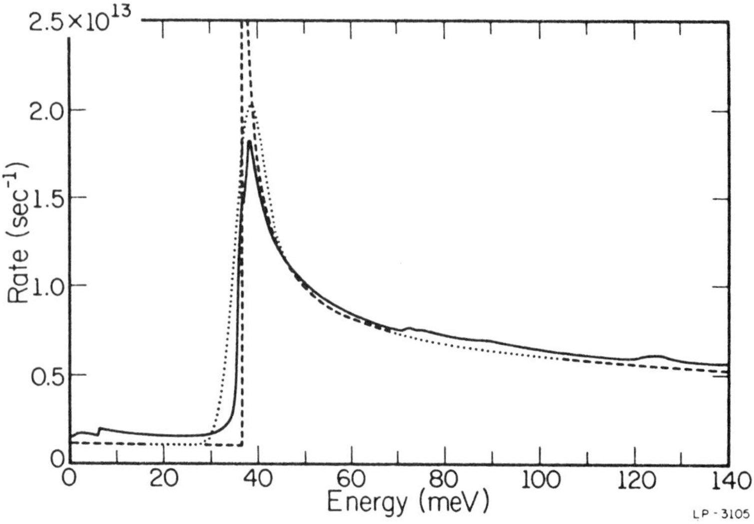

Figure 3: Comparison of POP scattering rates computed by the Fock approximation (solid), Fermi's golden rule using the classical unbroadened density of states (dashed), and Fermi's golden rule using the classical density of states convolved with a Gaussian broadening function fitted to the Fock result (dotted). The broadening of the fitted curve is conveyed in a constant broadening factor; $\Gamma = 2.5meV$ at 300K.

addition, Eq. (29) is computationally cumbersome owing to its self-consistent nature and further complicated by the introduction of round-off errors due to divergences in the denominator for the quasi-1D dispersion, $\epsilon(q_x)$. A solution by Briggs et al for the POP interaction (Fig. 3) demonstrates a substantial broadening in the emission peak which plays an important role in scattering processes at high temperatures. For practicality, the golden-rule formalism can be retained by convolving a Gaussian broadening function containing a constant broadening factor, Γ, with the bare 1D density of states so that[44]

$$D(E) = \frac{1}{2\hbar\Gamma}\sqrt{\frac{m^*}{\pi}} \int_{-E}^{\infty} e^{-E'^2/2\Gamma^2} \frac{1}{\sqrt{E'+E}} dE'. \tag{31}$$

This method, with a proper choice of Γ, gives a good representation of the quantum corrections to the scattering rates (Fig. 3) and substantiates using the golden rule with Eq. (31) replacing the classical density of states.

Computationally, the main differences between 1D simulations and 2D and bulk models is that quantum-scale confinement lifts some of the complexities associated with angular scattering. Due to the limited number of final scattering states, total rates can be stored in memory thereby eliminating lengthy computations during free-flight loops and final state selection. Quantized systems, however, introduce their own brand of difficulties in the treatment of subbands and singularities arising from the 1D density of states. The major computational bottlenecks of quasi-1D simulations occur during the numerical double integrations of the matrix elements over the transverse phonon wavevectors. Subbands add to the problem by requiring memory allocations that scale as the number of subbands squared. To alleviate these computational burdens, features unique to the confining potentials may be exploited as in the case of the harmonic oscillator confining potential; Hermite polynomials are generated recursively to save memory (as mentioned in Sec 2), the 3D matrix element coefficients are tabulated in an array, and finally an efficient table look-up algorithm is applied to expedite the double-integration routine and improve the overall speed performance. The vectorizability introduced by this method provides an order of magnitude speed up over the scalar code when run on a CRAY-2.

3 MONTE CARLO

Like the scattering rate computation, features specific to quasi-1D systems imply seeking for optimal Monte Carlo algorithms which may differ from those used in bulk or 2D simulations. For instance, the peaks in the quasi-1D scattering rates make constant-Γ methods inefficient because of the large percentage of self-scattering events which would occur. This effect coupled with the reduced degrees of freedom makes direct integration of the Monte Carlo free-flight equation extremely attractive for the quasi-1D problem. The probability that an electron in subband ν, which suffered a collision at time $t = 0$, scatters at time t is given by[44]

$$-\,ln(r) \;=\; \sum_{E=E_i}^{E_f} \lambda_\nu(E)\Delta t(E), \tag{32}$$

where r is a random number chosen from a uniform distribution. What makes this method particularly affordable is the restricted space of final 1D states. Since the outcome of a scattering event can be forward emission, forward absorption, backward emission, or backward absorption, and since there are only two possible wavevectors corresponding to each energy, all possible values of $\lambda_\nu(E)$ and $\Delta t(E)$ can be stored in look-up tables. This results in a significant integration speed-up of Eq. (32) over 2D and bulk simulations.

A typical single particle Monte Carlo program proceeds as outlined in Fig. 4. At invocation, the program requests the longitudinal electric field, F_x, and an initial state for the electron. The electron is placed in the initial state and allowed to undergo several scattering events to ensure independence of its initial conditions. Statistics are then collected for an appropriate number of scattering events to guarantee convergence of the electron distribution. After each free flight, a bin for the final wavevector, k'_x, is updated and the electron is allowed

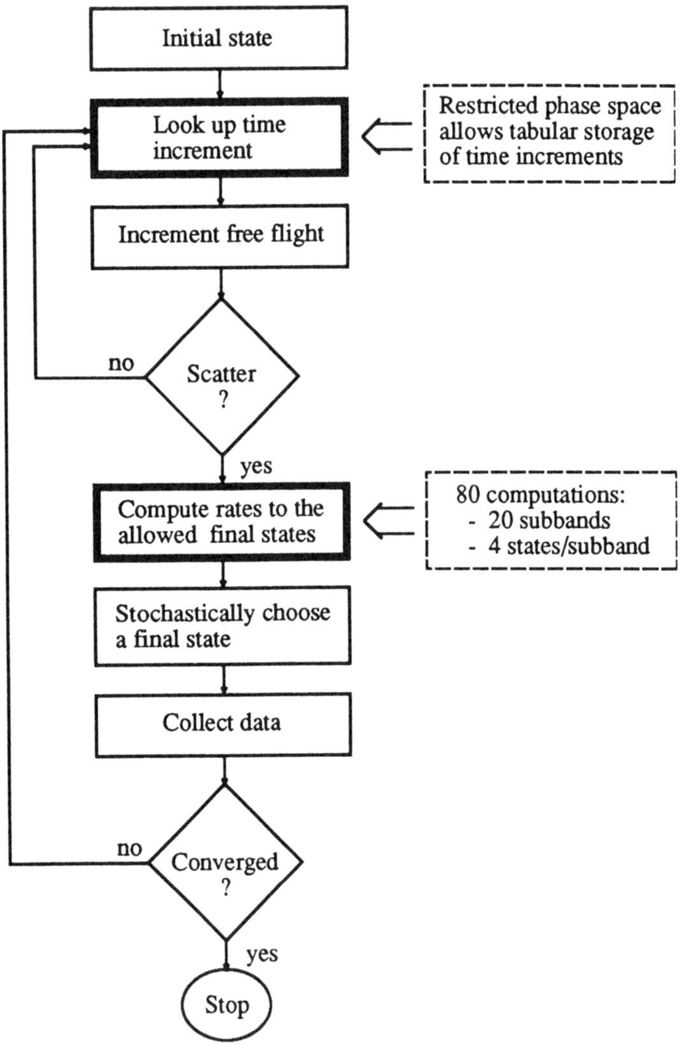

Figure 4: Flowchart for a typical single-particle multisubband Monte Carlo algorithm. The example pertains to a 20 subband model in which scattering rates are computed on-line for the stochastic determination of scattering species and final scattering states.

to scatter by randomly choosing a scattering agent (acoustic, piezoelectric, polar optic, or intervalley phonon). The final state information is then looked-up for the chosen scattering process, a new wavevector is stochastically chosen, and the electron is subsequently launched on another free flight.

The statistics compiled in this method are in the form of before scattering density functions ,$n_{b,\nu}(E)$. To recover the real distribution functions, $f_\nu(E)$, $n_{b,\nu}(E)$ must be normalized by the appropriate scattering rate, $D(E)$, and a normalization constant, τ_o[45]:

$$f_n(E) \; = \; \frac{1}{D(E)} \frac{1}{\lambda_\nu(E)\tau_o} n_{b,n}(E). \tag{33}$$

The affordability of the look-up method depends on the number of subbands necessary to achieve the desired level of accuracy in a particular simulation. Models near the extreme quantum limit (EQL) are well suited to the look-up method owing to the small number of subbands that need to be simulated and therefore don't tax the memory configurations of most contemporary computer systems. On the other hand, low confinement simulations require many subbands, especially at room temperature, to fully account for intersubband scattering effects. If memory constraints pose a problem, some of the look-up speed can be sacrificed by computing all possible final scattering rates for the initial electron state on-line after each scattering event. This speed-memory trade-off allows flexibility in tailoring code to the specific computer resources to which one has access.

In addition to single-particle Monte Carlo simulations, ensemble methods may be used to study the temporal or spatial variations of certain physical parameters. Instead of taking statistics after each scattering event, the time or space dimension is broken up into bins, and a statistical estimator for a particular bin is updated after each increment of the free-flight time. Spatial problems typically consist of establishing the 1D spatial grid, choosing the particle energies and momenta randomly out of some initial distribution, and drifting the particles through the spatial region under the influence of a longitudinal field in the background of scattering agents. Time-dependent problems follow in a similar manner but account for the temporal evolution of distribution functions and are therefore paramount in the study of diffusion phenomena. Statistics accumulated for both of these methods are space/time-averaged over each bin to obtain the physical properties of interest.

4 RESULTS

The Monte Carlo methods outlined in Sec. 3 have been used to study transport phenomena in quasi-1D systems at both low and high temperatures. Low temperature transport ($T \leq 77K$) is characterized by the absence of phonon absorption, an electronic distribution which occupies only the bottom few sub-

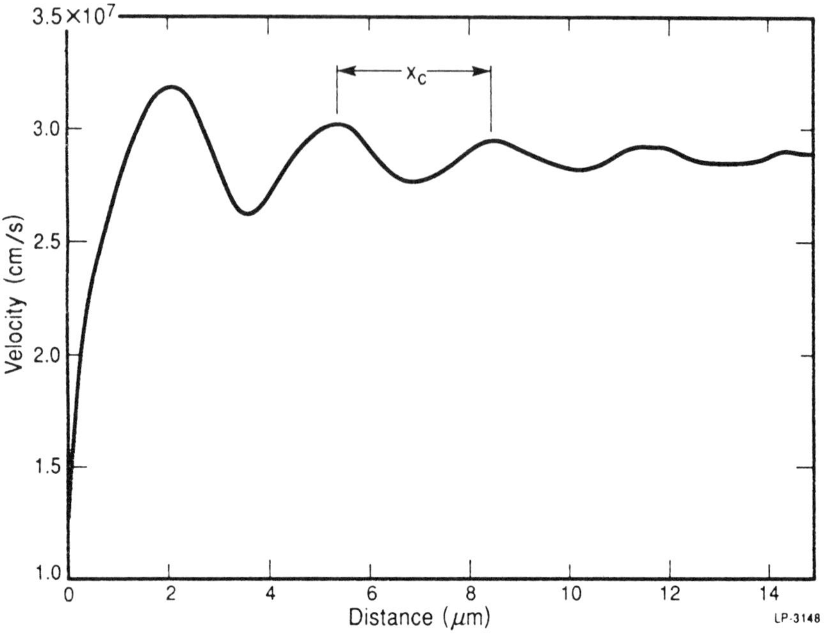

Figure 5: Electron velocity vs distance in a quasi-1D channel at 77K with $F_x = 100\text{V/cm}$. The initial velocity represents the average forward velocity of the initial distribution (Maxwell-Boltzmann).

bands, and evolution of the distribution that exhibits spatially dependent velocity oscillations. The contribution of phonon absorption at higher temperatures is predicted to lead to other phenomena such as anomalous carrier cooling, resonant intersubband optical phonon scattering (RISOPS), intersubband population inversions, and escape mechanisms similar to real-space transfer.

4.1 Low Temperature Transport

A spatially dependent Monte Carlo code was used to study the carrier evolution from an initial Maxwell-Boltzmann distribution under the sudden application of a longitudinal electric field F_x at low temperatures. This situation is analogous to the injection of carriers from a bulk region (source) at thermal equilibrium, into a quasi-1D channel. The low rate of phonon absorption implies that carrier transport will be essentially ballistic in the active region up to the POP emission threshold at which point carriers will emit a phonon and lose energy. This process is repeated for several periods until spatial dephasing and randomization by acoustic phonon scattering cause the electrons to converge onto a steady-state quasi-1D distribution. Fig. 5 shows the carrier velocity for a 100,000 electron ensemble at 77 K with a longitudinal field $F_x = 100V/cm$. The non-zero initial velocity reflects the forward component of the initial Maxwell-Boltzmann velocity distribution since these are the only carrier which are injected into the active region. As the electrons drift under the influence of the field they overshoot their steady state velocity several times in a manner analogous to an underdamped oscillator. The period of oscillation, x_c, can be derived classically by recognizing that quasi-ballistic transport occurs only for energies below the POP energy, $\hbar\omega_{pop}$, so that

$$x_c = \frac{\hbar\omega_{pop}}{eF_x} \tag{34}$$

For $F_x = 100\mathrm{V/cm}$, $x_c = 3.6\mu m$, which correlates well with the data in Fig. 5. This effect is similar to the velocity overshoot effect observed in short-channel MOSFET's in that the initial distribution is accelerated by the field causing the electron temperature to temporarily exceed the lattice temperature. In quasi-1D channels, this behavior persists over several periods due to the absence of angular randomization and the high peaks in the POP emission rates which permit relaxation to the subband minima in a quasi-coherent manner. The effect occurs for an intermediate range of F_x because at low fields, transport properties are dominated by acoustic deformation-potential and piezoelectric scattering which cause a strong randomization of the carrier distribution. Under high field conditions, the spatial coherence of the electron-phonon dynamics is destroyed by high-energy intersubband scattering. A similar effect has been predicted for bulk semiconductors although with a weaker intensity due to angular randomization and inherent background impurity scattering[46, 47, 48]. We believe that quantum wires would be ideal for observing this oscillatory behavior in a modulation-doped configuration.

4.2 High Temperature Transport

At high temperatures (T > 77 K), because of the absence of angular scattering, the electronic distribution shows significant deviations from the Maxwellian distribution at relatively low fields.

4.2.1 Carrier Cooling and Enhanced Mobility

Fig. 6 shows the profile of the energy distribution f(E) for F_x=100V/cm at T=300K compared with an approximate solution of the Boltzmann equation for POP scattering only. The quantum wire structure considered here consists of a QW with L_y=135 Å and a triangular potential with F_z=120 kV/cm. These high confinement conditions put the ground level at 138 meV above the GaAs band gap, and make the occupation of the upper subband practically negligible. The structures observed in f(E) are a direct consequence of the singularities in the 1D density of states and the emission and absorption of phonons with constant energy, $\hbar\omega_{pop}$, which favor energy occupation at $E = n\hbar\omega_{pop}$. The agreement between the numerical and analytical results is very good except near

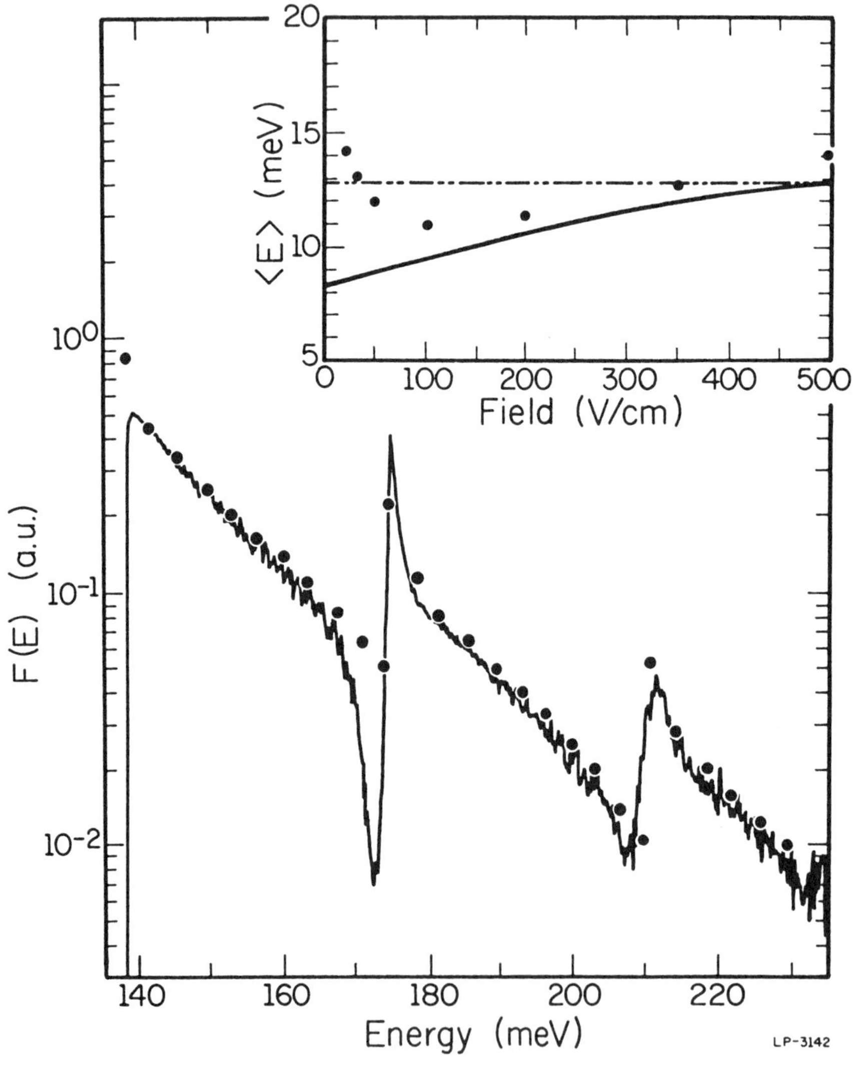

Figure 6: Energy distribution functions of a single band quasi-1D electron gas with F_x=100V/cm at T=300K. Both Monte Carlo (solid) and analytical (dots) results are plotted. The inset shows the average carrier energy as a function of F_x for the Monte Carlo (dots) and analytical (solid) models. The average thermal energy, kT/2, is also shown (dash-dot).

$E = n\hbar\omega_{pop}$ where the discrepancy is due to quantum broadening in the density of states accounted for in the Monte Carlo model (see Sec. 2). The structures manifest the onset of order in f(E) which arises from the disturbance of the Maxwellian equilibrium by the longitudinal electric field. The onset of order in f(E) is related to an important thermodynamical effect which is a cooling of the carriers below their thermal energy kT/2. The inset in Fig. 6 shows that for F_x <50V/cm the distribution is heated up due to acoustic phonon (AP) scattering; however, above 50 V/cm, the influence of AP's decreases rapidly and POP becomes the dominant scattering mechanism characterized by a minimum in the average energy $< E >$ around F_x=100V/cm. This minimum in $< E >$ corresponds to the onset of structures which result from the apparition of order in f(E). Increasing electric fields tend to compete with the POP-induced order and increase the carrier energy. As seen in the inset of Fig. 6, the analytical solution of the Boltzmann equation follows the same trend as the Monte Carlo simulation, except at low fields where AP scattering and other randomizing mechanisms have been neglected, and therefore shows a lower average energy. The decrease in $< E >$ below lattice temperature is associated with a mobility enhancement which, at 100V/cm, can be as high as $3 \times 10^4 cm^2/Vs$ and is thus much larger than the bulk value. The onset of order and carrier cooling are thus a consequence of the 1D forward absorption of dispersionless POP whereby under non-equilibrium high temperature conditions, thermal energy is converted in drift motion to enhance the transport response. The velocity-field relation is shown in Fig. 7.a, along with the equivalent relation for bulk GaAs. The velocity at $F_x = 500V/cm$ corresponds to a mobility of 16,160 cm^2/Vs, which is still twice the bulk value of 8000 cm^2/Vs. The differential mobility at low fields is over three times the bulk value and converges toward the latter at higher fields as electrons heat up and scatter to the upper subbands. The velocity-field curve for the same confinement conditions at 77 K is shown in Fig. 7.b. At low fields the differential mobility is higher in the 1D system than in the bulk, while at high fields the velocities again approach the same value.

The f(E) structures visible at multiples of the POP energy in Fig. 6 are relatively robust and persist in multi-subband systems, where the basic features of this effect contribute to enhanced POP-assisted intersubband resonances and onset of population inversion (see Sec. 4.2.3). We would like to point out that the absence of intrasubband (binary) e-e collisions limits the thermalization and provides grounds for the validity of our model.

4.2.2 Resonant Intersubband Optical Phonon Scattering

Two subbands are defined to be in resonance when the spacing between them approaches the POP energy $\hbar\omega_{pop}$. Of specific interest is high-order Resonant InterSubband Optic Phonon Scattering (RISOPS) for which one or more non-resonant subbands lie between two bands in resonance[49, 50]. In particular, as the spectrum of 1D subbands is passed through RISOPS, velocity oscillations similar to longitudinal magnetophonon effect occur over wide range of temper-

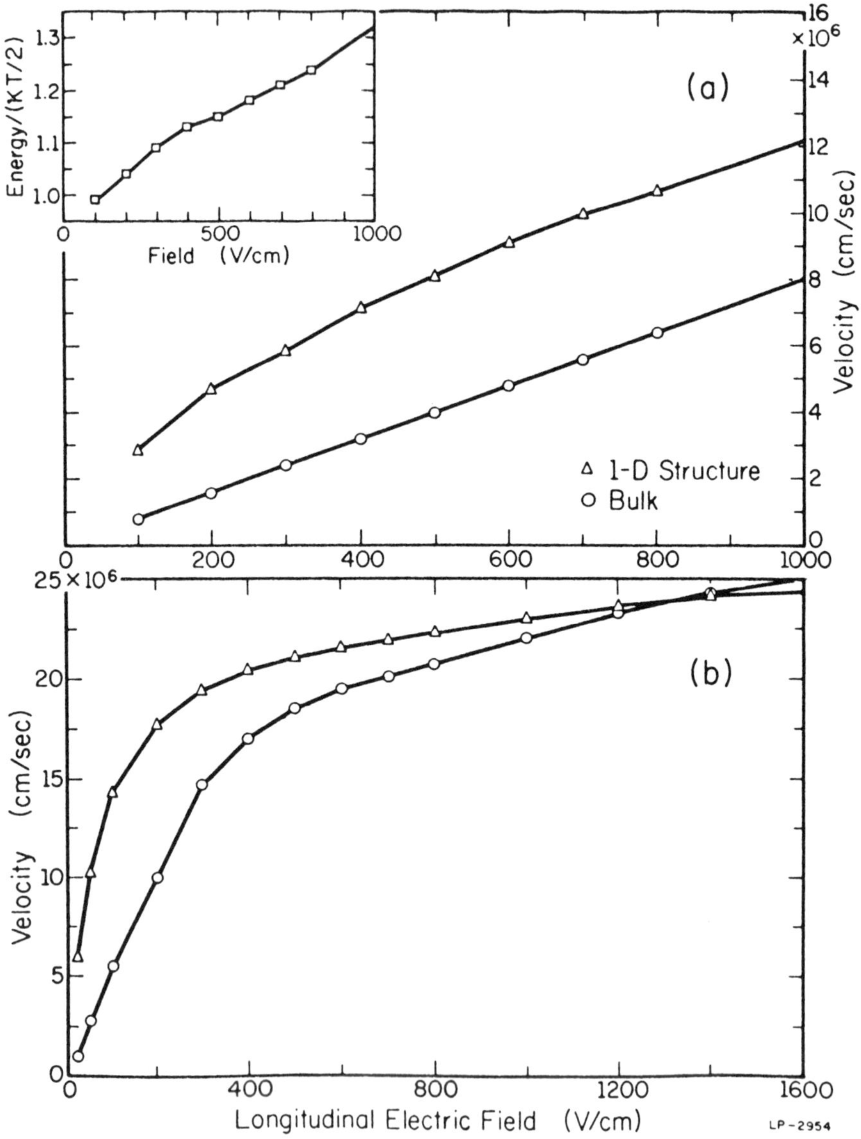

Figure 7: Velocity-field relation for a V-FET (L_y = 135Å and F_z = 120kV/cm) compared with bulk GaAs at (a) 300K and (b) 77K. The inset shows the average electron energy in units of kT as a function of the longitudinal field.

ature. Single carrier multisubband Monte Carlo simulation of a GaAs-AlGaAs structure with inelastic POP and acoustic phonon scattering characterized by appropriate broadening in the density of states shows velocity fluctuations as a function of confinement (Fig. 8.a). This effect which is modeled here for a quantum wire consisting of a infinite quantum well and an electrostatic parabolic potential, appears to be optimum at T = 150K since there is a trade-off between a significant absorption rate and small collision broadening[28]. Velocity minima occur during RISOPS, i.e., $n\Delta E = \hbar\omega_{pop}$ (n is an integer and ΔE constant energy separation of the harmonic oscillator) for which the intersubband POP scattering rate is maximum due to the peak in the joint 1D density of the states of the two interacting subbands. This phenomenon carries the potential for negative differential transconductance if the level separation can be modulated by an externally applied potential.

4.2.3 Intersubband Population Inversion

Another interesting feature in f(E) is the occurrence of intersubband population inversion under RISOPS conditions[51] with the potential for far infra-red (FIR) stimulated emission. Population inversion is found to occur for the 2nd order resonant configuration, at T = 150K, between the 3rd and 2nd subband with relative occupancies, O_{rel}, of 19.9% and 19.4%, respectively (Fig. 8.b). For a 3rd order resonance configuration, we found $O_{rel} = 14.0\%$ for the 4th subband and $O_{rel} = 11.8\%$ for the 3rd subband. It should be noted that although population inversions exist between harmonic oscillator subbands, they tend to be weaker than in quantum wires with irregularly spaced subbands. This is due to the propagation of RISOPS to the upper equally spaced harmonic oscillator subbands. As a result, the carrier population tends to be more dispersed in energy, thereby smoothing out any distribution anomalies. This is particularly noticeable at room temperature, where population inversion does not occur because the high absorption rate transfers electrons to the upper subbands very effectively.

Population inversion is be enhanced in quantum wire structures with unequally spaced energy levels, for instance those obtained with a triangular potential $eF_z z$ and a quantum well with width L_y. Two possible configurations leading to population inversion are realizable: a) $F_z = 23kV/cm$ and $L_y = 215\AA$ which places the first excited y-state ($n_y = 2, n_z = 1$) in resonance with the ground state ($n_y = 1, n_z = 1$), and the first excited z-state ($n_y = 1, n_z = 2$) just below the excited y-state (Fig. 9.a); b) $F_z = 10kV/cm$ and $L_y = 150\AA$ for which the second excited z-state ($n_y = 1, n_z = 3$) is in resonance with the ground state, and the first excited z-state ($n_y = 1, n_z = 2$) is the intermediate level (Fig. 9.b). However, stimulated emission is only possible in the latter case because the optical transition ($n_y = 2, n_z = 1 \rightarrow n_y = 1, n_z = 2$) in case (a) is forbidden due to the selection rules between independent y- and z- states. In case (b), Monte Carlo simulation for a longitudinal field F = 200 V/cm shows that 24 % of the carriers are in the upper band at T = 300K with only 4 %

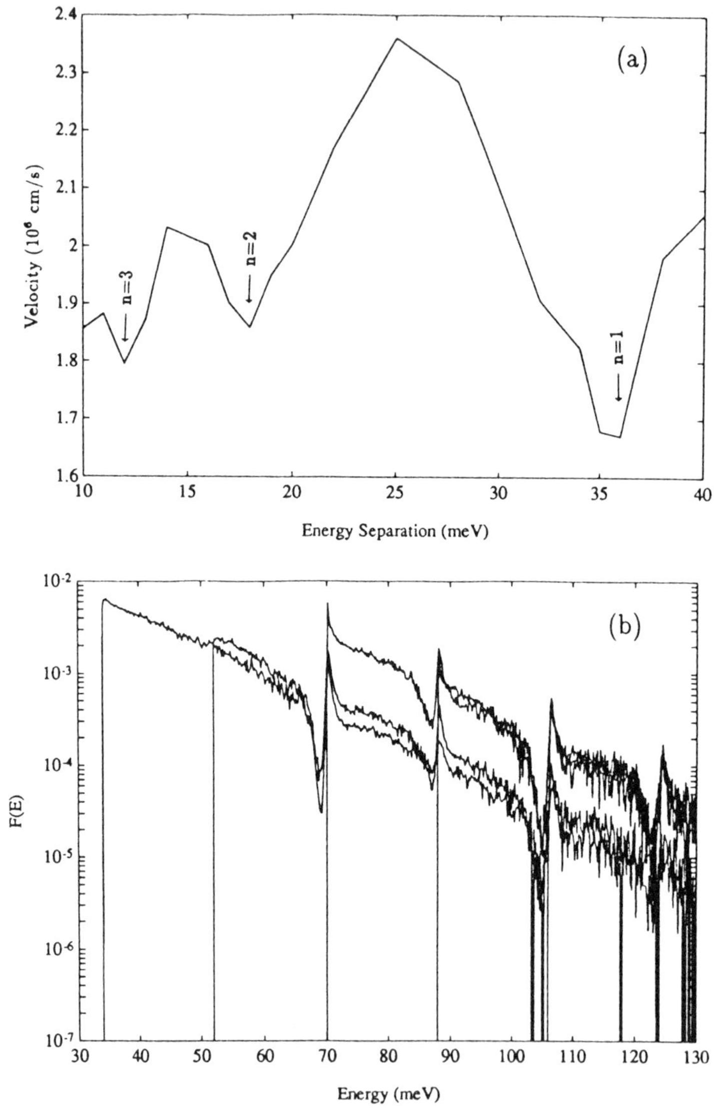

Figure 8: Electron velocity as a function of subband energy separation at T=150K. The data is taken for a system with L_y=150Å and F_x=100 V/cm. The occurrence and order, n, of RISOPS is indicated by the arrows. (b) Distribution functions for a 2nd-order resonant system with the same confinement as in a). Each curve represents the relative distribution of a particular subband with the energy origin lying at the Γ-valley minimum. Population inversion occur between the third (resonant) subband and the second.

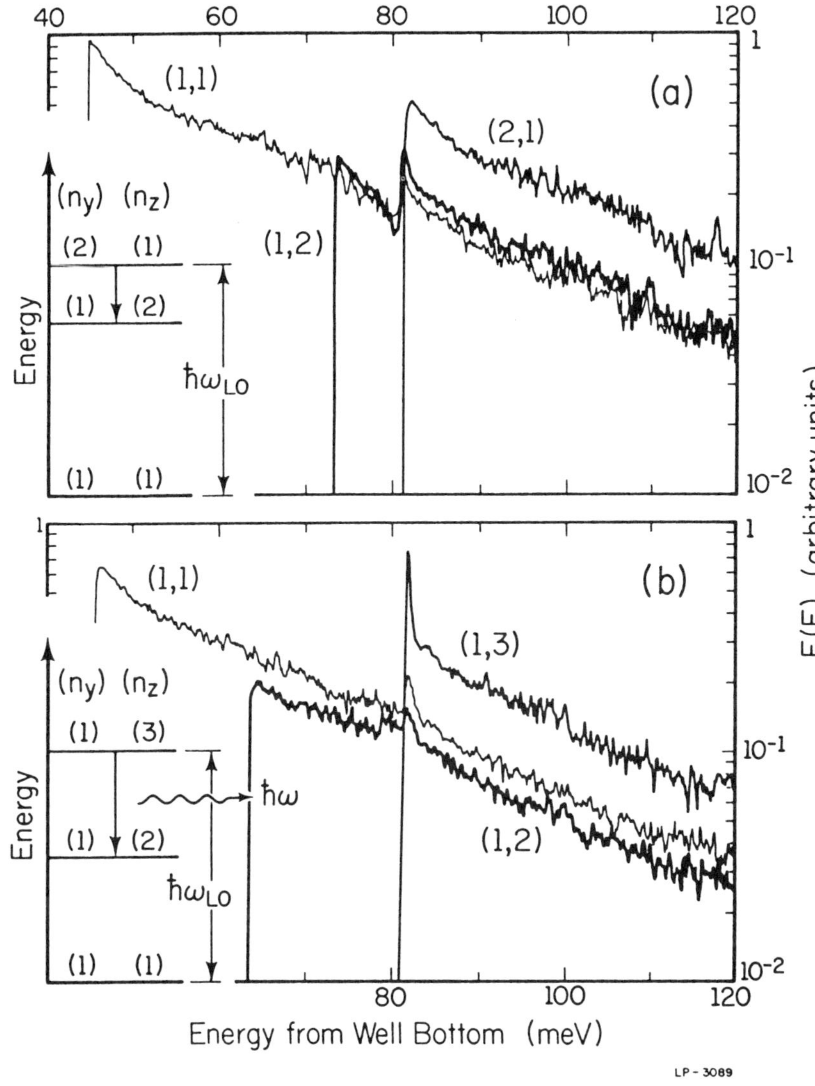

Figure 9: Distribution functions showing population inversion between (a) the (2,1) and (1,2) bands and (b) the (1,3) and (1,2) bands. The insets show the energy level diagrams under RISOPS conditions along with possible optical transitions. Because of selection rules in the optical matrix element, transitions between subbands cannot occur when both y- and z- quantum numbers change (case (a)).

in the intermediate subband. In order to assess the possibility for stimulated emission between inverted subbands, the theoretical gain given by[52]

$$g(\omega) = \frac{\Delta N}{A}\frac{e^2}{\bar{n}\epsilon_0 c}\frac{E_{ij}}{\hbar_2}|<j|\vec{e}\cdot\vec{r}|i>|^2\frac{T_2}{1+(\omega-\omega_{ij})^2T_2^2} \tag{35}$$

has been calculated to be $\simeq 900\ cm^{-1}$ at its peak value, for a population difference $\Delta N = 4\times 10^5/cm$, a wire cross section $A = 5\times 10^{-12}cm^2$, and an energy separation $E_{ij} = \hbar\omega_{ij} = 18meV$. Here, $\vec{e}$ is the optical polarization vector and we have assumed the intersubband scattering time to be $T_2 = 10^{-13}s$. The substantial gain value is encouraging and should make far FIR stimulated emission observable in highly packed 1D structures. In this respect, the novel techniques of quantum wire fabrication by direct growth on patterned vicinal substrates offer promising opportunities to realize FIR lasers operating by "phonon pumping" in quantum wire structures. We would like to emphasize that POP scattering is actually the pumping mechanism which creates population inversion here, while in other configurations such as tunneling injection in superlattices, it is the principal limitation to the quantum efficiency.

4.2.4 Intervalley Scattering and Carrier Escape

One serious issue affecting carrier dynamics is electron emission from the quantum wire at sufficiently high electric fields; the carriers gain enough energy from the field to overcome the confining barrier[32]. The escape mechanisms depend on the wire configuration and include real space transfer effects such as emission from a quantum well, or the drifting away of carriers from a heterojunction. Carrier losses due to scattering include intravalley as well as intervalley phonon scattering to 3D states. If the losses due to these mechanisms are significant, either in terms of electron lifetimes or mean paths in the wire, then the whole concept of an infinitely long, semi-classical quantum wire is invalid.

The influence of the longitudinal field on the lifetime is calculated by Monte Carlo techniques and shown in Fig. 10 for high and low confinement at both 77K and 300K; curve **a** (**b**) is for high (low) confinement at 300K while curve **c** (**d**) is high (low) confinement at 77K. In the high confinement cases, we have assumed that the initial electron energy is 140 meV while in the low confinement cases the initial electron energy is 50 meV. Varying the initial energy does not substantially alter the results except for cases with very high initial energies. All curves show a very strong dependence on the longitudinal field. At low fields, the effects of confinement and temperature are significant, but at higher fields these effects disappear and only the field is relevant in determining the lifetime. The escape times (for the same field) are larger at 77K because of the lack of phonon absorption, which tends to both lower the average energy and also raise the threshold for intervalley scattering. The 77K data do not go below 1500 V/cm because the lifetime is over 10 ns and too much CPU time is required to get an accurate estimate of the lifetime.

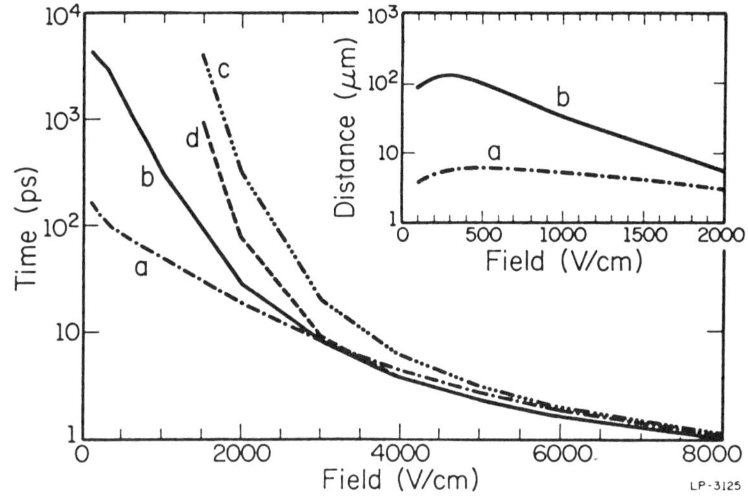

Figure 10: Electron lifetime vs. longitudinal field at 77K and 300K for high and low confinement conditions. **a** and **b** are the high and low confinement curves at 300K, respectively, while **c** and **d** are the corresponding curves at 77K. The inset shows the escape distance vs. longitudinal field.

The high confinement condition at 300K shows a much shorter lifetime than the low confinement case. This is due to the fact that under high confinement the average energy is high because of the location of the bottom subband relative to the L and X valley. However, at 77K the trend is reversed and under low confinement a large fraction of carriers is above the emission threshold. The inset in Fig. 10 shows an estimate of the distance an electron travels for **a**, high, and **b**, low confinement at 300K. The distance in the low confinement case is much larger than that for the high confinement case, even though electron velocities in the high confinement case are larger than in the low confinement case. The distance rises rapidly, peaks near 400 V/cm, and then drops off slowly.

The escape of carriers from quantum wires is a fundamental effect that can either be exploited or avoided by proper design of the 1D structure. With the aid of Monte Carlo simulation, we have obtained both length and time scales which establish a range of validity in treating the system as a closed 1D electron gas.

5 CONCLUSIONS

The Monte Carlo simulation is a powerful technique for investigating transport in highly confined semiconductor structures. Although of apparently greater complexity than 2D and bulk codes, realistic 1D Monte Carlo models can be handled by the capabilities of modern computers if efficient algorithms are implemented. The simulations have predicted several new effects which anticipate a wide range of applications for large-scale electronic as well as optoelectronic quasi-1D systems.

Acknowledgements

The Authors would like to thank Dr. Stephen Briggs for laying much of the foundation of this work. A debt of gratitude is also owed to Dr. Jack Higman for lending his expertise during times of need. This work was supported by JSEP under Grant No. N00014-90-J-1270. All of the computations were performed using the resources of the National Center for Supercomputing Applications (NCSA) and the National Center for Computational Electronics (NCCE) at the University of Illinois.

References

1. K. Kash, A. Scherer, J. Worlock, H. Craighead, and M. Tamargo, Appl. Phys. Lett., **49**, 1043 (1986).

2. A. C. Warren, I. Plotnick, E. H. Anderson, M. L. Shattenburg, D. A. Antoniadis, and H. I. Smith, J. Vac. Sci Technol., **B4**, 365 (1986).

3. P. M. Mankiewich, R. E. Howard, L. D. Jackel, W. J. Skocpol, and D. M. Tennant, J. Vac. Sci. Technol., **B4**, 380 (1986).

4. T.P. Smith III, H. Arnot, J.M. Hong, C.M. Knoedler, S.E. Laux, and H.Schmid, Phys. Rev. Lett., **59**, 2802 (1987).

5. K. Ishibashi, K. Nagata, K. Gamo, S Namba, S. Ishida, K. Murase, M. Kawabe, and Y. Aoyagi, Solid State Communications, **61**, 385 (1987).

6. K. Ismail, W. Chu, A. Yen, D. A. Antoniadis, and H. I. Smith, Appl. Phys. Lett., **54**, 460 (1989).

7. E. Colas, E. Kapon, S. Simhony, H.M. Cox, R. Bhat, K. Kash, and P.S.D. Lin, Appl. Phys. Lett., **55**, 867 (1989).

8. H. Sakaki, Japan J. Appl. Phys. **19**, L735 (1980).

9. H. Sakaki, *Proceedings of the International Symposium on GaAs and Related Compounds*, Osio, Japan, (1981).

10. For a review see e.g. *Nanostructure Physics and Fabrication*, edited by M.A. Reed and W.P. Kirk, Academic Press, Boston (1990).

11. W. J. Skocpol, L. D. Jackel, E. L. Hu, R. E. Howard, and L. A. Fetter, Phys. Rev. Lett., **49**, 951 (1982).

12. R. G. Wheeler, K. K. Choi, A. Goel, R. Wisnieff, and D. E. Prober, Phys. Rev. Lett., **49**, 1674 (1982).

13. G. Timp, A. M. Chang, P. Mankiewich, R. Behringer, J. E. Cunningham, T. Y. Chang, and R. E. Howard, Phys. Rev. Lett., **59**, 732 (1987).

14. B. J. van Wees, H. van Houten, C. W. J. Beenakker, J. G. Williamson, L. K. Kouwenhoven, D. van der Marel and C. T. Foxon, Phys. Rev. Lett., **60**. 848 (1988).

15. M. Roukes, A. Scherer, S. Allen, H. Craighead, R. Ruthen, E. Beebe, and J. Harbison, Phys. Rev. Lett. **59**, 3011 (1987).

16. C. J. B. Ford, T. J. Thornton, R. Newbury, M. Pepper, H. Ahmed, D. C. Peacock, D. A. Ritchie, J. E. F. Frost and G. A. C. Jones, Phys. Rev. Lett., **38**, 8518 (1988).

17. S. Datta, M. R. Melloch, S. Bandyopadhyay, R. Noren, M. Vaziri, M. Miller, and R. Reifenberger, Phys. Rev. Lett., **55**, 2344 (1985).

18. S. Washburn and R. A. Webb, Adv. Phys., **35**, 375 (1986).

19. M. Tsuchiya, J.M. Gaines, R. H. Yan, R. J. Simes, P. O. Holtz, L. A. Coldren, and P. M. Petroff,Phys. Rev. Lett., **62**, 466 (1989).

20. E. Kapon, D.M. Hwang, and R. Bhat, Phys. Rev. Lett., **63**, 430 (1989).

21. K. Ismail, D. A. Antoniadis, and H. I. Smith, Appl. Phys. Lett., **54**, 1130 (1989).

22. R. Stradling and R. Wood, J. of Physics C, **1**, 1711 (1968).

23. J. Lee and H. Spector, J. Appl. Phys., **57**, 366 (1985).

24. A. Ghosal, D. Chattopadhyay, and A. Bhattacharyya, J. Appl. Phys, **59**, 2511 (1986).

25. G. Fishman, Phys. Rev. B, **36**, 7448 (1987).

26. T. Hiramoto, K. Hirakawa, Y. Iye, and T. Ikoma, Appl. Phys. Lett., **51**, 1620 (1987).

27. T. Demel, D. Heitmann, P. Grambow, and K. Ploog, Appl. Phys. Lett., **53**, 2176 (1988).

28. D. Jovanovic, S. Briggs, and J.P. Leburton, Phys. Rev. B, **42**, 11108 (1990).

29. F. Stern, *CRC Critical Reviews in Solid State Sciences*, May 1974.

30. M. Abromowitz and I. A. Stegun, *Handbook of Mathematical Functions*, Dover Publications, New York, 782, (1972).

31. E.O. Kane, J. Phys. Chem. Solids, **1**, 249 (1957).

32. S. Briggs and J.P. Leburton, Phys. Rev. B, **43**, 4785, (1991).

33. H. Stormer, A. Gossard, and W. Wiegemann, Appl. Phys. Lett. **39**, 912 (1981).

34. S. Briggs and J. P. Leburton, Phys. Rev. B, **39**, 8025 (1989).

35. C. Kittel, *Quantum Theory of Solids*, John Wiley & Sons, (1987).

36. M. Stroscio, Phys. Rev. B **40**, 6428 (1989).

37. H. Fröhlich, H. Pelzer, and S. Zienau, Phil. Mag. **41**, 221 (1950).

38. B. K. Ridley, *Quantum Processes in Semiconductors*, Oxford Science Publications, 296 (1988).

39. K. Hirakawa and H. Sakaki, Phys. Rev. B, **33**, 8291 (1986).

40. W. Fawcett, A. D. Boardman, and S. Swain, J. Phys. Chem. Solids, **31**, 1963 (1970).

41. D. Pines in *Polarons and Excitons*, edited by G. Kupper and G. Whitfield (Scottish University Summer Schools, St. Andrews Scotland, 1962), p 155.

42. K. Kim, B. Mason, and K. Hess, Phys. Rev. B, **36**, 6547, (1987).

43. S. Briggs, B. Mason, and J.P. Leburton, Phys. Rev. B, **40**, 12001, (1989).

44. S. Briggs and J. P. Leburton, Phys. Rev. B, **38**, 8163 (1988).

45. C. Jacoboni and L. Reggiani, Rev. Mod. Phys., **55**, 645 (1983).

46. A. Matulionis, J. Pozela, and A. Reklaitis, Phys. Stat. Sol., (a) **31**, 83 (1975).

47. A. Matulionis, J. Pozela, and A. Reklaitis, Phys. Stat. Sol., (a) **35**, 43 (1976).

48. M. Brauer, Phys. Stat. Sol., (b) **81**, 147 (1977).

49. S. Briggs and J. P. Leburton, Superlattices Microstruct. **5**, 145 (1989).

50. S. Briggs, D. Jovanovic, and J. P. Leburton, Solid-State Electronics, **32** No.12, 1657 (1989).

51. S. Briggs, D. Jovanovic, and J.P. Leburton, Appl. Phys. Lett., **54**, 2012 (1989).

52. S. I. Borenstain and J. Katz, Appl. Phys. Lett., **55**, 654 (1989).

8

The Application of Monte Carlo Techniques in Advanced Hydrodynamic Transport Models

D. L. Woolard, H. Tian, M.A. Littlejohn, R. J. Trew, and K.W. Kim

Electrical and Computer Engineering Department
North Carolina State University
Raleigh, North Carolina 27695-7911

1 Introduction

The Monte Carlo method is a powerful technique for investigating electron transport physics in semiconductors and semiconductor devices[1, 2, 3]. Monte Carlo is a general statistical numerical method for solving mathematical and physical problems which was firmly established long before being introduced to semiconductor physics[4]. In the case of electron transport, the method provides the most efficiently implemented and physically adaptive technique currently available to determine *exact numerical* solutions of the electron motion in semiconducting materials and devices[5, and references therein].

The Monte Carlo method provides a relatively simple and accurate indirect approach to determine particle distribution functions and transport parameters resulting from the Boltzmann tranport equation subject to complex scattering mechanisms[6, 7]. In the method, a type of mathematical experiment is performed where the trajectory of an electron (or an ensemble of electrons) is simulated subject to applied fields and given scattering mechanisms. The duration of the carrier free flights and the particular collision events are selected stochastically (randomly) according to known scattering probabilities which describe the microscopic processes. When an electron is tracked over a *sufficiently long* history the dynamics can then be used to estimate time-independent properties of the entire electron gas (For time and/or space dependent problems, the sample ensemble must be sufficiently large to accurately represent the entire electron gas). This procedure permits the investigation and interpretation of physical phenomena unattainable solely by experimental methods. Thus, the Monte Carlo method is extremely valuable to solid state physics from both a theoretical and experimental point of view.

However, the approach does have important disadvantages. In general, the method does not *steadily* reduce the variation in average transport quantities from their true values and this convergence process is hard to monitor and estimate[2]. Therefore, the method is not well-suited for analyzing low-field transport, where the average electron velocity is small, due to the inherit statistical uncertainty present in most calculations[7]. Hence, the Monte Carlo method can be very computationally intensive when accurate results are desired.

An alternative approach to study electron device physics more accurately and efficiently is to develop macroscopic models which predict average transport quantities as opposed to the full electron distribution function. Combining the Monte Carlo technique with advanced hydrodynamic models, which include hot electron effects such as multiple nonparabolic conduction bands, is an excellent approach. These type of advanced hydrodynamic models represent a simplified mathematical approach to efficiently include realistic transport physics. A limited number of Monte Carlo simulations can then be utilized to verify the accuracy of the simplified models and to supply approximate parameteric dissipation data for the particular materials and/or devices under consideration.

In this presentation, the application of the Monte Carlo technique in advanced hydrodynamic models will be demonstrated in detail. In Section 2, the derivation of a new nonparabolic hydrodynamic transport model, which can model hot electron effects, will be presented. Also in Section 2, the Monte Carlo technique is used to extensively interrogate this macroscopic model under both stationary and nonstationary conditions. In Section 3, the issue of dissipation and how the Monte Carlo method can be used to generate approximated scattering terms is addressed. Finally in Section 4, an example of the tandem approach (Monte Carlo - Hydrodynamic) is used to study transit time enhancement in ballistic diode structures.

2 The Hydrodynamic Model Approach

2.1 Survey of Hydrodynamic Models

The last two decades have witnessed an increasing use of hydrodynamic conservation models to investigate nonstationary and nonequilibrium electron dynamics in submicron semiconductor devices[8]-[21]. These models, referred to by some[22, 23] as the hydrodynamic equations due to their similarity to the Euler equations of fluid dynamics[24, 25] used in classical hydrodynamics studies, are based upon higher moments of the Boltzmann transport equation (BTE). The popularity of hydrodynamic electron transport theory is due to it's physical and practical attributes. The hydrodynamic models have the capability to include nonstationary and hot electron effects[26] and hence are superior to simple drift-diffusion (extreme thermal equilibrium approximation (TEA)[27] and local electric field dependent case) models. While Monte Carlo methods which solve the full BTE can easily incorporate complicated band structures and detailed scattering rates, hydrodynamic models require much less computation time to generate solutions and possess macroscopic terms which offer important physical insight. Thus, the hydrodynamic approach offers much flexibility for future theoretical electron device studies.

Stratton[27] was first to introduce the general conservation or momentum-energy balance approach to investigate hot electron transport in semiconductors. The analysis performed by Stratton utilized a spherical harmonic expansion with the relaxation time approximation to define a nonequilibrium distribution which was only slightly anisotropic. This analysis replaced the nonequilibrium distribution by a Maxwellian distribution function in some terms of the BTE. Thus, some of the streaming terms in the final transport equations were suppressed. Blotekjaer[28] extended the theory, using a more general analysis which retained all terms of the moment equations and derived relations applicable to a two-valley semiconductor. These previous investigations were generally applicable to semi-classical transport and assumed homogeneous materials with parabolic conduction bands.

Since the early work of Stratton and Blotekjaer, the hydrodynamic approach has been utilized extensively to study various transport phenomena in many different Si[29, 16, 30], GaAs[8, 31, 32, 33] and heterostructure[18, 34, 35] electron devices. The particular material, device configuration and bias conditions under consideration have led to a variety of acceptable simplifications and to a hierarchy of approximate electron transport models (for example, see Sandborn *et al.*[36] and references therein). Also, much effort has been directed toward developing numerical techniques to solve the hydrodynamic transport model self-consistently with Poisson's equation for many semiconductor structures. This has included methods for solving the problem with and without time dependency and in multiple space dimensions [37]-[39] as well as an interest in special mathematical problems such as shock waves[40] and acceptable boundary conditions[41].

There has also been interest in improving the accuracy of hydrodynamic models for high field applications where electron energies may become very large and the electron gas may reside far from equilibrium. Stewart and Churchill[42] have recently introduced a fully nonparabolic hydrodynamic model for describing hot electron transport in GaAs. This model attempts to account for nonparabolicity and general distribution functions which become important for the extreme nonequilibrium or hot electron transport case. However, the analysis includes the introduction of heuristic effective mass parameters which are necessary to relate average momentum to average velocity as well as to relate an effective scalar temperature to thermal and displacement kinetic energy components. Furthermore, it is necessary to derive this effective mass from stationary Monte Carlo calculations. Also, Azoff[34] has developed a model for electron transport in nonparabolic degenerate semiconductor heterostructures. This analysis, which simultaneously addressed the issues of position-dependent energy bands, high-field energy band effects and degenerate statistics, is based on the first four moments of the relaxation time approximation Boltzmann transport equation. However, the results presented were primarily restricted to the thermal equilibrium limit (TEA) with many transport terms derived from approximate thermodynamic principles such as electronic heat capacity and the isothermal approximation. Also, the choice of moment operators and subsequent approximations for this nonparabolic analysis lead to a less than optimum form for comparing the results for the nonparabolic case to the general parabolic hydrodynamic transport equations.

Except for these examples, most of the current hydrodynamic transport models either directly or indirectly assume electron transport occurs in parabolic conduction bands and that the electron ensemble has the form of a displaced Maxwellian **k**-space distribution[36]. However, neither or these assumptions are valid for most high mobility compound semiconductors (such as GaAs) under moderately high electric fields[42]-[44]. Previously, we presented a hydrodynamic model [44] that accounted for both nonparabolicity and nondisplaced-Maxwellian distribution functions. This model made the extreme hueristic assumption that the effective mass, temperature tensors and heat flow vector could be described by stationary Monte Carlo generated parameters which only possessed energy dependence. Subsequently, simulations using this model[32] indicated that including a dependence on the displacement component[45] of the kinetic energy was important both physically and in regard to generating stable numerical solutions to the problem. In this presentation, a more general transport model will be developed which is less dependent on externally generated transport parameters and which approximates hot electron transport well in nonparabolic semiconductor conduction bands.

The sections that follow will present an alternate formalism for deriving a new nonparabolic hydrodynamic transport model which can model hot electron transport in semiconductors. Section 2.2 presents a novel approach for deriving this new model. The model equations are developed by applying a unique set of moment operators ($\mathbf{\Phi}_0(\mathbf{k}) = 1$; a constant, $\mathbf{\Phi}_1(\mathbf{k}) = \mathbf{u}(\mathbf{k})$; the nonparabolic velocity, $\mathbf{\Phi}_2(\mathbf{k}) = (m(\mathbf{k})/2)\mathbf{u}(\mathbf{k}) \cdot \mathbf{u}(\mathbf{k}) \approx E(\mathbf{k})$; an approximation to the particle energy) to the collisionless Boltzmann transport equation. The operators, used in this analysis and defined in Section 2.2, represent a significant improvement in the development of the hydrodynamic transport model since they lead to a more compact mathematical form. This resulting form can be manipulated more easily and reveals clearly the allowable simplifying approximations. Together, the operators and acceptable approximations lead to a model which is primarily expressed in terms of familiar averages: electron concentration, average electron velocity and average electron energy. The remaining integral terms involve only average velocity, momentum space velocity and a momentum space effective mass term. This compact nature is utilized later to develop an appropriate physical and mathematical closure for the system. The collisionless case developed in this section to permit a concise focus on the terms which evolve from the application of nonparabolic conduction bands. How nonparabolic bands and the particular scattering mechanisms involved affect the average loss terms will be addressed in a later section.

The system of equations, which are derived in Section 2.2, requires an additional mathematical closure before it can be used to study electron transport. In Section 2.3, a phenomenological-based distribution function is proposed as a constitutive (essential) relation to close the moment equations. This ansatz nonparabolic distribution function, which is derived in velocity space for mathematical convenience and approximates characteristics of the true nonparabolic distribution function, is then analyzed and used to develop a set of transport parameters. The resultant transport parameters are shown to depend only on the nonparabolicity parameter and physical quantities; electron concentration, average electron velocity and average electron energy.

The results from two independent Monte Carlo transport models[46, 47] are used in Section 2.4 to evaluate the nonparabolic transport parameters. These models were used to investigate electron transport in both stationary (uniform electric field) and nonstationary (ballistic diode) environments. The results of these studies show that there is excellent agreement between this new nonparabolic transport model and the Monte Carlo calculations. Hence, this new nonparabolic transport model presents a viable alternative to studying realistic electronic device structures operating under high bias conditions. Furthermore, this exercise demonstrates the utility of the Monte Carlo method in the verification of distribution function-based macroscopic transport models.

2.2 General Moment Equations

Consider the collisionless Boltzmann transport equation (BTE) expressed in terms of the phase-space variables $(\mathbf{r}, \mathbf{k}, t)$. The developments presented here are independent of the specific scattering phenomena which will be addressed later. Hence, the BTE is (see Section 3.2) is

$$\frac{\partial f}{\partial t} + \mathbf{u} \cdot \nabla_r f + \frac{\mathbf{F}}{\hbar} \cdot \nabla_{\mathbf{k}} f = 0 \tag{2.1}$$

where $f(\mathbf{r}, \mathbf{k}, t)$ is some arbitrary electron distribution function. Invoking the effective-mass theorem[48] (also called the quasi-free-particle approximation [49]) for electron motion defines the electron momentum as $\mathbf{p} = \hbar\mathbf{k}$ and identifies the electron group velocity as

$$\frac{\partial \mathbf{r}}{\partial t} = \mathbf{u}(\mathbf{k}) = \frac{1}{\hbar} \nabla_{\mathbf{k}} E(\mathbf{k}). \tag{2.2}$$

The driving force on the electron is $\mathbf{F} = \partial \mathbf{p}/\partial t$ and the final form of Eq. (2.1) is established.

The overall goal is to study moments of the BTE, Eq. (2.1), for the case of nonparabolic conduction bands. Thus, the first step is to develop a general moment equation. This is achieved by multiplying Eq. (2.1) by an arbitrary vector (or scalar) operator $\Phi(\mathbf{k})$ and integrating with respect to $\mathbf{k}$ to obtain

$$\int \Phi \frac{\partial f}{\partial t} d^3k + \int \Phi \mathbf{u} \cdot \nabla_r \, f d^3k + \int \Phi \frac{\mathbf{F}}{\hbar} \cdot \nabla_{\mathbf{k}} \, f d^3k = 0. \tag{2.3}$$

By noting that Φ is a function of $\mathbf{k}$ alone and invoking the assumption that f approaches zero rapidly enough in the limits of integration such that $\int \nabla_{\mathbf{k}}(\Phi \, f) d^3k$ is negligible[50], one obtains the general moment equation

$$\frac{\partial}{\partial t} \int \Phi f d^3k = -\nabla_r \cdot \int \Phi u f d^3k + \frac{\mathbf{F}}{\hbar} \cdot \int f \nabla_{\mathbf{k}} \Phi \, d^3k \tag{2.4}$$

in terms of the driving force $\mathbf{F}$, $\mathbf{k}$-space dependent electron velocity $\mathbf{u}(\mathbf{k})$ and an arbitrary vector moment operator $\Phi(\mathbf{k})$. Equation (2.4) represents the mathematical form required to develop the nonparabolic conservation or hydrodynamic transport equations.

For this nonparabolic model, the Kane energy band dispersion relation[42] for nonparabolic bands will be utilized. This relation is given by

$$\frac{\hbar^2 \mathbf{k}^2}{2m^*} = E(\mathbf{k})(1 + \alpha E(\mathbf{k})). \tag{2.5}$$

The specific choice of operators will determine what average physical quantity is conserved in each moment equation. For the first (or zero-th order) moment, let $\Phi_0(\mathbf{k})= 1$ in Eq. (2.4). This gives

$$\frac{\partial}{\partial t} \int f d^3k = -\nabla_r \cdot \int \mathbf{u} f d^3k. \tag{2.6}$$

This is the familiar continuity equation

$$\frac{\partial n}{\partial t} = -\nabla_r \cdot (n\mathbf{v}) \tag{2.7}$$

where $n = \int f d^3k$ is the conduction band electron density and $\mathbf{v} = (1/n) \int \mathbf{u} f d^3k$ is the average electron velocity. The continuity equation is unchanged by the nonparabolic band structure description.

Traditionally, to generate the second (or first order) moment of the BTE, one lets $\Phi_1(\mathbf{k}) = p = \hbar\mathbf{k}$ (or the parabolic band crystal momentum expression p raised to first order) to develop a conservation-of-momentum equation[31]. However, when considering nonparabolic bands, chosing another factor for the second moment leads to an improved, more compact, mathematical form. Since there is a clearly defined velocity from Eq. (2.2),

$$\Phi_1(\mathbf{k}) = \mathbf{u}(\mathbf{k}) \tag{2.8}$$

is chosen here as the second moment operator. The use of the nonparabolic velocity $\mathbf{u}(\mathbf{k})$ for the first order moment operator is a significant improvement in the development of the hydrodynamic transport model. Application of the Kane dispersion relation results in

$$\Phi_1(\mathbf{k}) = \mathbf{u}(\mathbf{k}) = \frac{\hbar \mathbf{k}}{m(\mathbf{k})}, \tag{2.9}$$

where from Eq. (2.5) and the definition for electron velocity

$$m(\mathbf{k}) = m^* \sqrt{1 + \frac{2\alpha\hbar^2\mathbf{k}^2}{m^*}}. \tag{2.10}$$

Using Eq. (2.9) $\Phi_1(\mathbf{k}) = \mathbf{u}(\mathbf{k})$ in Eq. (2.4) results in

$$\frac{\partial}{\partial t}\int \mathbf{u}(\mathbf{k}) f d^3k = -\nabla_r \cdot \int \mathbf{u}(\mathbf{k})\mathbf{u}(\mathbf{k}) f d^3k + \frac{\mathbf{F}}{\hbar}\cdot \int f \nabla_{\mathbf{k}} \mathbf{u}(\mathbf{k}) d^3k. \tag{2.11}$$

For typical upper limit values of electron velocity, $\mathbf{u} = 5.0\text{x}10^7 cm/sec$, Eq. (2.11) can be simplified to[52]

$$\frac{\partial \mathbf{v}}{\partial t} = -\mathbf{v}\cdot\nabla_r \mathbf{v} + \frac{\mathbf{F}}{m^{**}} - \frac{1}{nm^*}\nabla_r \cdot [\mathbf{P_v}], \tag{2.12}$$

which contains an average effective mass term defined by

$$\frac{n}{m^{**}} = \int \frac{f}{m(\mathbf{k})} d^3k \tag{2.13}$$

and an effective velocity pressure term

$$[\mathbf{P_v}]_{i,j} = m^* \int f(u_i(\mathbf{k}) - v_i)(u_j(\mathbf{k}) - v_j) d^3k. \tag{2.14}$$

The second moment of the BTE given by Eq. (2.12) is derived using the nonparabolic velocity $\mathbf{u}(\mathbf{k})$ and acceptable simplifications and has a form which is identical to the parabolic or constant effective mass case[28, 29], with the exceptions that m^{**} replaces m^* and the definition of $\mathbf{P_v}$ utilizes a more complicated definition of the particle velocity $\mathbf{u}(\mathbf{k})$. As in the parabolic case, additional requirements must be made to simplify $\mathbf{P_v}$ (and m^{**} for this nonparabolic model) and these will be discussed in the next section.

The third (or second order) moment equation is usually produced by chosing $\Phi_2(\mathbf{k}) = E(\mathbf{k}) = (m^*/2)|\mathbf{u}|^2$, (or $2m^*\mathbf{p}^2$ for parabolic bands: thus, second order in momentum $\mathbf{p}$) and results in an energy conservation equation. For nonparabolic bands, $E(\mathbf{k})$ is the positive root of $(\hbar^2\mathbf{k}^2)/(2m^*) = E(1+\alpha E)$. This leads to a very troublesome formulation when one attempts to define the moment equations in terms of familiar physical averages. Thus, a slightly more transparent path is taken. Energy is chosen as the third moment operator; however, the approximation $E_S(\mathbf{k}) \approx E(\mathbf{k})$ is chosen to once again produce an improved mathematical form more conducive to analysis and simplification. The chosen operator is

$$\Phi_2(\mathbf{k}) = E_S(\mathbf{k}) = \frac{1}{2} m(\mathbf{k})\mathbf{u}(\mathbf{k})\cdot\mathbf{u}(\mathbf{k}), \tag{2.15}$$

which has a familiar kinetic energy form $((m^*/2)\mathbf{u}^2)$ and is second order in $\mathbf{u}(\mathbf{k})$. Before continuing, the error in using the approximation $E_S(\mathbf{k}) \approx E(\mathbf{k})$ may be bounded by noting that

$$E_S(\mathbf{k}) = E(\mathbf{k})(\frac{1+\alpha E(\mathbf{k})}{1+2\alpha E(\mathbf{k})}). \tag{2.16}$$

For a typical value of $\alpha = 0.6eV^{-1}$, this approximation is better than 80% accurate up to energies of $0.5eV$. The agreement is sufficient since above $0.5eV$ the Kane dispersion relation does not accurately predict the true band structure[51].

Using Eq. (2.15) as an approximate definition of electron energy, the third moment of the BTE becomes

$$\frac{\partial}{\partial t}\int E_S(\mathbf{k}) f d^3k = -\nabla_r \cdot \int \mathbf{u}(\mathbf{k}) E_S(\mathbf{k}) f d^3k + \frac{\mathbf{F}}{\hbar} \cdot \int f \nabla_{\mathbf{k}} E_S(\mathbf{k}) d^3k. \tag{2.17}$$

For typical values of electron velocity Eq. (2.17) may be expressed as[52]

$$\frac{\partial w}{\partial t} = -\mathbf{v} \cdot \nabla_r\, w + \mathbf{v} \cdot \mathbf{F} - \frac{1}{n} \nabla_r \cdot (\mathbf{v} \cdot [\mathbf{P_w}] + \mathbf{q}) \tag{2.18}$$

with an energy pressure term of

$$[\mathbf{P_w}]_{i,j} = \int m(\mathbf{k})(u_i(\mathbf{k}) - v_i)(u_j(\mathbf{k}) - v_j) f d^3k \tag{2.19}$$

and effective heat flow vector term

$$\mathbf{q}_i = \int \frac{m(\mathbf{k})}{2}(u_i(\mathbf{k}) - v_i)(\mathbf{u}(\mathbf{k}) - \mathbf{v}) \cdot (\mathbf{u}(\mathbf{k}) - \mathbf{v}) f d^3k. \tag{2.20}$$

From Eqs. (2.7), (2.12) and (2.18), the nonparabolic hydrodynamic collisionless transport equations for conservation of particles, average momentum and average energy in Lagrangian form[24] are:

$$\frac{\partial n}{\partial t} = -\nabla_r \cdot (n\mathbf{v}), \tag{2.21}$$

$$\frac{\partial \mathbf{v}}{\partial t} = -\mathbf{v} \cdot \nabla_r \mathbf{v} + \frac{\mathbf{F}}{m^{**}} - \frac{1}{nm^*} \nabla_r \cdot [\mathbf{P_v}], \tag{2.22}$$

$$\frac{\partial w}{\partial t} = -\mathbf{v} \cdot \nabla_r\, w + \mathbf{F} \cdot \mathbf{v} - \frac{1}{n} \nabla_r \cdot (\mathbf{v} \cdot [\mathbf{P_w}] + \mathbf{q}) \tag{2.23}$$

where;

$$n = \int f d^3k \tag{2.24}$$

$$n\mathbf{v} = \int \mathbf{u}(\mathbf{k}) f d^3k \tag{2.25}$$

$$nw = \int \frac{m(\mathbf{k})}{2} \mathbf{u}(\mathbf{k}) \cdot \mathbf{u}(\mathbf{k}) f d^3k \tag{2.26}$$

$$\frac{n}{m^{**}} = \int \frac{f}{m(\mathbf{k})} d^3k \tag{2.27}$$

$$[\mathbf{P_v}] = m^* \int (\mathbf{u}(\mathbf{k}) - \mathbf{v})(\mathbf{u}(\mathbf{k}) - \mathbf{v}) f d^3k \tag{2.28}$$

$$[\mathbf{P_w}] = \int m(\mathbf{k})(\mathbf{u}(\mathbf{k}) - \mathbf{v})(\mathbf{u}(\mathbf{k}) - \mathbf{v}) f d^3k \tag{2.29}$$

$$\mathbf{q} = \int \frac{m(\mathbf{k})}{2} (\mathbf{u}(\mathbf{k}) - \mathbf{v})(\mathbf{u}(\mathbf{k}) - \mathbf{v}) \cdot (\mathbf{u}(\mathbf{k}) - \mathbf{v}) f d^3k. \tag{2.30}$$

Before this system of equations can be applied to the study of electron transport, additional assumptions must be invoked to express Eqs. (2.27) - (2.30) in terms n, $\mathbf{v}$ and w. This mathematical closure entails the use of specific fundamental physical properties to evaluate these parameters. Throughout this section efforts have been made to express these parameters in a mathematical form which is as compact as possible. It is clear that the integral definitions consist primarily of terms involving $\mathbf{u}(\mathbf{k})$ and differ from the definitions for the parabolic case only by the introduction of the nonparabolic mass $m(\mathbf{k})$ and the specific definition of the particle velocity $\mathbf{u}(\mathbf{k})$. This property will be utilized in the next section to develop approximate expressions so that this nonparabolic form of the hydrodynamic equations may be used to study electron dynamics in realistic electronic structures.

2.3 Nonparabolic Transport Model

The previous equations represent a general nonparabolic hydrodynamic transport model for electron transport in a single valley of GaAs. Additional assumptions or relationships are necessary to close the system of equations before they can be used to describe transport phenomenon. Specifically, one must either ignore one or more higher order terms in the formulation[28, 53], have some previous knowledge for the equations of state[54, 55], or assume some form of the electron distribution function in momentum space[56]. These are all acceptable methods of simplification and can lead to the same unique system of equations.

The use of a predefined distributional form is often viewed as being more restrictive and less general than other alternatives. Yet, many derivations, which seek to develop models applicable for more generalized distributions, partially implement closure relations with properties *proven* true only for specific distributional forms[16, 22]. Extending the electron transport model in this manner has merit and leads to results which are more physically accurate[16]. However, mixing assumptions in this manner can obsure regions of validity for the complete model. Defining a specific form of the distribution function is the most complete approach and is easiest to evaluate from both a physical and mathematical point of view. Therefore, this is the approach considered here.

The distribution most frequently chosen to describe nonequilibrium transport phenomena is the displaced Maxwellian function. For the case of parabolic conduction bands this distribution function, in momentum space, is[56]

$$f_{dm}(\mathbf{k}) = \frac{n\hbar^3}{(2\pi m^* k_B T_e)^{\frac{3}{2}}} \exp[\frac{-\hbar^2|\mathbf{k}-\mathbf{k_d}|^2}{2m^* k_B T_e}], \tag{2.31}$$

where k_B is the Boltzmann constant, T_e is the temperature of the electron gas representing the spread of the distribution, and $\mathbf{k_d}$ is the displacement from $\mathbf{k} = 0$ of the Maxwellian distribution. The previous distribution ignores any effect that Fermi-Dirac statistics may have on the total number of available electrons. Since this work is directed toward transport in conduction bands and it is a simple task to modify the normalization coefficient to include Fermi-Dirac statistics, they will be ignored throughout.

This Maxwellian approach[17] has been applied often in the analysis of transport problems; however, for the case of nonparabolic bands, Monte Carlo calculations indicate that such a distribution leads to a less than adequate treatment of the transport parameters of the hydrodynamic model[44].

In this derivation a unique treatment is proposed for closing the general moment equations for electron transport in GaAs. To articulate the basis of this approach, consider a form of the displaced Maxwellian distribution $\mathcal{F}_{dm}(\mathbf{u})$ transformed into velocity space such that $\mathcal{F}_{dm}(\mathbf{u})d^3u = f_{dm}(\mathbf{k})d^3k$ or specifically, $\mathcal{F}_{dm}(\mathbf{u}) = |J_{dm}|f(\mathbf{u}^{-1}(\mathbf{k}))$. Here, the inverse function represents $\mathbf{u}^{-1}(\mathbf{k}) = \mathbf{k} = \mathbf{g}(\mathbf{u})$ where $\mathbf{g}(\mathbf{u})$ is the correct mapping from momentum space to velocity space. Also, J_{dm} is the appropriate Jacobian transformation [57] from momentum space to velocity space. Defining this velocity space distribution in such a manner allows parametric definitions of the form $I = \int i(\mathbf{k})f_{dm}(\mathbf{k})d^3k$, similar to Eqs. (2.27) through (2.30), to be expressed in velocity space as $I = \int i(\mathbf{u}^{-1}(\mathbf{k}))\mathcal{F}_{dm}(\mathbf{u})d^3u$. Therefore, for the case of parabolic bands where the relationship between velocity space and momentum space is linear, $\mathbf{u}(\mathbf{k}) = (\hbar\mathbf{k})/m^*$ and $|J_{dm}| = (m^*/\hbar)^3$, this corresponding displaced Maxwellian distribution in velocity space is

$$\mathcal{F}_{dm}(\mathbf{u}) = n[\frac{m^*}{2\pi k_B T_e}]^{\frac{3}{2}} \exp[\frac{-m^*|\mathbf{u}-\mathbf{v}|^2}{2k_B T_e}], \tag{2.32}$$

where $\mathbf{v}$ now represents the average velocity as well as the displacement under consideration in velocity space. Both Eq. (2.31) and Eq. (2.32) are symmetric about their displacements within their respective spaces and the average velocity $\mathbf{v}$ is equal to $(\hbar\mathbf{k}_d)/m^*$; therefore, Eq. (2.32) does represent a completely consistent transformation of Eq. (2.31) for the case of parabolic bands.

One would expect that, under the conditions of nonparabolic bands where the effective mass depends on the electrons precise location in the conduction band, the use of constant m^* would be insufficient. Also, another dilemma exists in defining a single temperature of the electron gas since it has been shown that two distinctly different temperatures tensors $\mathbf{T_V}$ and $\mathbf{T_W}$ occur for the case of nonparabolic bands[44]. Thus, stationary Monte Carlo calculations and physical intuition suggests the following ansatz distribution

$$\mathcal{F}_{nonpar}(\mathbf{u}) = n[\frac{m_w(\mathbf{v},T_w)}{2\pi k_B T_w}]^{\frac{3}{2}} \exp[\frac{-m_w(\mathbf{v},T_w)|\mathbf{u}-\mathbf{v}|^2}{2k_B T_w}] \qquad (2.33)$$

as a constitutive relation to close the moment equations for the case of nonparabolic energy bands. Here a single effective temperature T_w is still utilized. However, T_w, with a distinct w subscript, has been chosen to replace T_e because $\frac{3}{2}k_B T_w$ (here T_w is defined from diagonal elements of the energy pressure $[\mathbf{P_W}]$) approximates the effective thermal energy well for the stationary transport case at moderate values of average energy[44]. In addition, $m_w(\mathbf{v},T_w)$ has been introduced because, based on physical considerations, we expect a nonconstant average effective mass which should depend strongly on the spread or thermal component T_w of the true electron distribution and on the displacement in velocity space $\mathbf{v}$.

In applying Eq. (2.33) to the integral definitions (2.24) - (2.30), it is correct to use $\mathcal{F}_{nonpar}(\mathbf{u})d^3u = f_{nonpar}(\mathbf{k})d^3k$, since $\mathcal{F}_{nonpar}$ was postulated in velocity space from a distribution which contained this transformation property. In general, the issue of transforming between momentum and velocity space for the case of nonparabolic bands involves very complicated nonlinear transformations. In fact, a careful analysis[52], using a momentum-space distribution (of Maxwellian form) and appropriate transformation factors, shows that the velocity-space distribution will have terms similar to those in Eq. (2.33).

Using those results as a basis for the $\mathcal{F}_{nonpar}(\mathbf{u})$ distribution function, some of it's properties will be presented before proceeding. The distribution is symmetric in velocity space about a displacement of the average velocity $\mathbf{v}$. Though the spread in velocity space is symmetric, the amount of spread depends in a nonlinear way on temperature T_w and displacement $\mathbf{v}$. Specifically, the dependence of the spread is partially determined by the free function $m_w(\mathbf{v},T_w)$. The distribution *does* meet the expected and necessary requirements of yielding the average electron density n upon evaluation of $\int \mathcal{F}_{nonpar} d^3u$ and the average electron velocity $\mathbf{v}$ upon evaluating $(1/n)\int \mathbf{u}\mathcal{F}_{nonpar} d^3u$. Also, $\mathcal{F}_{nonpar}(\mathbf{u})$ converges to $\mathcal{F}_{dm}(\mathbf{u})$ as $m_w \rightarrow m^*$. This last quality is desirable because the Maxwellian-based distribution $\mathcal{F}_{dm}(\mathbf{u})$ is considered to be a good approximation for the case of small perturbations from equilibrium and dominate electron-electron scattering[58].

At this point the first question to address is how to define $m_w(\mathbf{v},T_w)$. One way would be to study either experimentally-obtained or Monte Carlo-generated electron distributions in velocity space. However, since the postulation and inclusion of $m_w(\mathbf{v},T_w)$ in the distribution function was based on the idea of a nonconstant effective mass, a more interesting approach is to approximate $m_w(\mathbf{v},T_w)$ by the average effective mass of the entire distribution. Equating $m_w(\mathbf{v},T_w)$ to the definition of the inverse average effective mass of Eq. (2.27),

$$m^{**} = \left(\frac{1}{n}\int \frac{\mathcal{F}_{nonpar}}{m(\mathbf{u}^{-1}(\mathbf{k}))} d^3u\right)^{-1} \qquad (2.34)$$

yields the recursive definition

$$m_w(\mathbf{v}, T_w) = \left(\left[\frac{m_w(\mathbf{v}, T_w)}{2\pi k_B T_w} \right]^{\frac{3}{2}} \int \frac{\exp\left[\frac{-m_w(\mathbf{v},T_w)|\mathbf{u}-\mathbf{v}|^2}{2k_B T_w}\right]}{m(\mathbf{u}^{-1}(\mathbf{k}))} d^3u \right)^{-1} . \quad (2.35)$$

Here, $m(\mathbf{u}^{-1}(\mathbf{k}))$ is defined using Eq. (2.10) and Eq. (2.9), which to fourth order in $\mathbf{u}$ becomes

$$\frac{1}{m(\mathbf{u}^{-1}(\mathbf{k}))} = \frac{1}{m^*}(1 - \alpha m^* |\mathbf{u}|^2 - \frac{\alpha^2}{2}(m^*)^2 |\mathbf{u}|^4). \quad (2.36)$$

The above formulation has been written in velocity space to simplify the integration process required here and later in the paper. Performing the integration in Eq. (2.35) results in the approximate expression of

$$m_w(\mathbf{v}, T_w) = m^*(1 + 3\alpha k_B T_w + 7.5(\alpha k_B T_w)^2 + \alpha m^* |\mathbf{v}|^2). \quad (2.37)$$

The last formula resulted from the recursive definition of Eq. (2.35) and the method of defining $m_w(\mathbf{v}, T_w)$ merits further investigation. Within the framework of this nonparabolic model $m_w(\mathbf{v}, T_w)$ is a free parameter which one may arbitrarily chose. Consider the simple definition of

$$m_w(\mathbf{v}, T_w) = m^*(1 + \mathcal{A}_w \alpha k_B T_w + \mathcal{B}_w \alpha m^* |\mathbf{v}|^2) \quad (2.38)$$

where the result from Eq. (2.37) has been used to define an effective normalized lower order form for $m_w(\mathbf{v}, T_w)$ with $\mathcal{A}_w$ and $\mathcal{B}_w$ as fitting parameters. Using (2.38) in the definition for m^{**} results in the approximate formula

$$m^{**}(\mathbf{v}, T_w) = m^*(1 + 3\alpha k_B T_w + (7.5 - 3\mathcal{A}_w)(\alpha k_B T_w)^2 + \alpha m^* |\mathbf{v}|^2). \quad (2.39)$$

From this result one can observe that the first order result for m^{**} is independent of the definition of $m_w(\mathbf{v}, T_w)$ and represents a perturbational effect on the second order terms in T_w (here higher order terms have been excluded; hence, the disappearance of $\mathcal{B}_w$). Also, since $(1/n) \int m(\mathbf{u}^{-1}(\mathbf{k})) \mathcal{F}_{nonpar} d^3u \approx ((1/n) \int \mathcal{F}_{nonpar} / m(\mathbf{u}^{-1}(\mathbf{k})) d^3u)^{-1}$, to second order in T_w and $\mathbf{u}$, the last derivation has also resulted in an approximate expression for the transport parameter $m^{**}_{avg} = (1/n) \int m(\mathbf{u}^{-1}(\mathbf{k})) \mathcal{F}_{nonpar} d^3u$ (this parameter will be required later in the derivation). Hence, this work will utilize the definitions

$$m^{**}(\mathbf{v}, T_w) = m^{**}_{avg}(\mathbf{v}, T_w) = m_w(\mathbf{v}, T_w). \quad (2.40)$$

Combining the proposed distribution of Eq. (2.33) with the last result supplies the complete constitutive relation to close the moment equations of the last section. Applying these relations to the energy pressure tensor $[\mathbf{P_W}]$ results in

$$[\mathbf{P_W}] = n k_B T_w \mathcal{T}(\mathbf{v}, T_w)[\mathbf{I}] \quad (2.41)$$

where

$$\mathcal{T}(\mathbf{v}, T_w) = \frac{m^*}{m^{**}}(1 + 2\alpha m^* |\mathbf{v}|^2 + 5\alpha \frac{m^*}{m^{**}} k_B T_w + ...). \quad (2.42)$$

Approximating $\mathcal{T}(\mathbf{v}, T_w)$ to second order in $\mathbf{v}$ and T_w without cross-terms yields a simplified form of

$$\mathcal{T}(\mathbf{v}, T_w) = (1 + 2\alpha k_B T_w + \alpha m^* |\mathbf{v}|^2) \tag{2.43}$$

and an energy pressure tensor of

$$[\mathbf{P_W}] = n k_B T_w (1 + 2\alpha k_B T_w + \alpha m^* |\mathbf{v}|^2)[\mathbf{I}]. \tag{2.44}$$

Again applying Eq. (2.33) and the definition for the average distributional mass to the velocity pressure tensor $[\mathbf{P_V}]$ yields

$$[\mathbf{P_V}] = \frac{m^*}{m^{**}} n k_B T_w [\mathbf{I}]. \tag{2.45}$$

Using the same approach on the definition of heat flow vector $\mathbf{q}$ results in

$$\mathbf{q} = 5\alpha(\frac{m^*}{m^{**}})^2 (k_B T_w)^2 (1 + 21\alpha \frac{m^*}{m^{**}} k_B T_w) n \mathbf{v}. \tag{2.46}$$

A relation can now be developed to eliminate T_w from the formulation by studying $(1/n) \int (m(\mathbf{u}^{-1}(\mathbf{k}))/2) |\mathbf{u} - \mathbf{v}|^2 \mathcal{F}_{nonpar} d^3u$ which is the expression used to define the classical temperature of a gas except with $m(\mathbf{u}^{-1}(\mathbf{k})) =$ constant. Performing the same integration steps as used to simplify the energy pressure tensor results in

$$\frac{3}{2} k_B T_w \mathcal{T}(\mathbf{v}, T_w) = \frac{1}{n} \int \frac{m(\mathbf{u}^{-1}(\mathbf{k}))}{2} |\mathbf{u} - \mathbf{v}|^2 f d^3u. \tag{2.47}$$

The last expression can also be expanded to yield

$$\frac{1}{n} \int \frac{m(\mathbf{u}^{-1}(\mathbf{k}))}{2} |\mathbf{u} - \mathbf{v}|^2 f d^3u = w + \frac{m^{**}}{2} \mathbf{v} \cdot \mathbf{v} - m^{**}_{avg} \mathbf{v} \cdot \mathbf{v} \tag{2.48}$$

to second order in $\mathbf{v}$ and T_w. Using the definition of Eq. (2.40) leads to the final constitutive relation

$$w = \frac{3}{2} k_B T_{eff}(\mathbf{v}, T_w) + \frac{m^{**}(\mathbf{v}, T_w)}{2} |\mathbf{v}|^2 \tag{2.49}$$

with $T_{eff}(\mathbf{v}, T_w) = T_w \mathcal{T}(\mathbf{v}, T_w)$.

The last equation is a nonparabolic form relating the average energy of the electron gas to it's effective thermal and kinetic energy components. It is interesting that this equation is identical to the classical case except T_{eff} replaces T_e and m^{**} is substituted for the constant mass m^*. This equation provides the final relation necessary to close the general moment equations. To summarize the developments of this section, the following equations:

$$m^{**}(\mathbf{v}, T_w) = m^*(1 + 3\alpha k_B T_w + 7.5(\alpha k_B T_w)^2 + \alpha m^* |\mathbf{v}|^2) \tag{2.50}$$

$$[\mathbf{P_V}] = \frac{m^*}{m^{**}} n k_B T_w [\mathbf{I}] \tag{2.51}$$

$$[\mathbf{P_w}] = nk_BT_w(1 + 2\alpha k_BT_w + \alpha m^*|\mathbf{v}|^2)[\mathbf{I}] \tag{2.52}$$

$$\mathbf{q} = 5\alpha(\frac{m^*}{m^{**}})^2(k_BT_w)^2(1 + 21\alpha\frac{m^*}{m^{**}}k_BT_w)n\mathbf{v} \tag{2.53}$$

along with the rearranged form of Eq. (2.49)

$$T_w = \frac{2}{3k_B}(w - \frac{m^*}{2}|\mathbf{v}|^2)(\frac{1 - \frac{4}{3}\alpha w}{1 + \alpha m^*|\mathbf{v}|^2}) \tag{2.54}$$

supply all relations necessary to express the hydrodynamic transport model for nonparabolic conduction bands in terms of the quantities; electron concentration n, average electron velocity $\mathbf{v}$, average electron energy w and the space-times variable $(\mathbf{r}, t)$. Applying the last relations reveal the new collisionless nonparabolic hydrodynamic transport model equations to be :

$$\frac{\partial n}{\partial t} = -\nabla_r \cdot (n\mathbf{v}) \tag{2.55}$$

$$\frac{\partial \mathbf{v}}{\partial t} = -\mathbf{v} \cdot \nabla_r \mathbf{v} + \frac{\mathbf{F}}{m^*\mu} - \frac{2}{3nm^*}\nabla_r[n\nu(w - \frac{m^*}{2}|\mathbf{v}|^2)] \tag{2.56}$$

$$\frac{\partial w}{\partial t} = -\mathbf{v} \cdot \nabla_r w + \mathbf{F} \cdot \mathbf{v} - \frac{2}{3n}\nabla_r \cdot [n\mathbf{v}(\omega + \varphi)(w - \frac{m^*}{2}|\mathbf{v}|^2)] \tag{2.57}$$

with the nonparabolic correction terms

$$\begin{aligned} \mu(\mathbf{v}, w) &= 1 + 2\alpha\varepsilon(\mathbf{v}, w)(w - \frac{m^*}{2}|\mathbf{v}|^2) \\ &+ \frac{10\alpha^2}{3}\varepsilon(\mathbf{v}, w)^2(w - \frac{m^*}{2}|\mathbf{v}|^2)^2 + \alpha m^*|\mathbf{v}|^2 \end{aligned} \tag{2.58}$$

$$\nu(\mathbf{v}, w) = \frac{\varepsilon(\mathbf{v}, w)}{\mu(\mathbf{v}, w)} \tag{2.59}$$

$$\omega(\mathbf{v}, w) = \varepsilon(\mathbf{v}, w)[1 + \frac{4\alpha}{3}\varepsilon(\mathbf{v}, w)(w - \frac{m^*}{2}|\mathbf{v}|^2) + \alpha m^*|\mathbf{v}|^2] \tag{2.60}$$

$$\varphi(\mathbf{v}, w) = \frac{10\alpha}{3}(\nu(\mathbf{v}, w))^2(w - \frac{m^*}{2}\mathbf{v} \cdot \mathbf{v})[1 + 14\alpha\nu(\mathbf{v}, w)(w - \frac{m^*}{2}\mathbf{v} \cdot \mathbf{v})] \tag{2.61}$$

where

$$\varepsilon(\mathbf{v}, w) = \frac{1 - \frac{4\alpha}{3}w}{1 + \alpha m^*|\mathbf{v}|^2}. \tag{2.62}$$

The nonparabolic model equations above have been written in a form so that for the parabolic case, $\alpha = 0$, it is easy to observe that $\mu = \nu = \omega = 1$ and $\varphi = 0$. Thus, the nonparabolic model conveniently reduces to the more familiar classical hydrodynamic equations in Lagrangian form.

2.4 Monte Carlo Evaluation of the Transport Model

The nonparabolic hydrodynamic model derived in the last section ignored any specific form of scattering mechanisms. This was done primarily to simplify the analysis and to allow for a clear focus on the transport parameters which evolve from the collisionless BTE for nonparabolic bands. To evaluate the results of the previous section one could assume a form for the scattering mechanism and then solve the resulting system of equations with appropriate boundary conditions (The author has performed such a study[59]). Results for electron current could then be used to judge the macroscopic feasibility of the new nonparabolic model. However, the issue of chosing accurate forms for scattering terms is a complicated one and the specific form has a large influence on the results. Therefore, a different initial approach will be followed. The specific approach will be to calculate the resultant nonparabolic parameters m^{**}, $[\mathbf{P_V}]$, $[\mathbf{P_W}]$ and $\mathbf{q}$ of the hydrodynamic model for various structures using an elaborate Monte Carlo model. Since the results of the previous section presented formulas for each of the model parameters in terms of average velocity $\mathbf{v}$ and average energy w, the values predicted by these formulas can easily be compared to those from the more elaborate and detailed Monte Carlo model. Taking this approach will give an excellent gauge of the accuracy of the terms of the model without introducing errors by using incorrect forms for the loss mechanisms.

The first test was to generate the electron distribution and nonparabolic parameters, directly from their integral definitions (2.27) – (2.30), for a time steady-state constant field structure over a variety of electric field values using Monte Carlo techniques. For these stationary investigations, two independent Monte Carlo models[46, 47] were used to analyze electron transport in GaAs. Both of these models utilize single particle Monte Carlo procedures to improve computational accuracy and efficiency. The models incorporate three valley analytical band structures and included all the significant scattering mechanisms with polar optical intervally scattering. In addition, both models used the same material and physical parameters[60].

Under these stationary conditions, as can be observed directly from Eqs. (2.21) - (2.23) or indirectly in Eqs. (2.55) - (2.57), only the value of m^{**} will affect the final solution for transport. This is true because all the other terms are space differential in nature. However, as will be shown, these terms do have finite values which vary widely over the cases considered. Also, for these conditions, all the displacement terms (terms of form $\alpha m^{*}|\mathbf{v}|^2$) in Eqs. (2.50) - (2.53) are found to be negligible. Hence, this is an excellent test for the accuracy of the derived nonparabolic model parameters and their dependence on energy temperature T_w. The test also reveals how well T_w is determined by its dependence on total kinetic energy, w, and the displacement kinetic energy term, $(m^{*}/2)|\mathbf{v}|^2$ via Eq. (2.54). An investigation into the true electron distribution is also very important since $\mathcal{F}_{nonpar}$ is instrumental in determining the final formulas for the nonparabolic terms.

The results of these stationary investigations, with the two independent Monte Carlo models in agreement, are presented in Figs. 2.1 - 2.4. Fig. 2.1 exhibits the resulting Monte Carlo generated distribution, the ansatz distribution, $\mathcal{F}_{nonpar}$, and the displaced Maxwellian distribution $\mathcal{F}_{dm}$ along the direction of applied field, in velocity space, for two values of electric field. Here, the Monte Carlo distributions were obtained by sampling electrons in a differential velocity volume element from the three dimensional velocity space. In the calculations, a limiting velocity v_S is used to define the size of the element in directions other than that of the electric field. Also, v_S was chosen small enough to clearly and accurately show the velocity distribution and large enough for a practical simulation time.

Fig. 2.1 clearly indicates that the corrected electron distribution $\mathcal{F}_{nonpar}$ is a distinct improvement over the displaced Maxwellian. Using $\mathcal{F}_{dm}$ results in a general tendency, for any select average velocity and energy, to overspread and place too many electrons in the tail of the distribution. Conversely, the improved distribution, $\mathcal{F}_{nonpar}$, retains more electrons closer to equilibrium. This result seems correct for the situation of an energy dependent effective particle mass (or nonparabolic conduction band) and is verified by the Monte Carlo generated distribution function.

The stationary results for a variety of electric field values, corresponding to the nonparabolic transport parameters, are presented in Figs. 2.2 - 2.4 and are plotted versus the average electron energy for each case. In each instance, the nonparabolic transport parameters predicted by the hydrodynamic model were calculated using the Monte Carlo values for average velocity $\mathbf{v}$ and average energy w since they are fundamentally dependent on these variables.

In general, the predictions supplied by the nonparabolic parameter formulas are in excellent agreement with the results calculated by the Monte Carlo model. Fig. 2.2 shows that the values for the average effective mass m^{**} predicted by the hydrodynamic model are essentially equal (less than 1 percent difference) to those calculated by the Monte Carlo model for average particle energies up to $0.4eV$. Fig. 2.3, which gives both the momentum pressure $[\mathbf{P_V}]$ and the energy pressure $[\mathbf{P_W}]$, clearly shows that these tensors are approximated well by a diagonal matrix. Further, the diagonal elements for both are in very good agreement with the predictions made by the hydrodynamic model. Finally, Fig. 2.4 shows the comparison between the values from the hydrodynamic model for the heat flow vector $\mathbf{q}$ and those calculated by the Monte Carlo method. These results also agree well, which leads to a very important conclusion. The nonzero expression for the heat flow vector was derived here from first principles for the nonparabolic conduction band case. Our treatment represents an original *nonequilibrium* analysis for the heat flow vector which is usually based upon equilibrium thermodynamics arguments. However, it should also be noted that for highly nonstationary conditions this term may need to be determined by higher moments of the BTE before it is accurate for transport simulations[61].

While the previous stationary results are very positive, by themselves they are

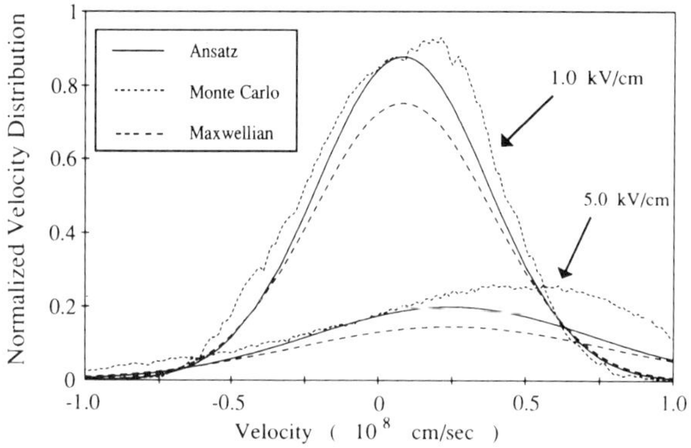

Fig. 2.1 Normalized Γ valley velocity distribution functions resulting from the constant electric field values of 1.0 kV/cm and 5.0 kV/cm. The profiles are exhibited along the direction of net average velocity (direction of applied field).

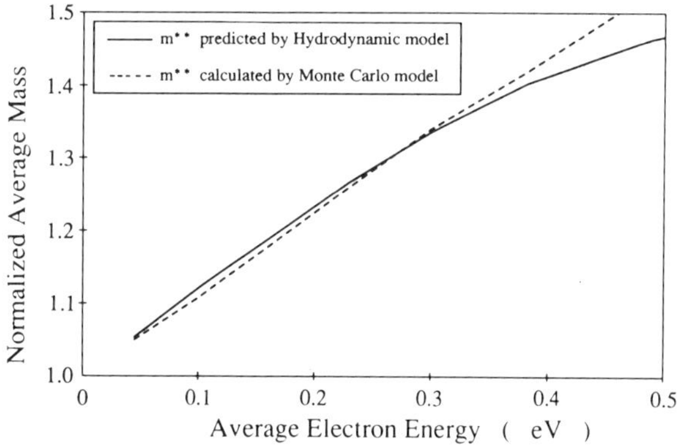

Fig. 2.2 Average Γ valley effective mass, m^{**}, normalized by near equilibrium mass value of $m^* = 0.57 \mathrm{x} 10^{-31} Kg$ and plotted versus average electron energy.

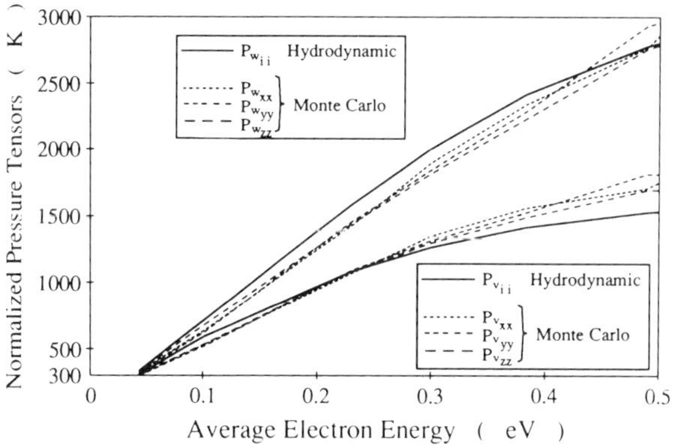

Fig. 2.3 Velocity and energy Γ valley pressure tensors, normalized by nk_B, versus average electron energy. The off-diagonal terms are omitted because they are negligible (an order of magnitude smaller) compared to diagonal terms.

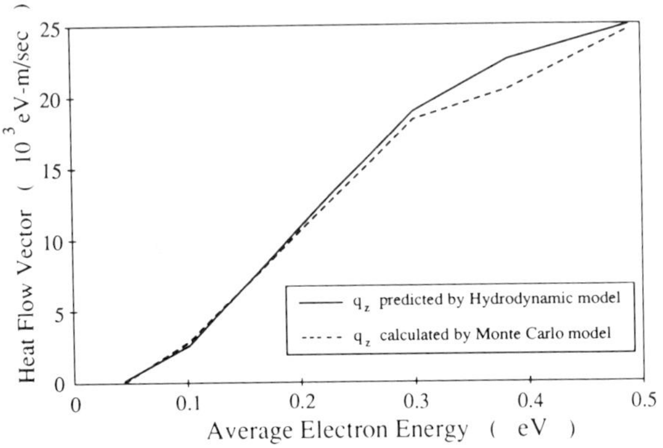

Fig. 2.4 Γ valley heat flow vector along the direction of applied electric field, normalized by n, versus average electron energy. The other vector components are omitted because they are negligible.

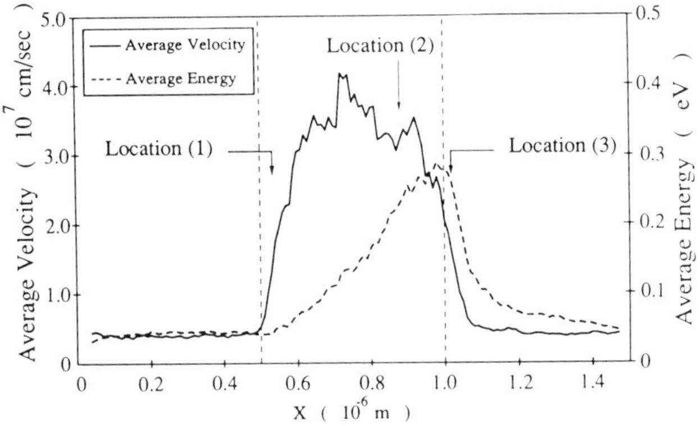

Fig. 2.5 Γ valley average velocity and energy profiles through a submicron GaAs $N^+ - N - N^+$ ballistic diode structure biased at 0.8 volts. The structure consists of 0.5 μm long source and collector regions doped 10^{17} cm^{-3} with a 0.5 μm active region doped 10^{16} cm^{-3}.

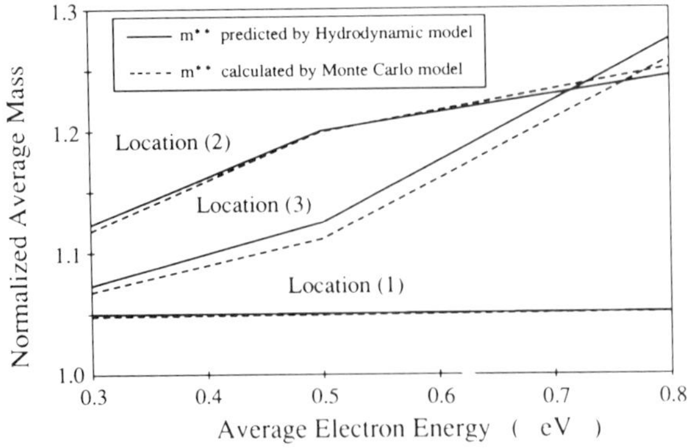

Fig. 2.6 Normalized average Γ valley effective mass, at three locations within the ballistic diode, plotted as a function of appled bias.

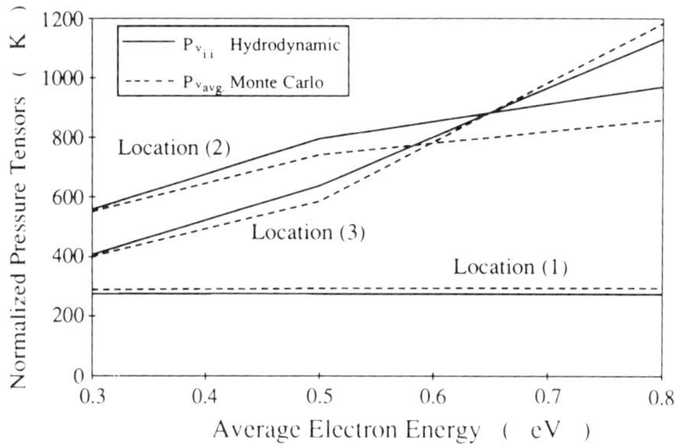

Fig. 2.7 Normalized Γ valley velocity pressure, at three locations within the ballistic diode. $P_{vavg.}$ represents the average of the three diagonal components which were found to be approximately equal. Off-diagonal terms were negligible.

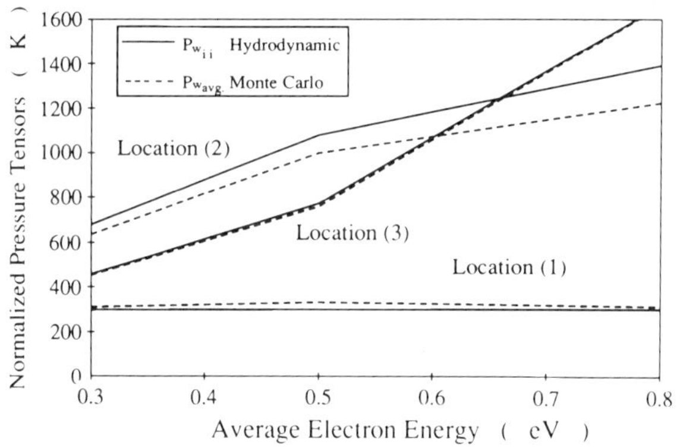

Fig. 2.8 Normalized Γ valley energy pressure, at three locations within the ballistic diode. $P_{wavg.}$ represents the average of the three diagonal components which were found to be approximately equal. Off-diagonal terms were negligible.

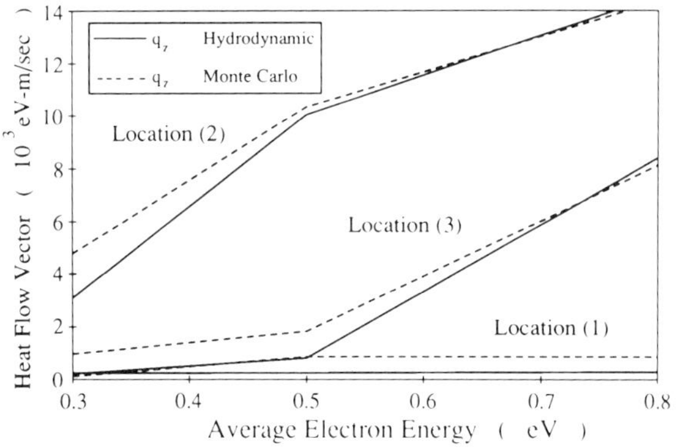

Fig. 2.9 Γ valley heat flow vector, along the direction of appled electric field, at three locations within the ballistic diode. The other components were found to be negligible.

not sufficient to insure the applicability of this new model to arbitrary electron device structures. Realistic device structures, operating under moderately high biases, always contain regions where transport occurs under nonstationary conditions. Hence, additional calculations were performed to determine the accuracy of the nonparabolic model subject to these conditions. The N^+-N-N^+ ballistic diode, a simple prototype submicron structure which has been extensively studied[62], was chosen as a test device.

A self-consistent ensemble Monte Carlo model[47] was used to calculate the nonparabolic parameters for several strategically selected locations within this device as illustrated in Fig. 2.5. Specifically, locations were selected which would exhibit different types of nonstationary transport. At location (1) the electron gas is relatively cool with a low average energy, while the average velocity is slightly ballistic. Conversely, at location (3) the average energy remains high and the average velocity is low. Finally, location (2) represents a case where both parameters of the electron gas are elevated and are experiencing significant spatial gradients.

The results of these nonstationary studies are presented in Figs. 2.6 - 2.9 where the various transport parameters are plotted as a function of applied bias. Summarizing the results, the transport parameters predicted by the hydrodynamic model are in very good agreement with the direct Monte Carlo calculations. The agreement for all parameters are within 10 percent for the majority of the biases studied. In fact, the difference only rises slightly higher to approximately 20 percent (for the heat flow vector) at low biases where the accuracy of the Monte Carlo statistics are themselves in question. The results agreed both qualitatively and quantitatively for all three locations and at all biases. Hence, all the results indicate this new nonparabolic hydrodynamic model can accurately approximate electron transport in realistic semiconductor devices which exhibit non-local electron transport effects.

In addition, these results have illustrated the applicability of the Monte Carlo method in verifying the accuracy of macroscopic transport models. This work has clearly shown how novel analytical approaches and Monte Carlo techniques can be combined to arrive at improved transport models (both physically and computationally) for investigating advanced semiconductor device behavior. However, before these advanced macroscopic models can be used to simulate practical devices, the effects of dissipation must be included. Fortunately, Monte Carlo methods can also be used to generate approximate transport parameters for including the effects of collisions during electron conduction in the macroscopic transport models.

3 Dissipation Effects in Hydrodynamic Models

3.1 General Discussion

The Monte Carlo method is an excellent tool to verify the accuracy of simplified transport models. The method also offers a convenient approach to supply average transport parameters arising due to physical dissipation mechanisms (particle relaxation rates, momentum relaxation rates, energy relaxation rates, etc.), for simplified electron transport models.

Any realistic model for electron transport must include terms to approximate the influence of collision processes. In semiconductors, there are many possible types of scattering mechanisms (ionized impurity, acoustic and optical phonon, piezoelectric, polar acoustic phonon, polar optical phonon and electron-electron[56, 63] which influence the operating characteristics of electron devices. These dissipation mechanisms directly determine the rate of microscopic scattering: and hence, the macroscopic momentum and energy losses and in what proportions electrons will reside among the various possible conduction band minima.

3.2 Scattering Effects on the BTE

Currently, the majority of semiconductor devices are modeled using variations of the Boltzmann transport equation (BTE)[7, 68, 64]. The BTE is an expression which describes the time evolution of an assumed electron distribution function, $f(\mathbf{r}, \mathbf{p}, t)$, in a six dimensional phase space consisting of real space $\mathbf{r}$ and momentum space $\mathbf{p}$. The BTE is a mathematical representation of the conservation of electrons in any elementary volume of phase space $\Delta\mathbf{r}\ \Delta\mathbf{p}$ as particles enter and exit due to internal and external processes. These processes include thermal motion, motion due to external fields and collisions with other particles. In the BTE description, the distribution function serves as a probability function which can be used to predict the average response of all the electrons as opposed to a kinetic approach which considers each electron individually. Hence, the BTE approach is a single-particle description of a multi-particle system which is only valid when the density of electrons is large enough to justify a statistical description, dilute enough to ignore electrical correlation effects and when the electrons can be treated as semiclassical particles which obey Newton's laws of motion.

The BTE can be eloquently derived by considering an arbitrary phase space volume and utilizing elemental calculus to describe how the distribution changes due to applied forces and scattering events[48, 6]. However, a shorter derivation can be achieved by applying the Liouville Theorem[65] which asserts that a distribution will follow a trajectory in phase space unchanged in the absence of collisions. Hence, the total time rate of change of the distribution $f(\mathbf{r}, \mathbf{p}, t)$ as it travels between two points can be written as[50]

$$\frac{Df}{Dt} = \left.\frac{\partial f}{\partial t}\right|_{coll.} + s(\mathbf{r}, \mathbf{p}, t) \tag{3.1}$$

where $\partial f/\partial t|_{coll.}$ is the change due to scattering events and $s(\mathbf{r}, \mathbf{p}, t)$ is a source term due to generation and recombination. The first term in Eq. (3.1) can be expanded to arrive at

$$\frac{\partial f}{\partial t} + \frac{\partial \mathbf{r}}{\partial t} \cdot \frac{\partial f}{\partial \mathbf{r}} + \frac{\partial \mathbf{p}}{\partial t} \cdot \frac{\partial f}{\partial \mathbf{p}} = \left.\frac{\partial f}{\partial t}\right|_{coll.} + s(\mathbf{r}, \mathbf{p}, t). \tag{3.2}$$

In dealing with conduction in solids, the semiclassical quantum mechanical description is used where the electron *crystal* momentum is described as $\mathbf{p} = \hbar\mathbf{k}$, where $\mathbf{k}$ evolves according to

$$\frac{\partial \mathbf{k}}{\partial t} = \frac{\mathbf{F}}{\hbar}. \tag{3.3}$$

Also, the electron is assumed to be a localized wave packet with a group velocity

$$\frac{\partial \mathbf{r}}{\partial t} = u(\mathbf{k}) = \frac{1}{\hbar} \nabla_{\mathbf{k}} E(\mathbf{k}). \tag{3.4}$$

Using these relations yields the BTE, in terms of the phase space variables $(\mathbf{r}, \mathbf{k}, t)$, as

$$\frac{\partial f}{\partial t} + \mathbf{u} \cdot \nabla_{\mathbf{r}} f + \frac{\mathbf{F}}{\hbar} \cdot \nabla_{\mathbf{k}} f = \left.\frac{\partial f}{\partial t}\right|_{coll.} + s(\mathbf{r}, \mathbf{p}, t). \tag{3.5}$$

The source term in Eq. (3.5) is usually assumed to be known in advance, independent of f, and does not introduce any additional difficulties in applying the BTE to electronic device structures. However, the scattering term does depend on f and dictates the need for additional assumptions in its treatment.

The scattering term in Eq. (3.5) is formally developed by considering the effects collision events have upon the electron distribution function f. In a single conduction band model, a collision event is defined as a process which changes a particle's *crystal* momentum. The event is assumed to be localized, position independent and to occur over a negligibly small time period[6]. Hence, the distribution can be altered at any momentum location (or state) $\mathbf{k}$ by two competitive processes. The distribution function $f(\mathbf{r}, \mathbf{k}, t)$ could be increased by particles which scatter in from a remote $\mathbf{k}'$ or particles at $\mathbf{k}$ could scatter out and decrease f. The net rate of increase in $f(\mathbf{r}, \mathbf{k}, t)$ is the difference between this in-scattering and out-scattering phenomena and is given by[66]

$$\left.\frac{\partial f}{\partial t}\right|_{coll.} = \sum_{all\ \mathbf{k}'} S(\mathbf{k}', \mathbf{k}) f(\mathbf{k}')[1 - f(\mathbf{k})] - \sum_{all\ \mathbf{k}'} S(\mathbf{k}, \mathbf{k}') f(\mathbf{k})[1 - f(\mathbf{k}')]. \tag{3.6}$$

The first term on the R.H.S. of Eq. (3.6) defines the rate which particles in-scatter from all possible states $\mathbf{k}'$. In this term, $S(\mathbf{k}', \mathbf{k})$ is the transition rate or probability per unit time that an arbitrary scattering mechanism will induce a particle to scatter from initial state $\mathbf{k}'$ to final state $\mathbf{k}$ (assuming that state $\mathbf{k}'$ is occupied and state $\mathbf{k}$ is empty). For a particle to scatter, it must exist in the initial state and the final state must be empty. Hence, the term must be multiplied by the probability that the particle resides in the initial state, $f(\mathbf{k}')$, and the probability that the final state is empty, $1 - f(\mathbf{k})$. Conversely, The second term on the R.H.S. of Eq. (3.6) defines the rate which particles out-scatter from the state $\mathbf{k}$. The explanation for this term is similar to the first, except now the scattering direction and existence probabilities are reversed.

Equation (3.6) can be cast as an integral equation directly[67] or transformed by noting that each discrete $\mathbf{k}'$ state occupies a $\mathbf{k}$-space volume of $\Omega_{\mathbf{k}'} = (2\pi)^3/V$ (V is an arbitrary volume) if spin states are neglected. Therefore, for a single conduction band and conservation of spin, any arbitrary sum can be approximated by an integral according to

$$\sum_{all\,\mathbf{k}} \mathcal{F}(\mathbf{k}) = \frac{V}{(2\pi)^3} \int_{\mathbf{k}_-}^{\mathbf{k}_+} \mathcal{F}(\mathbf{k}) d\mathbf{k} \tag{3.7}$$

where $\mathbf{k}_+$ and $\mathbf{k}_-$ are the symbolic upper and lower bounds on the entire domain of $\mathbf{k}$. Using Eq. (3.7), the rate of change in f due to scattering is

$$\begin{aligned} \left.\frac{\partial f}{\partial t}\right|_{coll.} &= -\frac{V}{(2\pi)^3} \int_{\mathbf{k}'_-}^{\mathbf{k}'_+} S(\mathbf{k},\mathbf{k}') f(\mathbf{k})[1 - f(\mathbf{k}')] d\mathbf{k}' \\ &\quad -\frac{V}{(2\pi)^3} \int_{\mathbf{k}'_-}^{\mathbf{k}'_+} S(\mathbf{k}',\mathbf{k}) f(\mathbf{k}')[1 - f(\mathbf{k})] d\mathbf{k}'. \end{aligned} \tag{3.8}$$

3.3 The Relaxation Time Approximation

When the nonlinear integral term in Eq. (3.8) is applied to Eq. (3.5) the resulting BTE is an integro-differential equation. Hence, the application of any realistic scattering mechanism (transition probability) results in a problem which is almost impossible to solve by direct analytical methods. Therefore, theoretical transport studies *always* involve the application of approximate methods. In practice, direct solutions of the full BTE are always generated using numerical methods[68]. The most powerful and popular approach is the statistical numerical Monte Carlo method[69, 70]. However, this technique is very computational intensive and suffers from statistical uncertainty when analyzing low field transport. To surmount these previous problems, most electron transport models incorporate dissipation using the strict relaxation time approximation (RTA) or some phenomenological-based ensemble relaxation time approximation (E-RTA).

The RTA is used to simplify the collision integral in such a manner that direct analytical solutions of the BTE may be obtained. Strictly, the RTA is valid only for low applied fields (or near equilibrium) so that the carriers in the symmetric part of the distribution are in thermal equilibrium with the lattice. The key element of the RTA approach is that one is able to express $\partial f/\partial t|_{coll.}$ in a form which is independent of the distribution function. Obviously, this reduces Eq. (3.5) to a partial differential equation which is much easier to solve.

While the RTA is often proposed from physical experience and intuition, it can be more formally developed[7] by writing the nonequilibrium distribution function as

$$f(\mathbf{r},\mathbf{k},t) = f_{\mathrm{s}}(\mathbf{r},\mathbf{k},t) + f_{\mathrm{a}}(\mathbf{r},\mathbf{k},t) \tag{3.9}$$

where the first term is symmetric and assumed to describe most of the particles in the ensemble. The second term, which is antisymmetric and is responsible for net transport of momentum and energy, is assumed to be small. Using Eq.(3.9), the collision integral for nondegenerate statistics (neglecting the Pauli Exclusion Principle) can be written as

$$\left.\frac{\partial f}{\partial t}\right|_{coll.} = \left.\frac{\partial f_{\mathrm{s}}}{\partial t}\right|_{coll.} + \left.\frac{\partial f_{\mathrm{a}}}{\partial t}\right|_{coll.}, \tag{3.10}$$

where

$$\left.\frac{\partial f_{\mathrm{s}}}{\partial t}\right|_{coll.} = -\frac{V}{(2\pi)^3}\int_{\mathbf{k}'_-}^{\mathbf{k}'_+}\left\{S(\mathbf{k},\mathbf{k}')f_{\mathrm{s}}(\mathbf{k}) - S(\mathbf{k}',\mathbf{k})f_{\mathrm{s}}(\mathbf{k}')\right\}d\mathbf{k}' \tag{3.11}$$

and

$$\left.\frac{\partial f_{\mathrm{a}}}{\partial t}\right|_{coll.} = -\frac{V}{(2\pi)^3}\int_{\mathbf{k}'_-}^{\mathbf{k}'_+}\left\{S(\mathbf{k},\mathbf{k}')f_{\mathrm{a}}(\mathbf{k}) - S(\mathbf{k}',\mathbf{k})f_{\mathrm{a}}(\mathbf{k}')\right\}d\mathbf{k}'. \tag{3.12}$$

It can be reasoned that in equilibrium, $f_{\mathrm{s}} = f_0$ (f_0 being the equilibrium function) and $f_{\mathrm{a}} = 0$. Therefore, $\partial f_{\mathrm{s}}/\partial t|_{coll.}$ must vanish in equilibrium since $f_{\mathrm{s}} = f_0$ is time invariant. For conditions where the applied electric fields are small, the symmetric component f_{s} should retain a form very nearly the same as f_0. Hence, the first component of the RTA is to neglect the effects due to the symmetric part of the distribution and set $\partial f_{\mathrm{s}}/\partial t|_{coll.} = 0$.

The second component of the RTA involves the manner used to treat the antisymmetric part of the distribution. This is achieved through a phenomenological approach. The application of applied fields disturbs the equilibrium distribution f_0 placing some of the particles in nonequilibrium states, f_{a}. The phenomenological approach reconizes that scattering processes will tend to restore these particles to equilibrium at a rate proportional to the perturbed distribution f_{a} and a characteristic relaxation time τ_{C}. Therefore, this results in the mathematical description

$$\left.\frac{\partial f}{\partial t}\right|_{coll.} \approx \left.\frac{\partial f_{\mathrm{a}}}{\partial t}\right|_{coll.} = -\frac{f_{\mathrm{a}}(\mathbf{r},\mathbf{k},t)}{\tau_{\mathrm{C}}(\mathbf{r},\mathbf{k},t)} = -\frac{f(\mathbf{r},\mathbf{k},t) - f_0(\mathbf{r},\mathbf{k},t)}{\tau_{\mathrm{C}}(\mathbf{r},\mathbf{k},t)} \tag{3.13}$$

where the approximation $f = f_{\mathrm{s}} + f_{\mathrm{a}} \approx f_0 + f_{\mathrm{a}}$ has been used in the last term. When Eq. (3.13) is compared to Eq. (3.12) it is not obvious under what conditions $\tau_{\mathrm{C}}(\mathbf{r},\mathbf{k},t)$ may be defined exactly. In fact, establishing a relaxation time independent of f can only be accomplished when the scattering mechanisms are either elastic or isotropic[56, 7].

When the RTA can be justified, it can be used to solve a number of simplified transport problems to gain fundamental physical insight. The RTA can be used to solve some problems analytically and to simplify the application of numerical methods for others. When the RTA fails, due to the scattering mechanisms or for the case of strong applied forces, other types of approximations are used. In some cases, precise analytical forms (e.g., spherical harmonics expansion or a displaced Maxwellian formulation[63]) are chosen for the nonequilibrium distribution function to facilitate the direct solution of the BTE. However, this type of approach also has limitations and can be very difficult to apply to practical electron devices.

3.4 The Ensemble RTA : Single-Valley

An alternative approach is to develop hydrodynamic (also called conservation, balance or moment) equations applicable for electron transport that include collision effects using an ensemble relaxation time approximation (E-RTA). The hydrodynamic transport model (HTM), as was demonstrated in Section 2.2, is developed by taking moments of the BTE and results in equations which predict the value of average physical quantities rather than an electron distribution function. The prescription for the hydrodynamic model can be used to generate an infinite number of moments. However, as discussed in Section 2.3 most transport models[36] utilize some subset of the first four moment terms (electron density, average momentum, average energy and heat flux) because higher order terms lack simple physical interpretation. The HTM is a simplified approach to solving the BTE in two ways. First, the truncation (or closure) of the collisionless HTM equations must include some type of constitutive relation (like assuming a distribution form or an equation of state). Second, approximate relations, with functional dependencies on the system variables (density, momentum, energy, etc), must be introduced to account for dissipation due to scattering processes. Hence, this type of model is a significant simplification and is much easier to solve than the full BTE.

The familiar drift-diffusion model (DDM) is a good example of this approach[64]. The DDM has been used extensively by applied physicists and engineers to develop a fundamental understanding of device principles since the inception of semiconductor electronics. Further, the method continues to be the pillar upon which most device analysis, design and optimization is based[71]. The DDM describes the velocity using the thermal (and time) equilibrium limit of the conservation of momentum equation found in the HTM[28, 64]. In the model, the average velocity is given by drift and diffusion components with coefficients that are determined only by local effects. The velocity equation may be written in simplest form, for electron transport, as

$$\mathbf{v} = -\mu_n \mathcal{E} + \frac{D_n}{n} \nabla_{\mathbf{r}}\, n, \tag{3.14}$$

where

$$\mu_n = \frac{q\tau_p}{m^*} \tag{3.15}$$

is the low field electron mobility and

$$D_n = \frac{k_B T_e \mu_n}{q}. \tag{3.16}$$

is the low field diffusion coefficient. This simple version of the DDM, given by Eq. (3.14), can be seen as the direct hydrodynamic analogy of the RTA used in solving the BTE. This is true because the transport parameters, Eqs. (3.15) and (3.16), are determined by a constant ensemble averaged loss rate for momentum of $1/\tau_p$. Hence, this specific version of the DDM is only valid for low field transport and cannot predict hot-electron effects. To extend the DDM to nonequilibrium, various methods of defining effective mobility and diffusion coefficients have been used[36]. These methods include defining effective transport parameters as functions of local electric field, local electron energy and local electron temperature (which is very similar to the local energy model). However, these *local* DDM approaches only successfully include some hot-electron effects such as velocity saturation. To effectively predict nonstationary effects, such as velocity overshoot, the energy equation and possibly higher moments of the BTE must be included in the HTM. Also, the treatment of dissipation mechanisms is crucial to how well hot-electron effects are modeled.

Therefore, to accurately model hot-electron effects one must combine a higher moment version of the HTM with physically appropriate collision terms. If electron transport is considered in a single conduction band minimum (or valley) of a uniform semiconductor material, each moment equation of the HTM will be of the form[7]

$$\frac{\partial \mathcal{A}_\Phi}{\partial t} = -\nabla_{\mathbf{r}} \cdot \mathcal{F}_\Phi + \mathcal{G}_\Phi + \mathcal{S}_\Phi - \mathcal{R}_\Phi \tag{3.17}$$

where $\mathcal{A}_\Phi$ is the average physical quantity, associated with the particular moment operator Φ, given by

$$\mathcal{A}_\Phi = \int \Phi(\mathbf{k}) f(\mathbf{r}, \mathbf{k}, t) d\mathbf{k}. \tag{3.18}$$

In equation (3.17), $\mathcal{F}_\Phi$ is the associated spatial flux of Φ, $\mathcal{G}_\Phi$ is a generation (of momentum) term due to the electric field, $\mathcal{S}_\Phi$ is a particle generation-recombination rate due to internal sources-sinks and $\mathcal{R}_\Phi$ is a relaxation term due to scattering. Examples for $\mathcal{F}_\Phi$ and $\mathcal{G}_\Phi$ have been given earlier in Section 2.2 and $\mathcal{S}_\Phi$ is usually treated with simple dependencies in bipolar devices[64]. The relaxation term, $\mathcal{R}_\Phi$, is usually included into the HTM formulation through phenomenological arguments.

Just as in the direct BTE approach, the way the collision term is included will directly dictate the accuracy of the model for nonequilibrium analysis. The dependency of $\mathcal{R}_\Phi$, on the average physical quantities in the HTM equations, is established using an *ensemble relaxation time approximation* (E-RTA) similar to the RTA. For any parameter $\mathcal{A}_\Phi$, the collisions are assumed to destroy any net deviation from equilibrium with rate of loss of $\mathcal{A}_\Phi$ defined by

$$\mathcal{R}_\Phi = -\left.\frac{\partial \mathcal{A}_\Phi}{\partial t}\right|_{coll.} = \frac{\mathcal{A}_\Phi - \mathcal{A}_\Phi^0}{\tau_{\mathcal{A}\Phi}} \tag{3.19}$$

where $\mathcal{A}_\Phi^0$ is the equilibrium value and $1/\tau_{\mathcal{A}\Phi}$ is an specially-defined *ensemble relaxation rate* with dependence on f. For the case of spatially uniform conditions, no applied fields and a constant $\tau_{\mathcal{A}\Phi}$ Eq. (3.19) leads to an arbitrary HTM equation of

$$\frac{\partial \mathcal{A}_\Phi}{\partial t} = -\frac{(\mathcal{A}_\Phi - \mathcal{A}_\Phi^0)}{\tau_{\mathcal{A}\Phi}} \tag{3.20}$$

with an exponentially decaying solution of

$$\mathcal{A}_\Phi(t) = \mathcal{A}_\Phi^0 + \left(\mathcal{A}_\Phi(0) - \mathcal{A}_\Phi^0\right) \exp[-\frac{t}{\tau_{\mathcal{A}\Phi}}]. \tag{3.21}$$

Hence, the particular form of Eq. (3.19) is reasonable since scattering is fundamentally a random process with a probability of scattering that decays exponentially for a constant scattering rate[5]. Further, when the RTA can be justified, one can easily prove that Eq. (3.19) is a valid approximation for the relaxation of momentum and energy[72]. However, it should be noted that Eq. (3.19) is an *assumed* form and may not be valid for electron transport far from equilibrium. In addition, when the approximation is applied, the *characteristic* relaxation time $\tau_{\mathcal{A}_\Phi}$ may have dependencies on one or more of the $\mathcal{A}\Phi$'s due to the dependence on f.

When the E-RTA is applied, it is necessary to supply characteristic relaxation times which are determined by the number and specific type of moment operators used. For the model equations of section 2.3, one will need to approximate the collision relaxation rates for particle density, $\partial n/\partial t|_{coll.}$, average velocity, $\partial \mathbf{v}/\partial t|_{coll.} = (1/n)\, \partial n\mathbf{v}/\partial t|_{coll.} - (\mathbf{v}/n)\, \partial n/\partial t|_{coll.}$, and average energy, $\partial w/\partial t|_{coll.} = (1/n)\, \partial nw/\partial t|_{coll.} - (w/n)\, \partial n/\partial t|_{coll.}$. For transport in a single valley and uniform semiconductor material the general relaxation rate may be written as

$$\left.\frac{\partial \mathcal{A}_\Phi}{\partial t}\right|_{coll.} = \int \Phi(\mathbf{k}) \left.\frac{\partial f}{\partial t}\right|_{coll.} d\mathbf{k} \tag{3.22}$$

where in this discussion $\Phi(\mathbf{k})$ will take on values

$$\Phi(\mathbf{k}) = 1 \quad \text{for electron density} \quad \mathcal{A}_1 = n, \tag{3.23}$$

$$\Phi(\mathbf{k}) = \mathbf{u}(\mathbf{k}) \quad \text{for velocity density} \quad \mathcal{A}_\mathbf{u} = n\mathbf{v} \tag{3.24}$$

and

$$\Phi(\mathbf{k}) = E(\mathbf{k}) \quad \text{for energy density} \quad \mathcal{A}_E = nw. \tag{3.25}$$

Equation (3.8) may now be used in Eq. (3.22) to develop a precise definition of the general relaxation rate. If degeneracy is neglected this results in

$$\left.\frac{\partial \mathcal{A}_\Phi}{\partial t}\right|_{coll.} = \left.\frac{\partial \mathcal{A}_\Phi}{\partial t}\right|_{coll.}^{out} + \left.\frac{\partial \mathcal{A}_\Phi}{\partial t}\right|_{coll.}^{in}, \tag{3.26}$$

where

$$\left.\frac{\partial \mathcal{A}_\Phi}{\partial t}\right|_{coll.}^{out} = -\int_{\mathbf{k}_-}^{\mathbf{k}_+} \Phi(\mathbf{k}) f(\mathbf{k}) \left[\frac{V}{(2\pi)^3}\int_{\mathbf{k}'_-}^{\mathbf{k}'_+} S(\mathbf{k},\mathbf{k}')d\mathbf{k}'\right] d\mathbf{k} \tag{3.27}$$

and

$$\left.\frac{\partial \mathcal{A}_\Phi}{\partial t}\right|_{coll.}^{in} = \int_{\mathbf{k}_-}^{\mathbf{k}_+} \Phi(\mathbf{k}) \left[\frac{V}{(2\pi)^3}\int_{\mathbf{k}'_-}^{\mathbf{k}'_+} S(\mathbf{k}',\mathbf{k}) f(\mathbf{k}')d\mathbf{k}'\right] d\mathbf{k}. \tag{3.28}$$

In Eq. (3.26), the first term may be viewed as the change in $\mathcal{A}_\Phi$ due to the rate of out-scattering from each $\mathbf{k}$ to all the $\mathbf{k}'$s *simultaneously*. Conversely, the second term represents the change in $\mathcal{A}_\Phi$ due to the rate of in-scattering to each $\mathbf{k}$ from all the $\mathbf{k}'$s *simultaneously*. This particular interpretation will become useful later when a model is developed for multi-valley electron transport where the *effective* regions of integration for $\mathbf{k}$ and $\mathbf{k}'$ are not the same. However, for conduction in a single valley the order of integration may be interchanged in Eq. (3.28) to simplify Eq. (3.26) to

$$\left.\frac{\partial \mathcal{A}_\Phi}{\partial t}\right|_{coll.} = -\int_{\mathbf{k}_-}^{\mathbf{k}_+} \Phi(\mathbf{k}) \frac{f(\mathbf{r},\mathbf{k}',t)}{\tau_\Phi(\mathbf{k})} d\mathbf{k} \tag{3.29}$$

where

$$\frac{1}{\tau_\Phi(\mathbf{k})} = \frac{V}{(2\pi)^3}\int_{\mathbf{k}'_-}^{\mathbf{k}'_+} \left[1 - \frac{\Phi(\mathbf{k}')}{\Phi(\mathbf{k})}\right] S(\mathbf{k},\mathbf{k}')d\mathbf{k}' \tag{3.30}$$

is an effective out-scattering rate associated with the moment operator Φ At this point it is not clear under what conditions Eq. (3.29) may be written in the E-RTA form of Eq. (3.19). Clearly, any specific alternative form will be strongly dependent on the physics of the scattering process $S(\mathbf{k},\mathbf{k}')$ and the form of f.

Thus, the development of the necessary $\tau_\mathcal{A}$'s for the HTM will require a careful investigation into S and f for particular semiconductor materials and device structures[63, 72]. Fortunately, such studies[28, 73] have shown the E-RTA, with average energy dependent characteristic relaxation times $\tau_{\mathcal{A}_*}$, is reasonably accurate for nonequilibrium transport in semiconductors. Specifically, this type of E-RTA approach will be accurate to second order in average velocity for electron transport in isotropic parabolic conduction bands provided f can be approximated by a displaced Maxwellian distribution function. While the displaced Maxwellian approach has been applied often to study transport physics[6], there is no reason for its global application to hot-electron transport. Also, there has been experimental[74] and theoretical[43] evidence that this can be a poor approximation in nonequilibrium. However, there are conditions (very large electron concentrations where electron-electron scattering dominates[56]) where the displaced Maxwellian is acceptable to model hot-electron transport. Hence, the E-RTA is an excellent tool to approximate scattering effects in HTM's and explore moderately nonequilibrium electron transport.

The E-RTA can now be used to approximate collision effects and complete the HTM. For the HTM equations developed in section 2.3 and electron transport in a single valley the scattering terms are

$$\left.\frac{\partial n}{\partial t}\right|_{coll.} = 0, \tag{3.31}$$

$$\left.\frac{\partial \mathbf{v}}{\partial t}\right|_{coll.} = \frac{-\mathbf{v}}{\tau_v(w)} \tag{3.32}$$

and

$$\left.\frac{\partial w}{\partial t}\right|_{coll.} = -\frac{(w - w_0)}{\tau_w(w)}. \tag{3.33}$$

where w_0 is the equilibrium value of average energy and the energy dependence of the characteristic times is assumed from the results for a displaced Maxwellian. Equation (3.31) follows simply from conservation of total electrons in the valley ($n = n_0$) and Eq. (3.32) is true since for any distribution the average velocity must be zero in equilibrium. At this point, the only remaining issue is how to accurately determine the functions $\tau_v(w)$ and $\tau_w(w)$.

The microscopic nature of electron transport prohibits one from probing experimentally to directly obtain values and energy dependence for the momentum and energy relaxation times. An exact determination of these parameters would obviously require evaluating Eq. (3.29) using the correct distribution function. As discussed earlier, assumptions about the nonequilibrium distribution may be used to theoretically obtain the needed transport parameters. However, this will always lead to inaccuracy for transport far from equilibrium since an exact nonequilibrium distribution function is not known.

Fortunately, if the forms given by Eqs. (3.31) - (3.33) are used (with strictly energy dependent relaxations times) an indirect approach is available. Due to the singular energy dependence, results from stationary (time steady-state and no spatial gradients) measurements or calculations (which directly solve the BTE) may be used to determine $\tau_v(w)$ and $\tau_w(w)$. Stationary conditions reduce the momentum and energy conservation equations of the HTM to

$$\frac{(-q)\mathcal{E}}{m^{**}(w)} = \frac{\mathbf{v}}{\tau_v(w)} \tag{3.34}$$

and

$$(-q)\mathbf{v} \cdot \mathcal{E} = \frac{(w - w_0)}{\tau_w(w)}. \tag{3.35}$$

Since the electron current density is $\mathbf{J} = (-q)n\mathbf{v}$ and the energy density is $W = nw$ one has

$$\tau_v(w) = \frac{m^* |\mathbf{J}|}{q^2 n |\mathcal{E}|} \tag{3.36}$$

and

$$\tau_w(w) = \frac{(W - W_0)}{\mathbf{J} \cdot \mathcal{E}}. \tag{3.37}$$

Equations (3.36) and (3.37) can now be used to determine the transport parameters if experimental values for the electron current and energy density can be obtained. However, while **J** is easy to obtain, the estimation of W can be difficult for hot-electron transport[75].

The necessary transport parameters, valid for extreme nonequilibrium conditions, may be obtained by utilizing numerical methods (like Monte Carlo) which directly solve the BTE. This is an effective approach because the parameters may be obtained from a limited number of stationary calculations due to the previous assumptions about the HTM scattering terms. Thus, computationally intensive mathematical methods are a viable choice to accurately establish these parameters. The Monte Carlo technique is an excellent approach because it can incorporate much transport physics and readily provide the needed solutions. In addition, this type of approach will become *even* more necessary when multi-valley HTM's are implemented. This is true because the number of relaxation rates and the experimental difficulty in obtaining them increases.

3.5 The Ensemble RTA : Multi-Valley

Thus far, only conduction in a single valley has been considered. When electronic transport is studied in materials where a significant number of electrons reside in multiple conduction band minima (valleys) additional issues must be addressed in defining the collision terms. Semiconducting materials always contain multiple momentum-space energy bands[43]. The electrons are usually confined to one or more of the valleys found in the first conduction band even under the influence of high electric fields. One of the prime factors in determining where the electrons will be found is the separation in energy and momentum between these minima. In a covalent semiconductor (Si for example), there are a set of equivalent minima in the first conduction band[7]. The energy separation between these lowest minima and all other minima is large. Since this large separation confines the electrons to these equivalent minima, transport can be adequately modeled by an effective single valley model[76].

In a polar material like GaAs this is not true and the effects of multiple valleys (or the entire band structure) must be considered for hot-electron transport[3]. Figure 3.1 illustrates the major features of the lowest conduction band and uppermost portion of the valence band in GaAs[51]. The conduction band represents the allowed energies for an electron in a perfectly periodic lattice.

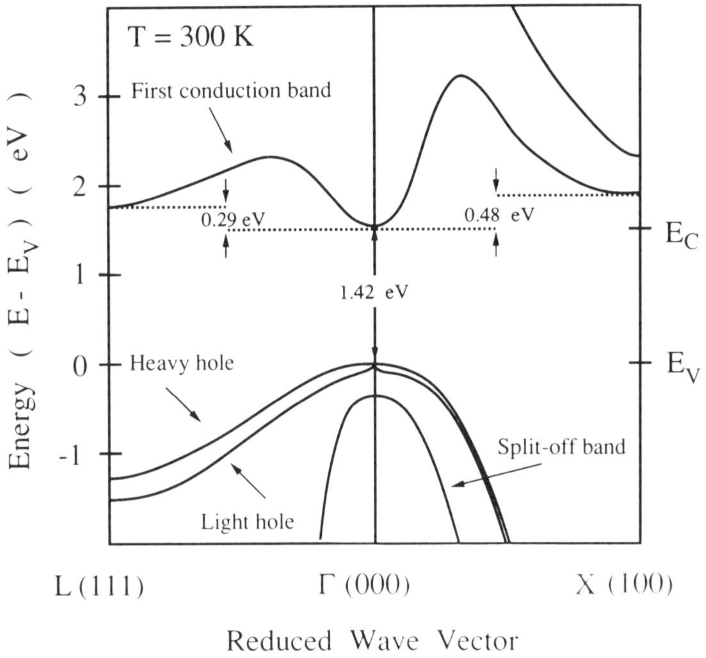

Fig. 3.1 Diagram illustrating the primary (lowest set) conduction band minima and uppermost part of the valence band system for GaAs. From Blakemore, J. Appl. Phys. **53**, 123 (1982).

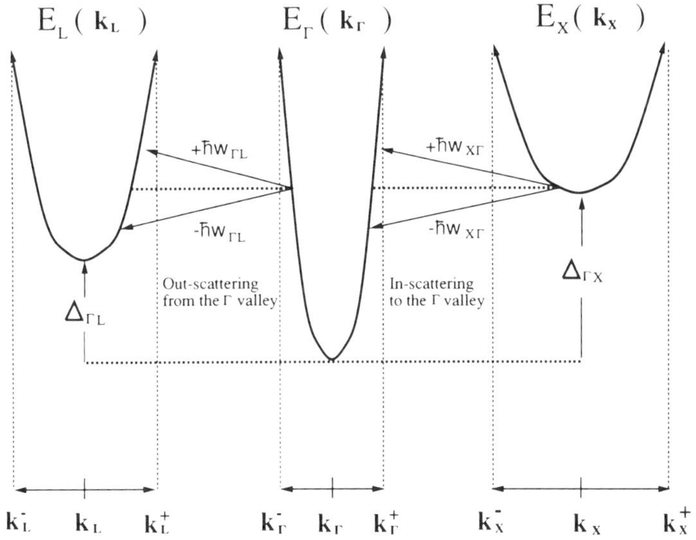

Fig. 3.2 Simplified (conduction) band model used to study electron transport in the lowest three minima (Γ, L and X) of GaAs.

Deriving such energy bands is a very difficult problem which requires a combination of complex theory and experimentation. Figure 3.1 shows the energy bands as a function of wave-vector $\mathbf{k}$ (or scaled crystal momentum $\mathbf{p} = \hbar\mathbf{k}$) for the first Brillouin zone in two different momentum directions. In GaAs[72], there is a single valley at the zone center ($\mathbf{k} = 0$), four equivalent valleys along the [111] direction and three equivalent valleys along [100] direction. To fully understand Fig. 3.1 and its usefulness a few fundamentals must be considered.

On the most fundamental level, an electron is a quasi-particle which exhibits both wave and particle properties. When electron transport is considered in a semiconductor the wave nature is apparent. This is due to the structure of the crystal which has microscopic variations (< 50 Å) in the lattice potential. Under these conditions, individual electrons must be treated *quantum mechanically* with the dynamics described by the single particle Schroëdinger equation[77],

$$i\hbar\frac{\partial}{\partial t}\Psi_0(\mathbf{r},t) = -\frac{\hbar^2}{2m_0}\nabla^2_{\mathbf{r}}\,\Psi_0(\mathbf{r},t) + U_T(\mathbf{r},t)\Psi_0(\mathbf{r},t). \tag{3.38}$$

Here, $\Psi_0(\mathbf{r},t)$ is the wave-function of the electron and determines all physical variables in the quantum formulation. In Eq. (3.38), $U_T(\mathbf{r},t)$ is the total potential energy of the particle. For electron transport in a crystal structure, this may be written as

$$U_T(\mathbf{r},t) = U_E(\mathbf{r},t) + U_L(\mathbf{r},t) + U_S(\mathbf{r},t) \tag{3.39}$$

where $U_E(\mathbf{r},t)$ is the potential energy due to externally applied electric fields, $U_L(\mathbf{r},t)$ is from a perfectly periodic *static* lattice and $U_S(\mathbf{r},t)$ is a scattering potential energy used to represent all the deviations from the ideal $U_L(\mathbf{r},t)$[77]. These deviations include defects (absence of regular ion or presence of impurity) and phonons (vibrations of lattice).

Solving Eq. (3.38) for an arbitrary U_T is *extremely* difficult. Still difficult, although a much simpler task, is to solve (3.38) for a single periodic function U_L. Solutions to the Schroëdinger equation, for such a periodic function, are known to be in the form of Bloch waves[78]. The Bloch solution for a particular conduction band v and $U_L(\mathbf{r})$ can be written as[77]

$$\Psi_0(\mathbf{r},t) = u_{v,\mathbf{k}}(\mathbf{r})\exp(i\mathbf{k}\cdot\mathbf{r})\exp\left[\frac{-iE_v(\mathbf{k})}{\hbar}\right] \tag{3.40}$$

where $u_{v,\mathbf{k}}(\mathbf{r})$ is a periodic function that is different for each conduction band v and $E_v(\mathbf{k})$ is a calculated energy band diagram like Fig. 3.1. These energy band diagrams can be used to simplify Eq. (3.38) by developing an *effective mass equation* which does not explicitly include $U_L(\mathbf{r})$. The simplified equation is

$$i\hbar\frac{\partial}{\partial t}\Psi(\mathbf{r},t) = E_v(-i\nabla_{\mathbf{r}})\Psi(\mathbf{r},t) + U(\mathbf{r},t)\Psi(\mathbf{r},t) \tag{3.41}$$

where $\Psi(\mathbf{r},t)$ is the effective mass approximation to the electron wave-function and $U(\mathbf{r},t)$ is the macroscopic potential equal to the sum of applied and scattering potentials. The effects of $U_L(\mathbf{r},t)$ have been incorporated in the known operator $E_v(-i\nabla_{\mathbf{r}})$. In applying Eq. (3.41), $E_v(-i\nabla_{\mathbf{r}})$ indicates an expansion of the known $E_v(\mathbf{k})$ function by Taylor's theorem and replacement of $\mathbf{k}$ by $-i\nabla_{\mathbf{r}}$. Equation (3.41) is known as the single-band effective mass equation and is suitable to model electron transport in periodic structures, with solutions of the form of Eq. (3.40), under certain conditions[77]. Equation (3.41) is accurate if the macroscopic potential, $U(\mathbf{r},t)$, is approximately constant over a unit cell and $u_{v,\mathbf{k}}(\mathbf{r})$ is not a strong function of $\mathbf{k}$.

If the electron is a highly localized wave packet, the correspondence principle can be used to equate the crystal momentum $\mathbf{p} = \hbar\mathbf{k}$ to the momentum operator $-i\hbar\nabla_{\mathbf{r}}$ to arrive at a very useful result. This allows the Hamiltonian function to be written as[72]

$$H(\mathbf{k},\mathbf{r}) = E_v(\mathbf{k}) + qV_E(\mathbf{r},t)\Psi_0(\mathbf{r},t) \tag{3.42}$$

where $V_E(\mathbf{r},t)$ is the applied electric potential. This resultant Hamiltonian now gives a center of motion (group) velocity for a localized electron packet of

$$\mathbf{v} = \frac{1}{\hbar}\nabla_{\mathbf{k}}\, E_v(\mathbf{k}) \tag{3.43}$$

and an evolution in the wave-vector of

$$\frac{\partial\mathbf{k}}{\partial t} = \frac{-q\mathcal{E}}{\hbar} \tag{3.44}$$

where the effective force on the electron is $\mathcal{E} = -\nabla_{\mathbf{r}}\, U_E$. The last two equations can now be used to define an effective mass of the electron wave packet[79]. Direct differentiation of Eq. (3.43) yields the tensor quantity

$$m_{i,j} = \left(\frac{1}{\hbar^2}\frac{\partial^2 E_v(\mathbf{k})}{\partial k_i \partial k_j}\right)^{-1}. \tag{3.45}$$

This relation can be used to define an effective *observable* momentum[80] of

$$\tilde{\mathbf{p}} = \mathbf{m}\cdot\mathbf{v} \tag{3.46}$$

for electron transport in a perfect crystal lattice. However, Eq. (3.45) is not absolutely necessary to describe the transport process. This is true because the crystal momentum $\mathbf{p}$ is the quantity which is seen to be conserved. For an isotropic parabolic energy band of the form $E_v(\mathbf{k}) = (\hbar^2\mathbf{k}^2)/(2m^*)$ that is centered about $\mathbf{k} = 0$, one can easily observe

$$\tilde{\mathbf{p}} = \hbar\mathbf{k} = \mathbf{p} \tag{3.47}$$

by direct application of Eqs. (3.43) and (3.46). Hence, the origin of the name crystal momentum for $\hbar\mathbf{k}$.

This previous discussion has outlined the *effective mass* approximation for transport in a single conduction band. Assuming that available electrons exist in the band, this approach consists of four separate components. First, one must obtain the energy band diagram for the electronic material. Second, the evolution of the crystal momentum (between scattering events) is described by Eq. (3.44). Third, the value for the actual velocity, at a particular moment and $\mathbf{k}(t)$, is obtained from Eq. (3.43) and $E_v(\mathbf{k})$. The final component deals with how the phenomenon of random scattering is integrated into the physical model.

For moderately doped semiconductors near equilibrium, statistical mechanics[78, 80] indicates that available electrons will seek to minimize their collective energy. Thus for GaAs, all conduction electrons will be found near the bottom of the Γ valley in Fig. 3.1. The application of external forces will induce the electrons to move and result in electron energies which increase as a function of time. Unchecked, the electrons would accelerate and leave the local valley. In reality, dissipative mechanisms do exist and even for high applied electric fields ($10 - 20$ keV/cm) intravalley scattering (collision processes which result in small changes in the electron's wave-vector) retains a large percentage of electrons well within the Γ valley[46] of bulk GaAs. However, even for field strengths of a few kilovolts per centimeter electrons can be transferred from the central Γ valley to higher energy (satellite) valleys[81]. When the Γ valley electrons become sufficiently *hot*, they are transferred to satellite valleys by scattering with short wavelength phonons (collisions which result in large changes in $\mathbf{k}$) Once in a satellite valley, they can be scattered between equivalent valleys (like L_1 to L_2 or X_1 to X_2 in Fig. 3.1) as well. These type of mechanisms have been verified both theoretically and experimentally as leading to dramatic transport characteristics (Gunn effect[1]).

The previous physical picture can be used to develop a simplified model for electron transport in multiple minima of a single conduction band. From this simplified transport model, an analogous multi-valley HTM and appropriate relaxation terms can be derived. Since electron conduction is primarily isolated to individual valleys except for those electrons which are transferred by intervalley mechanisms, the full band model is often replaced by the one shown in Fig. 3.2. In this simplified band model (SBM), each valley resides in a completely separate $\mathbf{k}$-space domain and with equilvalent minima (e.g, the four along the L direction) represented by a single effective valley. The electron transport is then described, in each valley i, by a local electron distribution function f_i and a corresponding BTE. Scattering, between any two valleys i and j, is described by appropriate terms which depend on f_i and f_j.

Similarly, the multi-valley HTM (M-HTM) for the SBM will have a set of moment equations, like Eqs. (2.55) - (2.57) in section 2.3, for each valley $i = \Gamma$, X or L. Therefore, the electron physics for each valley will be described by the quantities n_i, $\mathbf{v}_i$ and w_i. In this multi-valley HTM, each set of valley transport equations will contain scattering terms which can be divided into distinct parts. First, the intravalley collision terms which only influence, and depend on, local transport quantities. Second, the intervally terms which influence local transport quantities and depend on the average variables in other valleys.

In the M-HTM, all the dissipation terms can be approximated through the E-RTA using a superposition of intravalley and intervalley scattering. The total dissipation rate of a physical quantity $\mathcal{A}$ (where the subscript denoting the corresponding moment operator Ψ has now been dropped) in a valley i is given by

$$-\left.\frac{\partial \mathcal{A}_i}{\partial t}\right|_{coll.} = -\left.\frac{\partial \mathcal{A}_i}{\partial t}\right|_{coll.}^{intra.} - \left.\frac{\partial \mathcal{A}_i}{\partial t}\right|_{coll.}^{inter.} . \tag{3.48}$$

The form of the first term on the R.H.S of Eq. (3.48) is developed as before by assuming that the intravalley scattering will relax any deviations from equilibrium. Also, this equilibrium value $\mathcal{A}_i^0$ will be the one which would exist in the absence of the intervalley scattering. Therefore, the E-RTA intravalley scattering terms corresponding to the moment equations for electron density, average velocity and average energy are

$$\left.\frac{\partial n_i}{\partial t}\right|_{coll.}^{intra.} = 0, \tag{3.49}$$

$$\left.\frac{\partial \mathbf{v}_i}{\partial t}\right|_{coll.}^{intra.} = -\frac{\mathbf{v}_i}{\tau_{vii}(w)} \tag{3.50}$$

and

$$\left.\frac{\partial w_i}{\partial t}\right|_{coll.}^{intra.} = -\frac{w_i - w_0}{\tau_{wii}(w)}. \tag{3.51}$$

The intervalley scattering process, in the SBM as shown in Fig. (3.2), has features which differ from the previous intravalley discussion. When an i to j event occurs, the scattered electron *directly* removes a net amount of a physical quantity (like particle density, momentum density, energy density etc.) from valley i and delivers it to some other valley j. When out-scattering from a valley i is considered in the E-RTA, the process could remove all the electrons and result in an equilibrium value of zero for density n_i. Hence, the E-RTA out-scattering *loss* rate of density is[82] $n_i/\tau_{nij}(w_i)$ where $\tau_{nij}(w_i)$ is a time constant with an *assumed* energy dependence. The density of electrons is also affected by in-scattering from other valleys. The rate of density in-scattering gain in valley i is equal to the rate of density out-scattering loss in valley j. Therefore, the E-RTA in-scattering *gain* rate of density is just $n_j/\tau_{nji}(w_j)$. Then, the E-RTA intervalley scattering equation for electron density in a two valley SBM is

$$\left.\frac{\partial n_i}{\partial t}\right|_{coll.}^{inter.} = -\frac{n_i}{\tau_{nij}(w_i)} + \frac{n_j}{\tau_{nji}(w_j)}. \tag{3.52}$$

The process for intervalley scattering in semiconductors is known to belong to a class called randomizing collisions[1]. This means that the final vector momentum direction is not related to the initial direction and, on the average, in-scattering from j to i adds no net momentum to valley i. Thus, the corresponding E-RTA intervalley scattering equation is

$$\left.\frac{\partial \mathbf{v}_i}{\partial t}\right|_{coll.}^{inter.} = -\frac{\mathbf{v}_i}{\tau_{vij}(w_i)} - \left(\frac{\mathbf{v}_i}{n_i}\right) \left.\frac{\partial n_i}{\partial t}\right|_{coll.}^{inter.} \tag{3.53}$$

where the relation

$$\left.\frac{\partial \mathbf{v}_i}{\partial t}\right|_{coll.}^{inter.} = \left(\frac{1}{n_i}\right) \left.\frac{\partial n_i \mathbf{v}_i}{\partial t}\right|_{coll.}^{inter.} - \left(\frac{\mathbf{v}_i}{n_i}\right) \left.\frac{\partial n_i}{\partial t}\right|_{coll.}^{inter.} \tag{3.54}$$

has been used to include the effects of a nonzero $\partial n_i/\partial t|_{coll.}^{inter.}$.

The issues of energy exchange between valleys is similar to particle exchange. In the E-RTA, the loss rate of energy density in a valley i due to scattering from i to j is $n_i w_i/\tau_{wij}(w_i)$ since all the particles (energy) can be removed by intervalley scattering. On the other hand, the effects on the energy density due to in-scattering from j to i are more complex. Assume that the energy loss rate of out-scattering from some other valley j, which can be written as $n_j w_j/\tau_{wji}(w_j)$, is denoted by $\mathcal{R}_j^{out}$. The gain in energy density in valley i due to this out-scattering from j is not equal to $\mathcal{R}_j^{out}$. This is because the energy density nw is a kinetic energy and is measured from the bottom of the conduction band. Therefore, during a j to i transition each electron gains an additional kinetic energy of $\Delta_{i,j}$. Also, the dominant intervally scattering mechanism (phonons) will either absorb or emit an amount of energy $\hbar w_{ji}$ during the interaction. Therefore, to implement the E-RTA, an effective relaxation time $\tau_{eji}(w_j)$ is be defined so these effects are included. This type of approach is justified[28]. The rate of gain of energy density from j to i in-scattering is now approximated by $n_j w_j/\tau_{eji}(w_j)$. Thus, the corresponding E-RTA intervalley scattering equation for energy density is

$$\left.\frac{\partial w_i}{\partial t}\right|_{coll.}^{inter.} = -\frac{w_i}{\tau_{wij}(w_i)} + \left(\frac{n_j}{n_i}\right)\frac{w_j}{\tau_{eji}(w_j)} - \left(\frac{w_i}{n_i}\right) \left.\frac{\partial n_i}{\partial t}\right|_{coll.}^{inter.} \tag{3.55}$$

where

$$\left.\frac{\partial w_i}{\partial t}\right|_{coll.}^{inter.} = \left(\frac{1}{n_i}\right) \left.\frac{\partial n_i w_i}{\partial t}\right|_{coll.}^{inter.} - \left(\frac{w_i}{n_i}\right) \left.\frac{\partial n_i}{\partial t}\right|_{coll.}^{inter.} \tag{3.56}$$

is used to include the effects of a nonzero $\partial n_i/\partial t|_{coll.}^{inter.}$. When the scattering equations from the previous discussion are used in equation (3.48) one obtains

$$\left.\frac{\partial n_i}{\partial t}\right|_{coll.} = -\frac{n_i}{\tau_{nij}(w_i)} + \frac{n_j}{\tau_{nji}(w_j)}, \tag{3.57}$$

$$\left.\frac{\partial \mathbf{v}_i}{\partial t}\right|_{coll.} = -\frac{\mathbf{v}_i}{\tau_{vii}(w_i)} - \frac{\mathbf{v}_i}{\tau_{vij}(w_i)} - \frac{\mathbf{v}_i}{n_i}\left.\frac{\partial n_i}{\partial t}\right|_{coll.} \tag{3.58}$$

and

$$\left.\frac{\partial w_i}{\partial t}\right|_{coll.} = -\frac{w_i - w_0}{\tau_{wii}(w_i)} - \frac{w_i}{\tau_{wij}(w_i)} + \left(\frac{n_j}{n_i}\right)\frac{w_j}{\tau_{eji}(w_j)} - \frac{w_i}{n_i}\left.\frac{\partial n_i}{\partial t}\right|_{coll.} . \tag{3.59}$$

as the necessary dissipation terms for each valley i in the two-valley HTM. A similar derivation can be used to develop these terms for an arbitrary number of valleys. These collision terms have been previously established[28] as an adequate approximation for electron transport in isotropic (and parabolic) valleys if the valley distribution functions approximate displaced Maxwellians. While the actual conduction bands and distribution functions are quite different for extreme nonequilibrium conditions, this form should be qualitatively correct (with the true form containing only weak additional dependencies on moment variables). Therefore, the full M-HTM is an excellent tool to study hot-electron transport in semiconductors.

The M-HTM can only be utilized if quantitative approximations for the numerous characteristic τ's can be generated. Obviously, a direct experimental approach is not plausible. The Monte Carlo methods offers a powerful way to obtain the needed transport parameters. Since the Monte Carlo procedure can generate a distribution which solves the BTE, it can be used to evaluate the formal definitions of the τ's needed in the M-HTM.

In the SBM, *all* the intervalley scattering terms may be formally expressed as out-scattering integrals as given by Eq. (3.27). For an arbitrary parameter $\mathcal{A}_\Phi$, the scattering rate is

$$\left.\frac{\partial \mathcal{A}_i}{\partial t}\right|_{coll.}^{i\to j} = -\int_{i^{th}valley} \Phi_i(\mathbf{k})\frac{f_i(\mathbf{k})}{\tau_{Sij}(\mathbf{k})}d\mathbf{k} \tag{3.60}$$

where

$$\frac{1}{\tau_{Sij}(\mathbf{k})} = \frac{V}{(2\pi)^3}\int_{j^{th}valley} S_{ij}(\mathbf{k},\mathbf{k}')d\mathbf{k}' \tag{3.61}$$

is a $\mathbf{k}$-space dependent intervalley scattering rate determined by the intervalley transition rate S_{ij}.

Equation (3.60) can be used *directly* to determine an expression for all the particle ($\Phi_i = 1$) relaxation times. Applying Eq. (3.60) yields

$$\tau_{nij}(w_i) = \left(\int_{i^{th}valley} \frac{f_i'(\mathbf{k})}{\tau_{Sij}(\mathbf{k})}d\mathbf{k}\right)^{-1} \tag{3.62}$$

where f_i' is the normalized distribution $f_i' = f_i/n_i$. Equation (3.62) can be used to obtain all the needed τ_{nij}'s since the rates for density out-scattering (from valley i) and density in-scattering (to valley j.) are equal. Therefore, if $\tau_{nij}(w_i)$ and an occupancy ratio for n_i and n_j is known

$$\frac{n_i}{\tau_{nij}(w_i)} = \frac{n_j}{\tau_{nji}(w_j)} \tag{3.63}$$

can be used to derive $\tau_{nji}(w_j)$.

For intervalley momentum losses (but written directly in terms of velocity, $\Phi_i = \mathbf{u}_i(\mathbf{k})$), Eq. (3.60) results in

$$\tau_{vij}(w_i) = \left(\int_{i^{th}valley} \mathbf{u}_i(\mathbf{k})\frac{f_i'(\mathbf{k})}{\tau_{Sij}(\mathbf{k})}d\mathbf{k}\right)^{-1}. \tag{3.64}$$

In this case, Equation (3.62) is sufficient because in-scattering effects are random.

Since intervalley scattering in semiconductors is dominated by phonon collisions (which impart to or absorb energy from the electrons) and the bands are separated by Δ_{ij} the in-scattering and out-scattering must be addressed separately. For out-scattering, Equation (3.60) can be applied directly to obtain

$$\tau_{wij}(w_i) = \left(\int_{i^{th} valley} E_i(\mathbf{k}) \frac{f_i'(\mathbf{k})}{\tau_{Sij}(\mathbf{k})} d\mathbf{k} \right)^{-1} \tag{3.65}$$

which is valid for all the τ_{wij}'s. When an intervalley scattering event $j \rightarrow i$ occurs, an electron will leave valley j with energy E_j. However, the electron will either gain or lose an amount of kinetic energy Δ_{ji} (referenced from j) due to valley separation. In addition, the phonon will either absorb from or impart to the electron the phonon energy $\epsilon_{ji} = \hbar w_{ji}$. Therefore, the rate of change of *energy density* W_i due to in-scattering from j is

$$\left.\frac{\partial W_i}{\partial t}\right|_{coll.}^{j\rightarrow i} = \int_{j^{th} valley} \left(E_j(\mathbf{k}) + \Delta_{ji} + \epsilon_{ji}\right) \frac{f_j(\mathbf{k})}{\tau_{abji}(\mathbf{k})} d\mathbf{k} + \int_{j^{th} valley} \left(E_j(\mathbf{k}) + \Delta_{ji} - \epsilon_{ji}\right) \frac{f_j(\mathbf{k})}{\tau_{emji}(\mathbf{k})} d\mathbf{k} \tag{3.66}$$

where the relation

$$\frac{1}{\tau_{Sij}(\mathbf{k})} = \frac{1}{\tau_{abji}(\mathbf{k})} + \frac{1}{\tau_{emji}(\mathbf{k})} \tag{3.67}$$

has been used to divide the scattering rate into absorption and emission components. Equation (3.66) may be rearranged to yield

$$\left.\frac{\partial w_i}{\partial t}\right|_{coll.}^{j\rightarrow i} = \frac{1}{n_i}\left[-n_j \left.\frac{\partial w_j}{\partial t}\right|_{coll.}^{j\rightarrow i} - \left(\Delta_{ji} + w_j - w_i\right) \left.\frac{\partial n_j}{\partial t}\right|_{coll.}^{j\rightarrow i}\right] + \frac{\epsilon_{ji}}{n_i} \int_{j^{th} valley} \left(\frac{1}{\tau_{abji}(\mathbf{k})} - \frac{1}{\tau_{emji}(\mathbf{k})}\right) f_j(\mathbf{k}) d\mathbf{k} \tag{3.68}$$

where a constant phonon energy ϵ_{ji} has been assumed. If the difference between the average energies w_i and w_j is not large, then (3.62) - (3.65) and the phenomenological form for the rate of change of w_i due to in-scattering from j

$$\left.\frac{\partial w_i}{\partial t}\right|_{coll.}^{j\rightarrow i} = \left(\frac{n_j}{n_i}\right) \frac{w_j}{\tau_{eji}(w_j)} \tag{3.69}$$

can be used to obtain

$$\tau_{eji}(w_j) = \left(\frac{1}{\tau_{wji}} - \frac{\Delta_{ji}}{w_j \tau_{nji}} + \frac{\epsilon_{ji}}{w_j} \int_{j^{th} valley} \left[\frac{1}{\tau_{abji}(\mathbf{k})} - \frac{1}{\tau_{emji}(\mathbf{k})}\right] f_j'(\mathbf{k}) d\mathbf{k} \right)^{-1} \tag{3.70}$$

where $f_j' = f_j'/n_j$. Equations (3.62), (3.64), (3.65), and (3.70) can be evaluated directly from stationary Monte Carlo simulations and combined with electron density and average energy data to derive all necessary *intervalley* collision-related parameters for use in multivalley hydrodynamic transport models[44]. The *intravalley* parameters can be determined in a similar manner by applying Eq. (3.29) for each quantity $\mathcal{A}$ and valley in the model. However, once all the intervalley relaxation rates (or times) are determined the intravalley data is more simply obtained from the stationary hydrodynamic equations. Stationary conditions reduce each momentum and energy equation in the multi-valley model to

$$\frac{(-q)\mathcal{E}}{m_i^{**}(w_i)} = \frac{\mathbf{v}_i}{\tau_{vii}(w_i)} - \left.\frac{\partial \mathbf{v}_i}{\partial t}\right|_{coll.}^{inter.} \tag{3.71}$$

and

$$(-q)\mathbf{v}_i \cdot \mathcal{E} = \frac{(w_i - w_i^0)}{\tau_{wii}(w_i)} - \left.\frac{\partial w_i}{\partial t}\right|_{coll.}^{inter.} . \tag{3.72}$$

Eqs. (3.71) and (3.72) can now be used to determine all the remaining relaxation times.

4 Application of the Full M-HTM

When the collisionless hydrodynamic equations of section 2.3 are combined with the preceding multivalley dissipation terms (with the relaxation time data suppled by a limited number of Monte Carlo calculations), one has an advanced macroscopic transport model which can be used to accurately and efficiently study electron conduction in semiconductors. This model utilizes *acceptable* macroscopic properties to approximate realistic effects such as conduction band nonparabolicity and electron transfer between nonequivalent minima. By taking advantage of these macroscopic properties, one has a much simplified fundamental set of transport equations. Therefore, this will ensure less overall computational effort in arriving at transport solutions (for electron device structures) as compared to the full BTE approach, independent of the mathematical algorithms used.

Recently, a multi-valley hydrodynamic model (M-HTM) has been used to study the effects of velocity overshoot on transit time through submicron GaAs ballistic diode structures[52]. In these investigations, a M-HTM was used which contained a set of transport equations given by Eqs. (2.55) - (2.57) for each of the primary valleys (Γ, L and X) in GaAS. These equations were augmented by Poisson's equation so the simulation results would be valid for self-consistent potentials. For these initial studies, a simplified set of dissipation terms similar to those given by Eqs. (3.57) - (3.59), except extended to the three valley case, were used. In this simplied scattering model, full electron exchange was allowed between valleys with the scattering terms in each valley i given by

$$\left.\frac{\partial n_i}{\partial t}\right|_{coll.} = -\frac{n_i}{\tau_{nij}(w_i)} - \frac{n_i}{\tau_{nik}(w_i)} + \frac{n_j}{\tau_{nji}(w_j)} + \frac{n_k}{\tau_{nki}(w_k)} . \tag{4.1}$$

Momentum and energy losses were treated locally within each valley with the *multiple average-energy-dependent* exchange between valleys neglected. This was done to simplify the process of determining the necessary 14 boundary conditions so that they were mathematically *sufficient* for the entire system of transport equations. Thus, the momentum and energy loss terms used were

$$\left.\frac{\partial \mathbf{v}_i}{\partial t}\right|_{coll.} = -\frac{\mathbf{v}_i}{\tau_v(w_i)} \tag{4.2}$$

and

$$\left.\frac{\partial w_i}{\partial t}\right|_{coll.} = -\frac{w_i - w_0}{\tau_w(w_i)}. \tag{4.3}$$

where $\tau_v(w_i)$ and $\tau_w(w_i)$ are effective relaxation rates that reflect both intravally and intervalley dissipation. In Eqs. (4.1) - (4.3), the indices i, j and k take on permutations of Γ, L and X depending on which particular valley (i) is being considered. The values of the required relaxation times were obtained from previous Monte Carlo simulations[44]. The boundary conditions were selected to simulate ideal ohmic contacts operating under perfectly stationary conditions.

Transit time studies were performed on an $N^+ - N - N^+$ structure consisting of a 1.0 μm long source and collector regions doped 10^{17} cm^{-3} and a 0.2 μm long active region doped 10^{16} cm^{-3}. The very long source and collector regions were chosen to minimize the effects of boundary conditions applied at the ideal source and drain. An outline of the study and results is given in Fig. 4.1. In the study, a variety of different doping spikes and/or electric field spikes were strategically placed within the active region in an effort to reduce the average electron energy near the collector and retain more electrons in the high mobility Γ valley. This was performed in hopes that the the region of velocity overshoot could be extended to result in transit time enhancement through the active region and/or collector region.

The general results of the study, regardless of type of spike, location, width, intensity etc., were that the interval average velocity speedup was minimal in the neighborhood of the classical active region (inside the region defined by the sharp $N^+ - N$ spatial transitions). However, for some spikes the velocity overshoot could be extended far into the collector region with a significant enhancement of transit time from ideal source and drain. In order to illustrate this effect clearly, a summary of the numerical results for a relatively wide electric field spike (0.07 μm in extent and front edge located just past the midpoint of the active region) will be presented. The calculated results for *classical* active region transit time τ_a and total device transit time τ_d for the case with and without the spike is given in the lower portion of Fig. 4.1. The introduction of the spike produces an initial degradation in both transit times (τ_a and τ_d). However, there is a distinct enhancement in τ_d at sufficient biases to over come the retarding potential of the spike.

Important specific details can be obtained from Figs. 4.2 - 4.4 which reveal spatial profile data given for both cases well into the saturation region (1.5

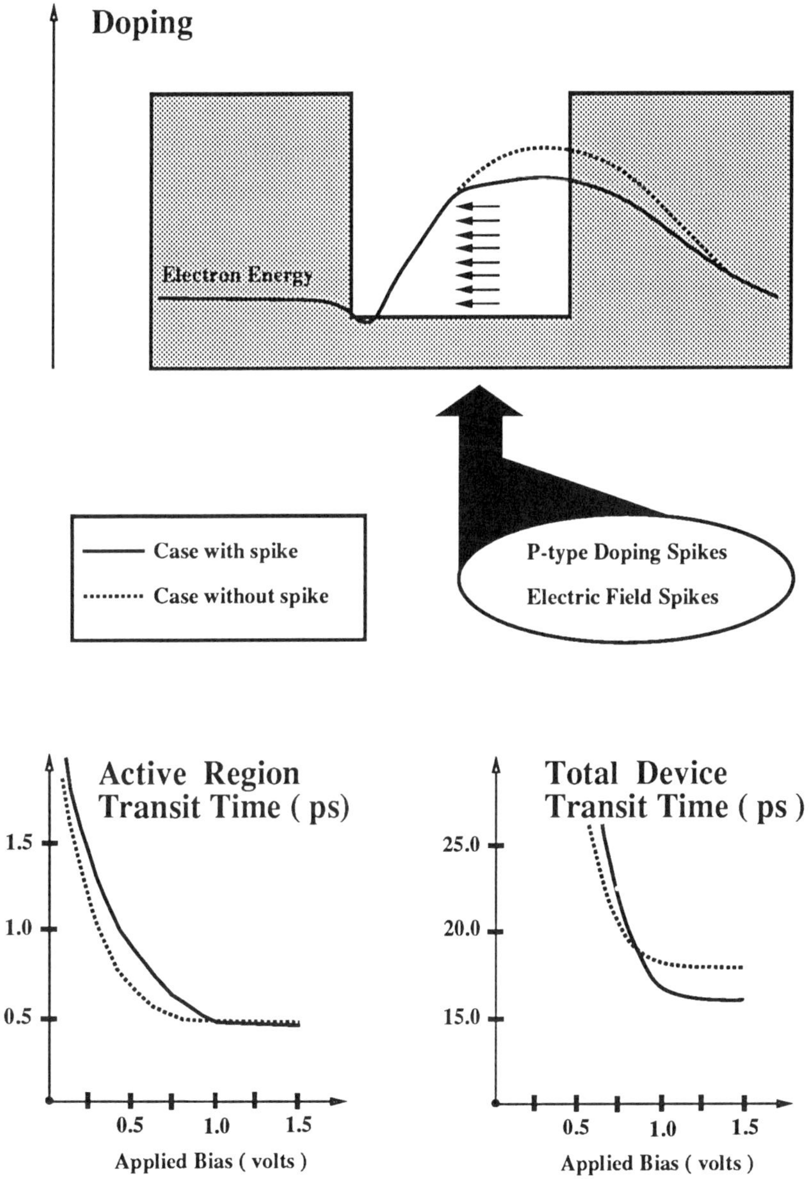

Fig. 4.1 Outline of M-HTM study and transit time results.

volts applied bias). Fig. 4.2 shows no significant difference in electric field profile (as compared to the uniform material in active region case) before the leading edge of the spike. Fig. 4.3 presents a comparison of the fractional Γ valley occupancy of electrons for the two cases. This figure clearly shows a reduction in fractional occupancy of n_Γ before the leading edge of the spike and an increase within and after the spike. This increase is shown to persist well into the collector region. Fig. 4.4 gives a comparison of the total average velocity

$$v_{avg} = \frac{n_\Gamma v_\Gamma + n_L v_L + n_X v_X}{n_\Gamma + n_L + n_X} \tag{4.4}$$

for the two cases. Fig. 4.4 shows the additional effect of the velocity directly reacting to the presence of the electric field spike and being reduced near the midpoint of the active region. The major observations, from Figs 4.3 and 4.4, are that the changes (relative increases and decrease in the spatial profile) to n_Γ and v_{avg} are nearly balanced within the active region. However, the increase in n_Γ and v_{avg} persists well into the collector region. Hence, this is the underlying explanation for the *considerable* transit time enhancement in collector region and minor improvement inside the active region.

These general results of minimal velocity speed-up in the classical active region and extension of velocity overshoot into the collector region with overall device transit time enhancement have been previously observed in transport modeling. Specifically, these results agree qualitatively with previously reported self-consistent Monte Carlo simulations[83] and by assumed electric field profile studies[84].

The solutions of the M-HTM (14th order system of ODE's) applied to the ballistic diode structure were generated by a *non-optimized* numerical approach. In this M-HTM study, as many as ten different bias point solutions could be generated on a VAXstation 3200 in only a few hours ($<$ 3 hours). A corresponding Monte Carlo study of the ballistic diode structure with an active region field spike, even with an *assumed electric field profile* and an optimized computer code, could take over a day. In general, optimizing computer codes for the Monte Carlo method is very difficult and solving the full BTE using other mathematical techniques is currently even more computationally demanding. On the other hand, there is much promise for further improvements in the numerical methods used to solve the M-HTM. Hence, these results have demonstrated that the Monte Carlo technique in combination with advanced hydrodynamic methods can be used to more efficiently study hot-electron transport in novel electronic structures. The tandem approach, while sacrificing some physical detail, leads to a simpler mathematics problem with the potential for rapid generation of solutions using numerical methods.

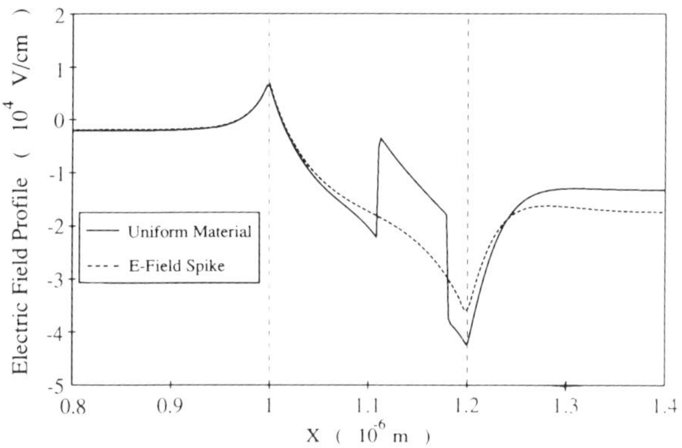

Fig. 4.2 Resultant electric field profiles for $N^+ - N - N^+$ structure with and without electric field spike.

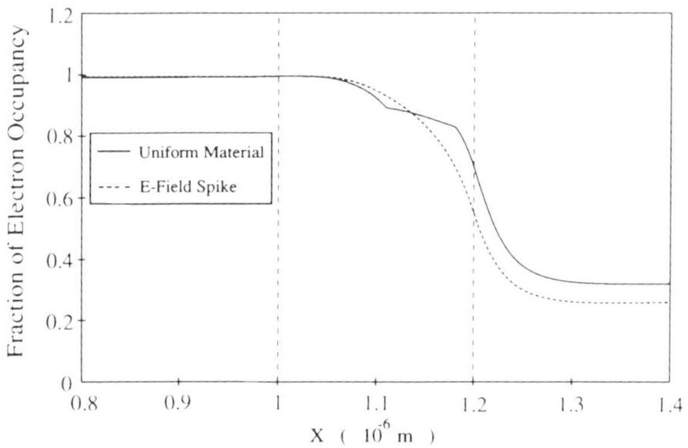

Fig. 4.3 Comparision of fractional occupancy n_Γ for the $N^+ - N - N^+$ structure with and without electric field spike.

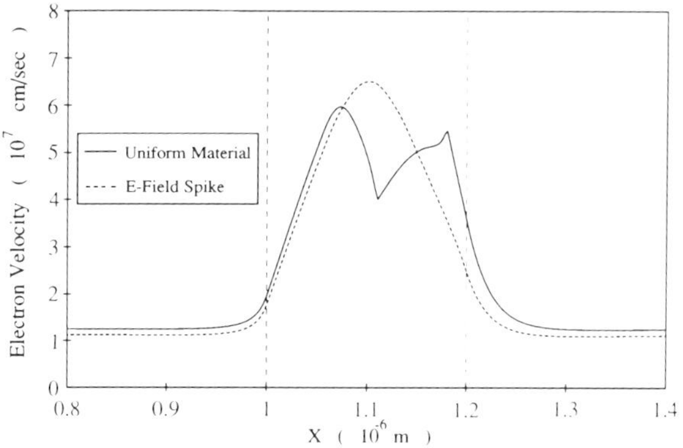

Fig. 4.4 Comparision of total average velocity v_{avg} for the $N^+ - N - N^+$ structure with and without electric field spike.

5 Conclusion

In this paper, a new approach has been presented for the study of hot electron transport in semiconductors and semiconductor devices. The approach utilizes an advanced macroscopic transport model in combination with Monte Carlo techniques to implement a more efficient model which is *almost* equal in accuracy to a full Monte Carlo analysis. This paper has demonstrated the valuable use of Monte Carlo techniques in the development and verification of this tandem transport model. In addition, the development and generation of the necessary dissipation terms has been given. Finally, the utility of the Monte Carlo - Hydrodynamic model has been illustrated through transit time studies of ballistic diode structures at high applied biases. Therefore, this new approach has been shown to offer much potential for future theoretical and experimental investigations of electronic devices.

References

[1] A.A. Quaranta, C. Jacoboni and G. Ottaviani, *Rivista Del Nuovo Cimento* **4**, 445 (1971).

[2] P.J. Price in *Semiconductor and Semimetals, Vol. 14*, edited by R.K. Willardson and A.C. Beer (Academic, New York, 1979).

[3] C. Jacoboni and L. Reggiani, *Advances in Physics* **28**, 493 (1979).

[4] H.A. Meyer, Editor of *Symposium on Monte Carlo Methods* (Wiley, New York, 1956).

[5] C. Jacoboni and L. Reggiani, *Reviews of Modern Physics* **55**, 645 (1983).

[6] B.R. Nag, *Electron Transport in Compound Semiconductors* (Springer-Verlag, Berlin, 1980).

[7] M. Lundstrom, *Fundamentals of Carrier Transport, Volume X. Modular Series on Solid State Devices* (Addison-Wesley, Reading, Mass., 1990).

[8] J.G. Ruch, IEEE Trans. Electron Devices **ED-19**, 652 (1972).

[9] R. Bosch and H.W. Thim, IEEE Trans. Electron Devices **ED-21**, 16 (1974).

[10] M. Shur, Electronics Letters **12**, 615 (1976).

[11] M.S. Shur and L.F. Eastman, Solid-State Electron. **24**, 11 (1981).

[12] J.P. Nougier, J.C. Vaissiere and D. Gasquet, J. Appl Phys. **52**, 825 (1981).

[13] R.K. Cook and J. Frey, IEEE Trans. Electron Devices **ED-29**, 970 (1982).

[14] F.A. Bout and J. Frey, Solid-State Electron. **26**, 617 (1983).

[15] H.L. Grubin and J.P. Kreskovsky, J. Vac. Sci. Technol. B **2**, 527 (1984).

[16] G. Baccarani and M.R. Wordeman, Solid-State Electron. **28**, 407 (1985).

[17] C.M. Snowden and D. Loret, IEEE Trans. Electron Devices **ED-34**, 212 (1987).

[18] R.B. Darling, Solid-State Electron. **31**, 1031 (1988).

[19] M.O. Vassell and J. Lee, J. Appl. Phys. **63**, 5380 (1988).

[20] P.A. Sandborn, J.R. East and G.I. Haddad, Solid-State Electron. **32**, 191 (1989).

[21] T. Shawki, G. Salmer and O. El-Sayed, IEEE Trans. Electron Devices **37**, 21 (1990).

[22] E.M. Azoff, IEEE Trans. Electron Devices **36**, 609 (1989).

[23] H.L. Grubin and J.P. Kreskovsky, Solid-State Electron. **32**, 1071 (1989).

[24] D. Potter, *Computational Physics* (Wiley, London/New York, 1973) Ch. 9.

[25] S. Chapman and T.G. Cowling, *The Mathematical Theory of Non-uniform Gases* (Cambridge, London/New York, 1973).

[26] K. Hess and G.J. Iafrate, Proc. IEEE **76**, 519 (1988).

[27] R. Stratton, Phys. Rev. **126**, 2002 (1962).

[28] K. Blotekjaer, IEEE Trans. Electron Devices **ED-17**, 38 (1970).

[29] R.K. Cook and J. Frey, COMPEL **1**, 65 (1982).

[30] N. Goldsman and J. Frey, IEEE Trans. Electron Devices **35**, 1524 (1988).

[31] H.L. Grubin and J.P. Kreskovsky, in *VLSI Electronics: Microstructure Science, Vol. 10*, edited by N.G. Einspruch (Academic, New York, 1985).

[32] D.L. Woolard J-L Pelouard, R.J. Trew, M.A. Littlejohn and C.T. Kelly, Solid-State Electron. **32**, 1347 (1989).

[33] D.L. Woolard, M.A. Stroscio, M.A. Littlejohn, R.J. Trew and H.L. Grubin, in *Computational Electronics* edited by K. Hess, J.P. Leburton and U. Ravaioli (Kluwer, Boston, 1991), p. 59.

[34] E.M. Azoff, J. Appl. Phys. **64**, 2439 (1985).

[35] D.J. Widiger, I.C. Kizilyalli, K. Hess and J.J. Coleman, IEEE Trans. Electron Devices **ED-32**, 1092 (1985).

[36] P.A. Sandborn, A. Rao and P.A. Blakey, IEEE Trans. Electron Devices **36**, 1244 (1989).

[37] R.K. Mains, M.A. El-Gabaly and G.I. Haddad, J. Comput. Phys. **59**, 456 (1985).

[38] J.P. Kreskovsky and M. Meyyappan and H.L. Grubin, COMPEL **6**, 99 (1987).

[39] M. Rudan, F. Odeh and J. White, COMPEL **6**, 151 (1987).

[40] C.L. Gardner, J.W. Jerome and D. Rose, IEEE Trans. CAD **8**, 501 (1989).

[41] E. Thomann and F. Odeh, COMPEL **9**, 45 (1990).

[42] R.A. Stewart and J.N. Churchill, Solid-State Electron. **33**, 819 (1990).

[43] W. Fawcett, A.D. Boardman and S. Swain, J. Phys. Chem. Solids **31**, 1963 (1970).

[44] D.L. Woolard, R.J. Trew and M.A. Littlejohn, Solid-State Electron. **31**, 571 (1988).

[45] An example of this is the $\frac{m^*}{2}|\mathbf{v}|^2$ in the expression $T_e = \frac{2}{3k_B}(w - \frac{m^*}{2}|\mathbf{v}|^2)$ which defines the temperature for the case of parabolic bands and a displaced-Maxwellian distribution.

[46] T.H. Glisson, C.K. Williams, J.R. Hauser and M.A. Littlejohn, in *VLSI Electronics : Microstructure Science, Vol. 4*, edited by N.G. Einspruch (Academic, New York, 1982), p. 99.

[47] K. Kim and K. Hess, J. Appl. Phys. **64**, 3057 (1988).

[48] A.C. Smith, J.F. Janak and R.B. Adler, *Electronic Conduction in Solids* (McGraw-Hill, New York, 1967).

[49] P.A. Blakey, J.R. East and G.I. Haddad, *VLSI Electronics: Microstructures Science, Vol. 2*, edited by N.G. Einspruch (Academic Press, New York, 1981), p. 105.

[50] J.J. Duderstadt and N.R. Martin, *Transport Theory* (Wiley, New York, 1979).

[51] J.S. Blakemore, J Appl. Phys. **53**, 123 (1982).

[52] D.L. Woolard, PH.D. Thesis, North Carolina State University (1991).

[53] E.M. Azoff, *Solid-State Electronics* **30**, 913 (1987).

[54] N.A. Krall and A.W. Trivelpiece, *Principles of Plasma Physics* (McGraw-Hill, New York, 1973), p. 96.

[55] S.R. Seshadri, *Fundamentals of Plasma Physics* (American Elsevier, New York, 1973), Ch. 1.

[56] B.R. Nag, *Theory of Electrical Transport in Semiconductors*, (Pergamon Press, New York, 1972).

[57] A. E. Taylor and W. R. Mann, *Advanced Calculus* (Wiley, New York, 1983), Ch. 8.

[58] D.K. Ferry, in *Handbook on Semiconductors, Vol. 1*, edited by W. Paul (North-Holland, New York, 1982), p. 563.

[59] D.L. Woolard, R.J. Trew and M.A. Littlejohn to be published.

[60] M.A. Littlejohn, J.R. Hauser and T.H. Glisson, J. Appl. Phys. **48**, 4587 (1977).

[61] S. Lee, T. Tang and D.H. Navon, in *Proceedings of the Sixth International Nasecode Conference*, edited by J.J.H. Miller (Boole, Dublin, 1989), p. 478.

[62] H.U. Baranger and J.W. Wilkins, Phys. Rev. **B 36**, 1487 (1987).

[63] E.M. Conwell, *High Field Transport in Semiconductors* (Academic Press, New York, 1967).

[64] C.M. Snowden, *Semiconductor Device Modelling* (Peter Peregrinus Ltd., London, 1988).

[65] K. Haung, *Statistical Mechanics* (Wiley, New York, 1963).

[66] C.M. Wolfe, N. Holonyak JR., and G.E. Stillman, *Physical Properties of Semiconductors* (Prentice Hall, New Jersey, 1989).

[67] N.W. Ashcroft and N.D. Mermin, *Solid State Physics* (Holt, Rinehart and Winston, New York, 1976).

[68] H.L. Grubin, D.K. Ferry, G.J. Iafrate and J.R. Barker, in *VLSI Electronics : Microstructure Science, Vol. 3*, edited by N.G. Einspruch (Academic, New York, 1982), P. 197.

[69] W. Fawcett and H.D. Rees, *Phys. Lett.* **29A**, 643 (1969)

[70] W. Fawcett, in *Electronic Transport in Crystalline Solids* edited by A. Salam (International Atomic Energy Agency, Vienna, 1973).

[71] W. Fichtner, D.J. Rose and R.E. Bank, *IEEE Trans. Electron Devices* **ED-30**, 1018 (1983).

[72] K. Hess, *Advanced Theory of Semiconductor Devices* (Prentice Hall, New Jersey, 1988).

[73] K. Blotekjaer and E.B. Lunde, *Physica Status Solidi* **35**, 581 (1969).

[74] G. Bauer, in *Springer Tracts in Modern Physics, Vol 74* (Springer, Berlin, 1978).

[75] K. Seeger, *Semiconductor Physics, Springer Series in Solid-State Sciences 40* (Springer-Verlag, Berlin, 1985).

[76] C.M. Snowden, *Rep. Prog. Phys* **48**, 223 (1985).

[77] S. Datta, *Quantum Phenomena, Volume VIII. Modular Series on Solid State Devices* (Addison-Wesley, Reading, Mass., 1989).

[78] R.F. Pierret, *Advanced Semiconductor Fundamentals, Volume VI. Modular Series on Solid State Devices* (Addison-Wesley, Reading, Mass., 1987).

[79] S. Wang, *Fundamentals of Semiconductor Theory and Device Physics* (Prentice Hall, New Jersey, 1989).

[80] J.P. McKelvey, *Solid State and Semiconductor Physics* (Harper and Row, New York, 1966).

[81] I.B. Bott and W. Fawcett, *Adv. Microwaves* **3**, 223 (1968).

[82] J.E. Carroll, *Rate Equations in Semiconductor Equations* (Cambridge University Press, Cambridge, 1985).

[83] K. Tomizawa, Y. Awano and N. Hasizume, *IEEE Electron. Device Lett.* **EDL-5**, 362 (1984).

[84] C.M. Maziar, M.E. Klausmeier-Brown and M.S. Lundstrom, *IEEE Electron. Device Lett.* **EDL-6**, 483 (1986).

9

VECTORIZATION OF MONTE CARLO ALGORITHMS FOR SEMICONDUCTOR SIMULATION

Umberto Ravaioli

Beckman Institute and Coordinated Science Laboratory
University of Illinois at Urbana-Champaign
Urbana, IL 61801

ABSTRACT

The improving capabilities of computers have deeply influenced the progress of large scale semiconductor device simulations. Supercomputing techniques can greatly benefit Ensemble Monte Carlo methods, provided that the history of a large number of particles is treated simultaneously. The intrinsic randomness of the electronic transport is a major obstacle to the efficient vectorization of self-consistent algorithms which require synchrony of the ensemble at the end of periodic time intervals, when Poisson's equation is solved to update the electric field distribution. Here, a discussion of optimization/vectorization schemes is presented, with particularly attention to strategies suitable for self-consistent Monte Carlo device simulations. Benchmarks and comparisons obtained on the CRAY X-MP/48 and the CRAY Y-MP4/464 supercomputers are also provided.

INTRODUCTION

Semi-classical electronic transport in semiconductors is described by the Boltzmann Transport Equation (BTE), given here in non-degenrate form,

$$\frac{\partial f}{\partial t} + \vec{v}\cdot\nabla f + \vec{F}\cdot\nabla_{\vec{k}} f = \Sigma_{\vec{k}'}\{S(\vec{k}', \vec{k})f(\vec{r}, \vec{k}', t) - S(\vec{k}, \vec{k}')f(\vec{r}, \vec{k}, t)\} \quad (1)$$

where $f(\vec{r}, \vec{k}, t)$ is the electron distribution function; $\vec{r}$ and $\vec{k}$ represent the position and crystal momentum coordinates in phase space; $\vec{F}$ is the electric field, $\vec{v}$ the velocity, and $S(\vec{k}, \vec{k}')$ are the second order perturbation theory scattering rates. In this formulation, the BTE treats the electrons as a classical gas, while the scattering rates in the collision integral on the right hand side are calculated using quantum mechanical techniques. In addition, the structure of the semiconductor conduction band is included through the dispersion relation for the kinetic energy $E(\vec{k})$ which enters the BTE through the velocity, defined as

$\vec{v} = \nabla_{\vec{k}} E(\vec{k})/\hbar$. From the dispersion relation $E(\vec{k})$ one can deduce an effective mass for the electron motion [1].

The BTE is an integro-differential equation, and for the majority of real systems, exact analytical solutions are not viable. Transport quantities are derived from the averages over a number of processes whose relative importance is not clearly known beforehand. This makes the formulation of a reliable microscopic model very difficult. Even greater difficulties arise in nonlinear response conditions, and analytical solutions of Boltzmann's equation, without linearization with respect to external forces, are only available for special cases. Approximate analytical solutions usually involve assumptions about the distribution function or approximations on the form of the scattering rates. The most used techniques are the displaced Maxwellian approximation [2] and the Legendre polynomial expansion [3]. In many cases of interest, such approximations are too drastic, and it is not always clear when differences in the results are due to the microscopic model itself or to the mathematical model.

The Monte Carlo method was developed to solve Boltzmann's equation directly, with a stochastic procedure. In a typical Monte Carlo simulation a single particle or an ensemble of particles are followed by computer simulation. In the frame of the semi-classical approximation, the motion of the particles consists of free flights, due to drift in an electric field, interrupted by instantaneous scattering events. The time of flight, the type of scattering process, and the final state after scattering, are random variables with probability distributions which can be expressed in terms of the strength of the electric field and of the transition rates due to the various processes. The probability distributions may be quite complex, while the generation by computer of random number sequences with uniform probability over some finite range is relatively simple. The key point in the method is to use techniques to convert easily generated uniform probability distributions into the complicated distributions required.

For the steady-state study of semiconductor bulk or of structures with a fixed known potential, the one-particle Monte Carlo technique is often used. The evolution of one single particle is followed in time and the particle properties (e.g. velocity, energy) are time-averaged. In steady-state conditions, according to the ergodic theorem, the properties so calculated represent also the properties of the whole electron ensemble. A natural development of the technique involves the simultaneous simulation of an ensemble of electrons. Obviously, only in this situation one can take advantage of parallel and vector machines for the computation. The ensemble Monte Carlo technique is also the only one suitable for transient calculations and for self-consistent device simulations.

In a self-consistent Ensemble Monte Carlo procedure, Poisson's equation is solved numerically at regular time intervals, to update the space distribution of the electric field, used the calculation of the particle acceleration. In this way, the simulation adapts itself to the charge movements and redistribution, until a steady-state is reached. Typically, a mesh is defined over the domain under study and a numerical procedure is used to solve for the potential on the nodes of the mesh. The charge on the mesh points, necessary to evaluate the right hand side of Poisson's equation, is obtained from the particle distribution inside the meshes, using techniques reviewed, for instance, in the book by Hockney and Eastwood [4]. An important simulation parameter is the time interval between two consecutive solutions of Poisson's equation. If such an interval is too long, spurious plasma oscillations arise in the simulated gas of electrons, making the reaching of a steady-state problematic. In typical semiconductor applications, Poisson's equation is solved at intervals of $10^{-14}s$ or less.

FREE-FLIGHT TIME GENERATION

The duration of the free-flights between collisions depends on the scattering probability associated with the particle. There are many sources of scattering for an electron moving inside a crystal. The most important mechanisms are related to collisions with phonons and ionized impurities, and at high energies one has to consider also impact ionization. The main difficulty in determining the distribution of the free-flight times lies in the strong momentum dependence of the scattering rates, so that during the accelerated motion of the particles, the associated scattering rates vary continuously according to the momentum evolution of the single particles. One can write the density of probability for a free-flight to have a duration t as

$$P(t) = \lambda(\vec{k}(t))\, p(t) \tag{2}$$

where $\lambda(\vec{k}(t))$ is the instantaneous total scattering rate and $p(t)$ is the probability for a particle to fly without scattering for a time t, after the previous collision. The total scattering rate is simply the sum of the rates for all the possible scattering processes, which are in general quite different and strongly dependent on the momentum as well. One can show [5] that $p(t)$ is related to the evoultion of $\vec{k}(t)$ according to

$$p(t) = \exp\left(-\int_0^t \lambda(\vec{k}(\tau))\, d\tau\right) \tag{3}$$

In order to derive the random duration of the free-flight, according to the density of probability given in (3), one can resort to random numbers uniformly distributed between 0 and 1, which are normally available on digital computers through a function call. A selected random number r can be mapped into a corresponding random time of flight t, through the equation

$$\int_0^r P(r)\,dr = \int_0^t P(\tau)\,d\tau \tag{4}$$

where $P(r)$ is the density of probability for the computer generated random numbers. Since $P(r) = 1$ for a uniform distribution, one can show that using (3) the following equation holds [5]

$$-\ln r = \int_0^t \lambda\,(\vec{k}(\tau))\,d\,\tau \tag{5}$$

Therefore, the determination of the random time of free-flight t reduces to the evaluation of the integral on the right hand side of (5) as the flight progresses, until the equality in (5) is satisfied.

The efficiency of the computer implementation of the Monte Carlo method for semiconductors is largely determined by the procedure used to solve (5). A complete review and a comparison of the techniques appearing in the literature can be found in the paper by Yorston [6].

A very simple procedure by Rees [7] introduces a virtual collision mechanism called *self-scattering*. The associated self-scattering rate $\lambda_{ss}(\vec{k})$, is chosen so that the resulting total scattering rate

$$\Gamma = \lambda\,(\vec{k}) + \lambda_{ss}(\vec{k}) \tag{6}$$

is constant. The self-scattering does not affect the state of the particle, so, whenever it is selected the trajectory continues unperturbed. It can be shown that the statistical distribution of the free-flight times obtained in a Monte Carlo computer experiment is not affected by the introduction of the self-scattering [5]. At this point, one can trivially solve (5) which yields the time of flight

$$t = -\frac{\ln r}{\Gamma} \tag{7}$$

This procedure is often called Constant Γ technique. Although it is quite appealing for its simplicity, the Constant Γ technique has computational drawbacks. In fact, since the total scattering rate is kept artificially high, in most actual simulations the vast majority of events is constituted by self-scatterings. Even if no calculations are required to update the electron state, every selection of a self-scattering entails two additional function calls to generate random numbers: one to select the self-scattering itself among the other possible events and a second one to restart the time of flight procedure. As shown in [6], alternative techniques for the determination of the time of flight may reduce the computation time by an order of magnitude. The concept of self-scattering remains however very useful, since it allows one to pick any total scattering rate which is more convenient for the calculations, provided that it is never smaller than the real total rate. The difference between the total rate used and the actual one is accounted for by self-scattering events.

Free-flight time generation procedures, in alterantive to the Constant Γ technique, follow three approaches:

(a) During the whole simulation a fixed total scattering rate is choosen over the momentum (or energy) range, in a way that reduces as much as possible the probability of selecting a self-scattering event. Examples in the literature use a piecewise-constant [8] or a polynomial [9] curve for the total scattering rate. The self-scattering levels for these techniques are illustrated in Fig. 1, where the total scattering rate is plotted as a function of energy.

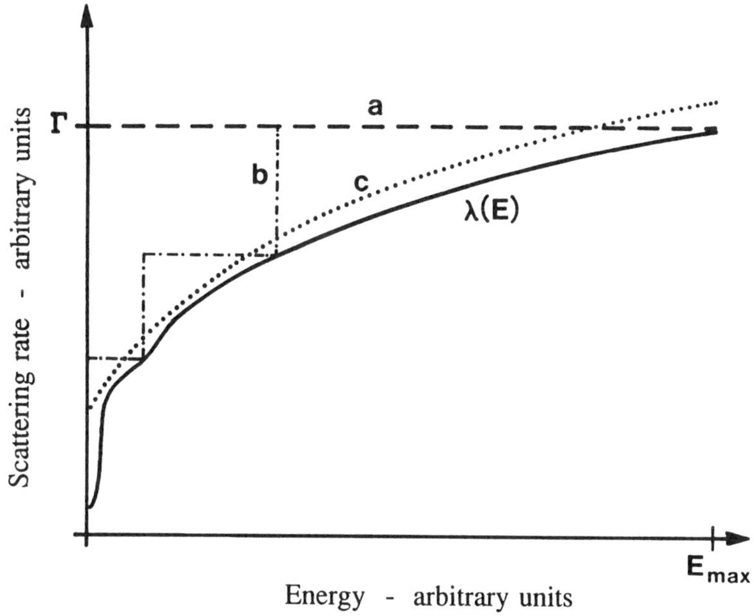

Fig. 1 - Examples of self-scattering levels determined with respect to particle energy at the beginning of a simulation for: (a) Constant Γ technique; (b) Piecewise-constant Γ technique; (c) Polynomial Γ technique.

(b) A variable Constant Γ level is associated to each particle, so that the total rate assumed is barely higher than the actual maximum scattering rate during each flight. The total rate should not be changed after a self-scattering is selected, but can be varied after a real scattering. An example is the Iterative Γ technique [4] illustrated in Fig. 2. A level Γ_1 is chosen initially, equal to the total scattering rate corresponding to the particle energy. The time of flight t_1 is calculated using (7) and is accepted if the scattering rate for the final energy is lower than Γ_1. Obviously, in the cases when the scattering rate increases with time, the value t_1 must be discarded and a slightly larger value Γ_2 is selected. The procedure is iterated until the current value of Γ is higher than the scattering rate at the end of the selected time of flight. Another example is the Constant Time technique [6] illustrated in Fig. 3. The flight is subdivided into subflights of constant duration τ, and the self-scattering level Γ_i is taken to be the maximum scattering rate within each subflight. The free flight is terminated at the time t which satisfies the equality

$$- \ln r = \sum_{i=1}^{n-1} \Gamma_i \, \tau_i + \Gamma_n \, \tau^* \, . \qquad (8)$$

The procedure is restarted after the occurrence of a scattering, and in the case of self-scattering one should begin the new free flight without changing any of the flight parameters, and selecting a new random number.

(c) The integral on the right hand side of (5) is evaluated directly. This approach obviously does not need the introduction of self-scattering.

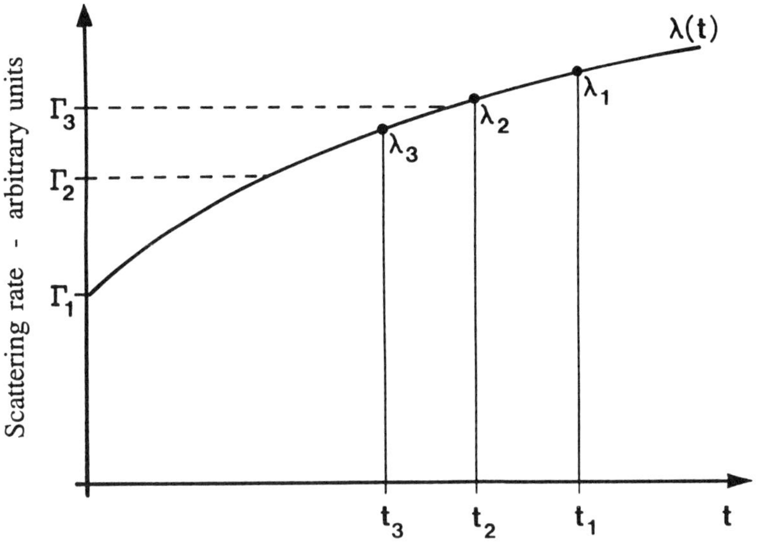

Fig. 2 - Example of self-scattering level determined for a particular free flight, using the Iterative Γ technique. In the example the third attempt is successful, since the total scattering rate λ_3, determined at $t_3 = -\,ln(r)/\Gamma_3$, is lower than Γ_3. Note that the evolution of the total scattering rate λ for the particle considered is plotted as a function of time.

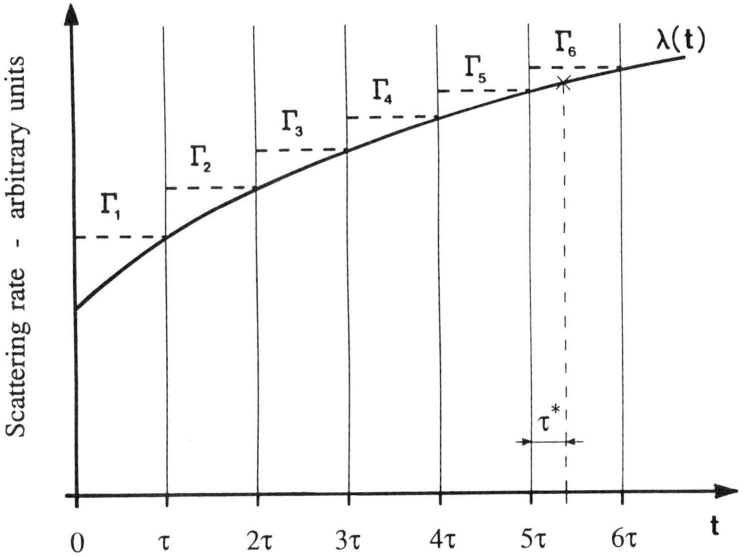

Fig. 3 - Example of self-scattering levels adjusted as a function of time during a particular free flight time, using the Constant Time technique. In the example, The flight is terminated with a scattering at time $t = 5\tau + \tau^*$.

VECTORIZATION OF THE MONTE CARLO METHOD

The Ensemble Monte Carlo method is quite appealing for supercomputing applications, since the simulation may require thousands of individual particles moving at the same time. In the following, we will concentrate mainly on the issue of vectorization, since this process gives a true gain in terms of CPU time. During vectorization, data vectors are pipelined into a single processor. Parallelization instead divides the tasks between a number of identical processors and the only gain is in the real time of the computation, while there is no reduction, rather an increase, of the overall CPU time required.

The randomness in the Monte Carlo algorithm, which is the key ingredient to provide statistically good answers, is also the main hindrance for an efficient vectorization. The standard Monte Carlo approach treats an ensemble of particles synchronously. The history of each particle is followed within a given time step, leading to a succession of free flights and scattering events selected through the generation of uniform random numbers. This so called *history-based* approach is difficult to vectorize, since the single electrons may undergo quite a different history, and the computing operations cannot proceed in parallel. The situation is exemplified in Fig. 4. For vectorization to maintain synchrony, a considerable number of additional instructions must be added. The timestep should be large enough to reduce the overhead involved with restarting each iteration, but the longer the timestep the more random will be the particle behavior, creating serious bookkeeping problems. On the other hand, one cannot simply choose very short time steps, to have the electrons in the ensemble behave in a more uniform fashion, with any free flight generation strategy, because the overhead introduced may just wipe out any vectorization advantage. A reasonable trade-off must be achieved with a Monte Carlo algorithm which is efficiently implemented with short timesteps.

In the case of a steady-state bulk simulation, synchrony is not an important requirement if steady-state properties of the ensemble are investigated. It has been shown [10] that one could obtain accurate estimates just from the knowledge of the ensemble properties *before scattering*. In this *event-based* approach [11], one can follow the ensemble between two different scattering events, treating the particles simultaneously during free-flights and the scatterings. The situation is shown in Fig. 5. A bulk calculation can be simply limited to momentum space, since the actual real space is assumed to be infinite and uniform. Obviously, for long enough simulations, the ensemble ends up being completely asynchronous. As mentioned above, the total scattering rates may differ by an order of magnitude in different energy ranges, leading to extremely dissimilar average times of flight. In addition, the distribution function of the simulated particles will differ widely when different electric fields are applied.

The event-based Monte Carlo algorithm has been successfully applied to a number of transport problems in physics, notably neutron and photon transport. A review is found in [11]. Of course, it can also be applied to situations where other events, besides scattering, are involved. These events may be the interaction with the boundaries or generation-recombination processes.

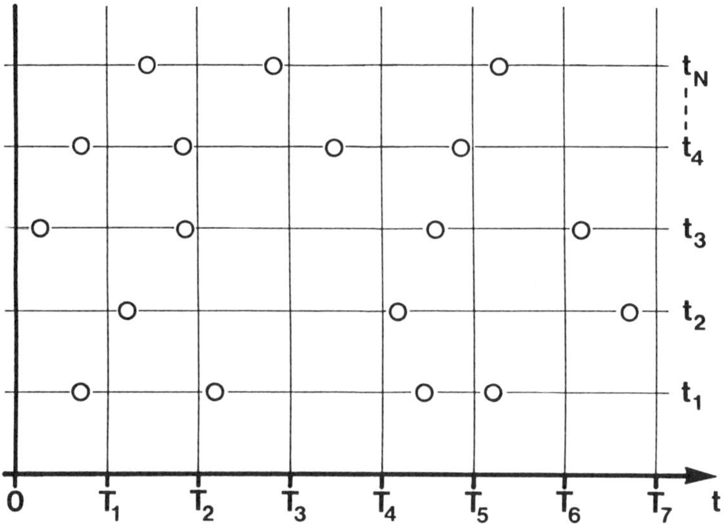

Fig. 4 - Time evolution of free flights for an ensemble of particles, simulated using a *history based* Monte Carlo procedure. The circles indicate scattering events. The procedure is suitable for self-consistent treatment, since the ensemble is synchronous at the time steps indicated.

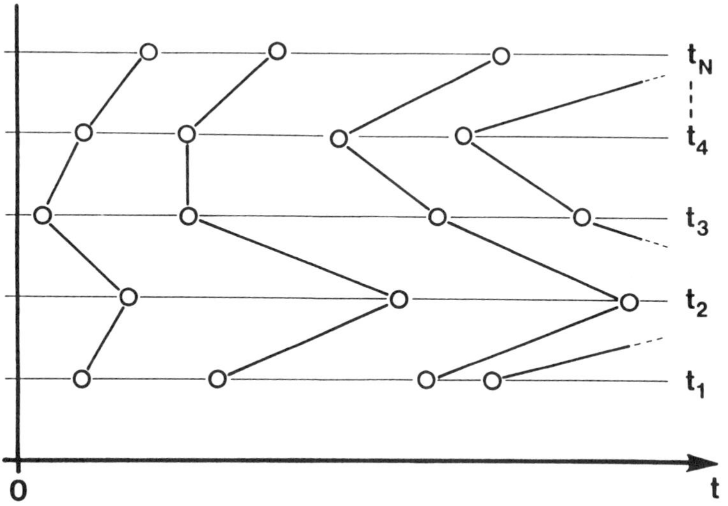

Fig. 5 - Time evolution of free flights for an ensemble of particles, simulated using an *event based* Monte Carlo procedure. The circles indicate scattering events. The procedure is not suitable for self-consistent treatment, since the ensemble is kept asynchronous.

In a broad sense, an event is any occurrence which terminates a free electron flight as such.

The event based algorithm retains some limitations of the one-particle Monte Carlo. Although the particles of the ensemble are studied at the same time, the history of the particles are totally uncorrelated in the simulation because they cannot be synchronous. The main conclusion is that, although this technique can be very efficient for some specific purpose, it is not the technique of choice for more complicated simulations which involve carrier-carrier interaction or need a self-consistent treatment. Therefore, an event-based Monte Carlo approach appears to be very efficient only for the investigation of steady-state bulk semiconductors or device structures with a predetermined potential (non-self-consistent).

The event-based procedure could be modified to yield a synchronous ensemble with some code additions to perform checks on the particles and stop the flights at given simulation times. Besides being cumbersome, the efficiency may degrade rapidly when one wants to apply the algorithm to a highly non-homogeneous situation, where particles may have quite different histories.

Synchrony of the ensemble is always necessary when studying time-dependent phenomena or when a self-consistent simulation solves Poisson's equation periodically. As mentioned before, the interval between consecutive solutions of Poisson's equation should not be larger than $10^{-14}s$ in typical semiconductors, to avoid the onset of spurious plasma oscillations in the ensemble. This is a major constraint which must be considered when a Monte Carlo simulation algorithm is designed. In the development of Monte Carlo codes, one usually starts from a bulk simulation, to builds on it a device, by introducing geometry data and real space evolution of the particles, and then charge density evaluation on mesh points and Poisson's equation for a self-consistent treatment. We see the need for a vectorization approach which can perform well in all situations and eventually perform best in the most complicated device simulation. Such a vectorization approach should be optimal in terms of self-scattering reduction and maintain the synchrony of the particle ensemble (i.e. make available a synchronous ensemble at prefixed times). For an efficient vectorization/parallelization and for future code development, one has to settle on a technique in which the calculation steps flow in a naturally parallel fashion so that the intrinsic randomness in the process can be represented with simplicity.

The Constant Γ technique is suitable for vectorization of an event-based algorithm, for instance, because the times of flight are always obtained in a one-step calculation for all particles. The overwhelming presence of self-scattering events in most practical situations, however, remains a drawback for the overall efficiency of the procedure. The free-flight time generation techniques mentioned earlier, may present various advantages or disadvantages when vectorization is attempted for use with the event- or history-based algorithms. The piecewise Γ technique is probably quite suitable for an efficient event-based vectorization, since the majority of the self-scattering can be eliminated and relatively few extra computations are needed to check when a particle moves to an energy range with different Γ during a flight. The iterative Γ technique can be vectorized efficiently if the procedure is calibrated so that an acceptable time of flight is selected at the first or second iteration for most particles. A good degree of experimentation is necessary to tune up such a vectorized procedure.

The application of the Constant Time technique to the event-based algorithm is somewhat difficult, since one has to split the actual flight into many sub-flights, whose number is not known *a priori*. However, with simple modifications the Constant Time algorithm is suited for an efficient vectorized synchronous treatment of the ensemble, as shown in the next section.

THE ENSEMBLE CONSTANT TIME TECHNIQUE

The Constant Time technique for the determination of the times of flight, as introduced by Yorston [6], is quite competitive with all the other self-scattering reduction techniques. With simple modifications, it becomes suitable for the vectorization of an Ensemble Monte Carlo. In the original formulation, the subdivision of the flight into sub-flights of equal duration is restarted immediately after a real scattering, since the new constant Γ level can be changed. This cannot be done after a self-scattering, in which case the new flight is started by completing the last subflight interrupted by the scattering event, keeping the previous value of Γ. To maintain synchrony in an ensemble Monte Carlo, it is better to use one single time reference for all the particles. The simulation proceeds as follows:

a) Initially, a synchronous ensemble of particles is considered. A random number $r(i)$ is calculated for each particle i, to obtain the quantities $\beta(i) = -\ln r(i)$ necessary to determine the random free-flight times $t(i)$.

b) The particle trajectories in phase space are calculated for a subflight of length τ.

c) The known values of the momentum at the beginning and at the end of the subflight allow to calculate the maximum of the scattering rate $\lambda(k(t))$ experienced by the particles. These maximum scattering values $\Gamma(i)$ are considered to be the constant Γ for the particles during the subflight. The $\beta(i)$ values are updated

$$\beta(i)_{new} = \beta(i)_{old} - \Gamma(i)\,\tau. \tag{9}$$

d) The results obtained at step b) are checked to detect negative $\beta(i)$ values. When this occurs, the free-flight should terminate before the end of the timestep τ just considered.

e) The last portion of the time of flight for the particles which must scatter is calculated from

$$\tau^*(i) = \frac{\beta(i)_{old}}{\Gamma(i)} \tag{10}$$

and the particle trajectories are recalculated up to the exact scattering point. The scattering type is randomly selected and the final state of the particles is determined.

f) The quantities $\beta(i) = -\ln r(i)$ are recalculated for the scattered particles and trajectories are generated for free-flights of length $\tau(i)_{start} = \tau - \tau^*(i)$, necessary to make the scattered particles synchronous with the ensemble again.

g) The maximum scattering rates $\Gamma(i)_{start}$ for the partial subflights are determined and the update

$$\beta(i)_{new} = \beta(i)_{old} - \Gamma(i)_{start}\, \tau(i)_{start}. \tag{11}$$

is performed for the particles restarted after scattering.

h) If $\beta(i) < 0$, the steps d) to g) are repeated to accomodate additional scattering events, until the end of the timestep is reached by all the particles. The iteration is now complete and the procedure can be repeated starting from b). For a self-consistent simulation, Poisson's equation is solved before b) at predetermined iterations.

In device simulations, any other event which interrupts or alters a free-flight (e.g. absorption by a contact, crossing of internal boundaries, reflection at a non-contact boundary) is treated similarly to a scattering event. When very small timesteps are used, the algorithm can be considerably simplified by assuming that only one event of any kind occurs during the timestep τ. The size of τ is usually so small, in order to reduce the self-scattering efficiently, that the occurrence of two scatterings during the same sub-flight is rare. One scattering and one boundary interaction may occasionally occur. These situations may be handled with simple code modifications or may even be forbidden by having the second event occur right at the beginning of the next subflight, if this is physically acceptable.

The algorithm presented is well optimized for semiconductor modeling besides being suitable for vectorization. This particular formulation with very short timesteps makes the technique efficient and at the same time it allows the vast majority of particles to move from one timestep to the following one without scattering. These calculations can be easily vectorized. The particles which scatter within the timestep can be grouped according to the various events and the calculations at steps e), f), g) and h) may be vectorized to some degree as well. Obviously, if the number of particles associated to a given event is much smaller than the vector length, vectorized operations cannot be very efficient.

The use of a very short timestep τ, on the order of $10^{-15}\ s$ for typical semiconductors, is very advantageous to simplify the implementation of the Monte Carlo algorithm for device simulation. Since the maximum possible velocity is $v_{\max} \approx 10^8\ cm/s$ in such materials, a particle can travel at most a distance $\Delta x = v_{\max}\tau = 10$ Å during the timestep. The average distance travelled will be considerably lower, since only very few particles in a device will be able to approach such a velocity. The size of the mesh used to discretize the device domain will be in general larger. Under these conditions, the bookkeping related to trajectory calculation and interaction with boundaries can be simplified by using less rigorous algorithms without loss of accuracy. Of course, it is advantageous to increase τ to reduce the overall CPU time, as long as the self-scattering is tolerable and the possibility of multiple events within the timestep is properly taken care of.

COMPUTER EXPERIMENTS

As an example, we present here computer benchmarks for ensemble Monte Carlo calculations using the Constant Γ and the Constant Time Technique, obtained with a vectorized code simulating electron transport in III-V semiconductor compounds. The model includes three non-parabolic valleys in the conduction band and includes all the important scattering rates, along the lines of the work by Littlejohn *et al.* [12]. Codes have been developed for the simulation of bulk material, 1-D and 2-D devices. According to our experience, the intrinsic simplicity of the Constant Time Technique makes code development easier and helps in the design of a modular structure. We report here typical results for simulations on a CRAY X-MP/48 and a CRAY Y-MP4/464. The code was written in standard FORTRAN 77, to be completely portable to a UNIX environment, therefore any machine specific programming was avoided. In particular, the gather-scatter procedure, necessary to operate vectorially on a random subset of the particle ensemble, subject to events during the timestep, was implemented using solely indirect addressing. This assures portability, but the efficiency of the calculations may vary with different compilers. The vectorization efficiency for loops containing IF statements and IF ... THEN constructs is also usually dependent on the compiler. We used the internal random number generator of the CRAY supercomputers, which allows full vectorization. Absolute performance evaluations are quite difficult on a complex vector machine, and various performance indicators should be considered to get a complete picture. As a general reference, the maximum vectorization speed-up achieved on fully vectorized DO loops is approximately 10 for the supercomputers used.

In order to isolate the behavior of the Monte Carlo algorithm itself, we performed simulations with the code designed for self-consistent operation, but in conditions of bulk transport with a uniform field. Comparisons were made between identical ensemble simulations using the Constant Γ and the Constant Time techniques in scalar conditions, by inhibiting vectorization. Rrepresentative results obtained for GaAs at room temperature are shown in Tables I and II for the CRAY X-MP, and on Tables III and IV for the CRAY Y-MP. Here, ECTT indicates the vectorized code implementing the Ensemble Constant Time technique and CGAM indicates the code implementing the Constant Γ technique. The experiments included 5,000 electrons and were carried on for a total simulation time of $5\times10^{-12}\,s$. This was done starting from an initial condition already at steady-state for the electric field used, to avoid the influence of transients. The Ensemble Constant Time technique performs consistently better. One should note that, since here a realistic semiconductor was simulated and very short timesteps were necessary, the Constant Γ technique performs better than one would have expected considering the results in [6], when compared to the Constant Time Technique, both in scalar and in vector mode. The code ECTT was run in scalar mode *as is* without changes to optimize the scalar performance, therefore the comparisons presented are quite conservative. It also appears that in [6] the experiment was carried on by generating a given number of flights for a single particle, without the restrictive situation of dealing with an ensemble within a selfconsitent framework. This allows one to increase the constant time τ arbitrarily (Yorston [6] used $\tau = 10^{-13}$) and achieve shorter running times. In addition, when comparing vector and scalar performance, one should notice that when the code is run on the CRAY supercomputers with inhibition of the vectorization, the computer does not operate as a purely scalar machine but tries to take advantage of the extra hardware available whenever possible. The exact influence of this scalar optimization is extremely difficult to determine, and one should take the vectorization speed-up obtained as slightly underestimated.

The analysis of the computer experiments was performed using the HPM timing tool available on both the CRAY X-MP and the CRAY Y-MP. The option "novector" was invoked during compilation to run in scalar mode the codes. In all cases, scalar optimization was allowed. The results show a vectorization speed-up between 3 and 4.5 at a computation speed between 30 and 40 Mflops on the CRAY X-MP for the ECTT code, as shown in Table I. The results obtained on the CRAY Y-MP, given Table III, not only show a higher Mflop rate between 48 and 66, expected because of the faster clock, but also show a somewhat better vectorization speed up, probably due to both architecture and compiler improvements. It can be noticed that when the timestep is doubled from 10^{-15} to 2×10^{-15}, the performance in terms of Mflops declines somewhat and the duration of the computer experiment is reduced by roughly 20-30%, not 50%. This is due to the fact that while the number of actual timesteps is halved, the number of scatterings occurring within each timestep increases. Since scattering calculations are based on indirect addressing operations with lower vectorization speed-up, one lowers the ratio between instructions with high and low speed-up. Also, at higher electric field a slightly lower vectorization speed-up is recorded because more scattering events occur.

Benchmark results for the CGAM code are shown in Tables II and IV. The required CPU time is normally about double, with respect to the same simulations run with the ECTT code. In scalar mode, a slightly better performance is comparatively observed on the CRAY Y-MP, although the vectorization speed-up is not substantially altered.

TABLE I

Results of computer experiments for the code ECTT in vector and scalar mode, on a CRAY X-MP/48 supercomputer. An ensemble of 5,000 electrons in bulk GaAs is simulated for $5\times10^{-12}\ s$, at an electric field of $1kV/cm$ and $5kV/cm$.

		ECTT	vector	ECTT	scalar	
Field (kV/cm)	Timestep (s)	CPU time (s)	Mflops	CPU time (s)	Mflops	Speed up
1.0	10^{-15}	33.22	41.90	147.37	8.6	4.44
1.0	2×10^{-15}	21.67	37.05	80.53	8.92	3.71
5.0	10^{-15}	37.24	37.01	139.81	8.98	3.75
5.0	2×10^{-15}	25.47	30.84	81.33	8.97	3.19

TABLE II

Results of computer experiments for the code CGAM in vector and scalar mode, on a CRAY X-MP/48 supercomputer. An ensemble of 5,000 electrons in bulk GaAs is simulated for $5\times10^{-12}\ s$, at an electric field of $1kV/cm$ and $5kV/cm$.

		CGAM	vector	CGAM	scalar	
Field (kV/cm)	Timestep (s)	CPU time (s)	Mflops	CPU time (s)	Mflops	Speed up
1.0	10^{-15}	72.03	26.6	287.1	7.45	3.99
1.0	2×10^{-15}	60.81	21.49	160.41	7.73	2.64
5.0	10^{-15}	75.11	26.72	290.72	7.59	3.87
5.0	2×10^{-15}	61.3	21.59	162.36	7.78	2.65

TABLE III

Results of computer experiments for the code ECTT in vector and scalar mode, on a CRAY Y-MP4/464 supercomputer. An ensemble of 5,000 electrons in bulk GaAs is simulated for $5\times10^{-12}\ s$, at an electric field of $1kV/cm$ and $5kV/cm$.

		ECTT	**vector**	**ECTT**	**scalar**	
Field (kV/cm)	Timestep (s)	CPU time (s)	Mflops	CPU time (s)	Mflops	Speed up
1.0	10^{-15}	20.42	66.47	105.9	11.69	5.18
1.0	2×10^{-15}	16.01	48.62	61.95	11.54	3.87
5.0	10^{-15}	20.75	65.97	106.84	11.73	5.14
5.0	2×10^{-15}	16.37	48.54	62.60	11.60	3.82

TABLE IV

Results of computer experiments for the code CGAM in vector and scalar mode, on a CRAY Y-MP4/464 supercomputer. An ensemble of 5,000 electrons in bulk GaAs is simulated for $5\times10^{-12}\ s$, at an electric field of $1kV/cm$ and $5kV/cm$.

		CGAM	**vector**	**CGAM**	**scalar**	
Field (kV/cm)	Timestep (s)	CPU time (s)	Mflops	CPU time (s)	Mflops	Speed up
1.0	10^{-15}	43.10	42.23	160.52	10.95	3.72
1.0	2×10^{-15}	36.90	34.03	109.68	10.54	2.97
5.0	10^{-15}	44.12	42.90	162.33	11.01	3.68
5.0	2×10^{-15}	37.74	37.29	111.03	10.71	2.94

A NOTE ON PARALLELIZATION

The efficient implementation of Monte Carlo algorithms on parallel supercomputers depends both on the type of simulation and on the architecture involved. For multitasking applications on supercomputers with several vector processors, the first priority is vectorization, to take fully advantage of the architectures. The techniques described in the previous sections can be adapted easily, with the vector computations split equally between the various processors. Problems will obviously arise when short vectors are treated, since the vector capability of the single processors may not be fully exploited.

Computations on highly parallel computers (Hypercube, Connection Machine) present quite different issues, due to Amdahl's law, which prescribes a very high degree of parallelization for efficient computations on a large number of processors. Bulk simulations, which do not require self-consistency, perform well, as reported for instance in [13], where a 64-processor Hypercube was used. The main problem that is encountered in simulations requiring absolute synchrony, is the cross-communication between processors. In principle, a number of particles can be assigned to each processor, but while the bulk of the computations can proceed in perfect parallelism, the necessity to share information at given time intervals can create serious bottlenecks. A degradation in the parallelism of just a few percents may have enormous repercussions on the overall performance. The development of hardware and algorithms for highly parallel computers is still in the early stages of development, and future improvements may circumvent the current limitations.

CONCLUSION

Vectorized procedures for Monte Carlo simulation of semiconductors on supercomputers have been examined. Benchmark calculations have been performed on the CRAY X-MP and the CRAY Y-MP supercomputers for code structures suitable for self-consistent device simulation, using two different free-flight time generation approaches. The results show that computation rates up to 60 Mflops and vectorization speed up between 4 and 5 can be achieved in realistic conditions. This is obtained with a fully transportable code which does not use any machine-specific utilities or constructs, with exception of the internal totally vectorized random number generator. It is expected that the basic algorithm developed with the Ensemble Constant Time Technique can be extended effectively to multitasking applications.

ACKNOWLEDGEMENTS

This work was supported by the National Center for Computational Electronics (NCCE) of the University of Illinois and by the National Science Foundation. The computations were performed on the CRAY X-MP/48 and the CRAY Y-MP4/464 of the National Center for Supercomputing Applications (NCSA) of the University of Illinois.

REFERENCES

[1] K. Hess, *Advanced theory of Semiconductor Devices* (Prentice Hall, Englewood Cliffs, NJ, 1988).

[2] E.M. Conwell, "High Field Transport in Semiconductors", *Solid State Physics*, suppl. 9 (Academic Press, New York, 1967).

[3] H.G. Reik and H. Risken, *Phys. Rev.* **126**, p. 1737 (1962).

[4] R.W. Hockney and J.W. Eastwood, *Computer Simulation Using Particles* (McGraw-Hill, New York, 1981).

[5] W. Fawcett, A.D. Boardman, and S. Swain, *J. Phys. Chem. Solids* **31**, p. 1963 (1970).

[6] R.M. Yorston, *J. Comp. Phys.* **64**, 177 (1986).

[7] H.D. Rees, *Phys. Lett. A* **26**, 416 (1968).

[8] V. Borsari and C. Jacoboni, *Phys. Status Solidi B* **54**, 649 (1972).

[9] M. Cheng and E.E. Kunhardt, *J. Appl. Phys.* **63**, 2322 (1988).

[10] C. Jacoboni and L. Reggiani, *Reviews of Mod. Phys.* **55**, p. 645 (1983).

[11] W.R. Martin and F.B. Brown, *Int. J. of Supercomp. Applic.* **1**, p. 11 (1987).

[12] M.A. Littlejohn, J.R. Hauser and T.H. Glisson, *J. Appl. Phys.* **48**, 4587 (1977).

[13] U.A. Ranawake, P. Lenders, and S.M. Goodnick, in *Computational Electronics*, K. Hess, J.P. Leburton and U. Ravaioli eds., p. 137 (Kluwer Academic, Norwell, Mass., 1990).

10

FULL BAND MONTE CARLO PROGRAM FOR ELECTRONS IN SILICON

H. Shichijo[1], J.Y. Tang[2], J. Bude and D. Yoder

Beckman Institute for Advanced Science and Technology and ECE

University of Illinois at Urbana-Champaign, Illinois 61801

This program has been developed by J.Y. Tang from the first version of a full band Monte Carlo program written by H. Shichijo. The program has then been extensively commented, structured, and improved by J. Bude and D. Yoder. All persons involved have performed their work during graduate studies with K. Hess.

This version of the program represents a compromise between the most recent and most correct developments. Large numerical codes such as this can only be created in an evolutionary way in a University environment. At the point when one is finally sure that all "bugs" have been removed over the course of several Ph.D. theses, one is also sure that the physical content and engineering content are obsolete. (For example, what is called X-L scattering in this code is actually treated as a randomizing scattering mechanism.) Meanwhile, one has developed other programs with better physics and more efficient numerics. The program presented here is reasonably modern by the standards of this book, and hopefully mostly free of errors.

To make use of this program, input files are necessary. These input files contain mainly the band structure, i.e. the E(k) relation, as calculated by the empirical pseudopotential method. Should the reader wish to obtain these files, it may be possible to transfer them (and the program) by e-mail after a request letter is written to the editor of this book.

We would like to acknowledge the help which we have received from Prof. M. Littlejohn during the work on the first version (Shichijo) of the full band Monte Carlo program. Prof. Littlejohn has generously provided his expertise and time to teach us about important basic elements of Monte Carlo simulation. All of the contributors to the program have received financial support from the Office of Naval Research and the Army Research Office, which is gratefully acknowledged.

1. Texas Instruments, Central Research, Dallas, TX.
2. 1315 Miraveille Ave, Los Altos, CA, 94024.

```
c*****************************************************************************************
c
c       PROGRAM SIMC
c       June 15, 1988
c       June 15, 1990
c
c       RECENT AUTHORS:
c       Jeff Bude  - University of Illinois, Urbana-Champaign
c                  - Semiconductor Physics Group
c       Doug Yoder - University of Illinois, Urbana-Champaign
c                  - Semiconductor Physics Group
c
c       PROGRAM DESCRIPTION:
c       This program performs a Monte Carlo Simulation for Bulk Silicon using 2-band
c       local empirical pseudopotential band structure and first order scattering rates.
c       The scattering mechanisms included in this simulation are acoustic intravalley
c       absorption and emission, optical intervalley absorption and emission and
c       impact ionization.  Scattering rates are calculated for a two valley (X,L)
c       non-parabolic analytic band structure model.  Total scattering rates are
c       normalized to the full density of states at high energy, and the electron
c       drifts are calculated according to the pseudopotential band structure.
c
c       KNOWN LIMITATIONS:
c       X->L scattering rates used for L->X scatterings, and mechanism
c       is treated as randomizing.
c
c
c       INPUT FILES:  DATA, EK_DER2.DAT, EORD.SI
c       OUTPUT FILES: SIMC.OUT
c
c*****************************************************************************************
c
        IMPLICIT DOUBLE PRECISION (A-H)
        IMPLICIT DOUBLE PRECISION (O-Z)
        DOUBLE PRECISION KBZ(3),KRZ(3),R(3),EXT(2,0:41,0:41,0:41)
     2      ,DEXT(3,2,0:41,0:41,0:41), DEN(3), KPOINT(3), KB, KT
     3      ,MTX, MLX, MTL, MLL, M0, KORD(3,5584), LATTICE, K0(3)
     4      ,INTCP(499), KTMP(3), KTEST(3), ETEST
        DIMENSION RATE(24,500), TPHXX(3,2), TPHXL(4), XXIVD(3,2)
     2      ,XLIVD(4), Z(3), EORD(5584), EPH(22), DV(3), SLOPE(499)
     3      ,ELOSSRATE(23), NE(23), TRATE(500), DIR(3)
     4      ,POS(3), POLD(3), ELS(23), EDR1(4), EDR2(4)
        INTEGER EVENT, KPER(3), BAND(5584), IDV(3)
        INTEGER ICOUNT
        DOUBLE PRECISION DSEED
        COMMON/BAND/EXT, DEXT
        COMMON/BAND2/EORD, KORD, BAND
        COMMON/CONSTANT1/KB, PI, T, HBAR, RHO, VS, Q, ES
        COMMON/CONSTANT2/LATTICE, MTX, MLX, MTL, MLL, M0, ACD, XXIVD
     2      ,XLIVD, ALPHA, Z, P, ETH, XLDIFF, TPHXX, TPHXL
        COMMON/CONSTANT3/EPH, SRPA
        COMMON/CONSTANT4/TSTEP, PMFDT, FIELD, DV, NTOT, NLMT, NLOOK
        COMMON/DISSIPATION/SEAC1, SEAC2
        COMMON/KVECTOR/KBZ, KPER, INDEX
        COMMON/RANDOM/DSEED
        COMMON/II/RTOT
        DOUBLE PRECISION C(4),A,B,Y,EV
        DSEED=123457.D0


        DATA K0, KB, PI, HBAR, RHO, VS, M0, Q/0.85, 0.0, 0.0, 8.61573E-5
     2      ,3.1415926, 6.5821829E-16 ,2.329, 9.04E5, 9.109E-31
     3      ,1.6021E-19/
        DATA Z/4.0, 1.0, 6.0/,XLDIFF/1.0/

c       Units:
c              HBAR:  eV sec
c              KB  :  eV / Kelvin
c*****************************************************************************************
c
c       VARIABLE DESCRIPTIONS:
c
c       KBZ - current K-vector.  Value is return in 1st Brilluoin zone after
c             subroutine ENERGY is called.
c       EN - current energy of the electron
c       TSTEP - coarse time step in seconds
c       DTIME - multiplies TSTEP to give a drift time consistent with
c               the electon's energy
c       RATE - 23 by 500 array, lookup table for scattering rates.  For first
```

```
c             index designates the scattering mechanism, the second, carrier
c             energy in 0.01 eV.  RATE(23,I) holds the total scattering rate
c             for all mechanisms.  See subroutine TAU for complete description.
c       POS - current position in real space
c       INDEX - current band index
c       DEN - gradient of E(k) for current value of EN
c       K0 - initial K vector
c       SLOPE, INTCP - slope and intercept values used to interpolate the
c             total scattering rate for energies between two known energies
c       TRATE - total scattering rate normalized by the coarse time step
c       DELK - the change in electron wave-vector k per coarse time step,
c             normalized to 2*pi/a where a is the lattice constant.
c
c*******************************************************************************
c
c
c       Read in all necessary data for the simulation.
        CALL INDAT
c
c       Calculate the scattering rates.
        CALL TAU(RATE,TRATE)

c       Multiply the total rate by the time step TSTEP.
c       Calculate slope/intercept for 'trate.'
        DO 20 I=1,499
           SLOPE(I)=(TRATE(I+1)-TRATE(I))/0.01
           INTCP(I)=TRATE(I)-SLOPE(I)*0.01*DBLE(I)
 20     CONTINUE

c       Calculate directional cosines.
        DMAG=SQRT(DV(1)*DV(1)+DV(2)*DV(2)+DV(3)*DV(3))
        DO 30 I=1,3
           DIR(I)=DV(I)/DMAG
 30     CONTINUE

c       Assign DELK once, since electric field is uniform.
        DELK=(LATTICE/(2.0*PI))*(FIELD/HBAR)*TSTEP

c       Initialize accumulators
        SDED=0.0                     {sum of delta E between drifts
        SVEL=0.0                     {sum of velocity in field direction
        svelfsq=0.                   {sum of the sqare of the vel. in fld dir.
        svelpsq=0.                   {sum of the sqare of the vel. in perpendicular dir.
        sev=0.                       {sum of velocity in field direction times energy
        EAV=0.0                      {average electron energy
        SCL=0.0                      {total distance electron travels along its path
        STIME=0.0                    {total time of simulation (in units of TSTEP)
        STIME2=0.0                   {total time spent in 2nd band (in units of TSTEP)
        SDFD=0.0                     {total distance traveled in field direction
        SEAC1=0.0                    {total acoustic phonon energy, abs. (intravalley)
        SEAC2=0.0                    {total acoustic phonon energy, emi. (intravalley)
        SUMRATEII=0.                 {weighted sum of imp. ion. rate, used for alpha estimator.
        NIMP=0                       {number of impact ionizations
        NBAL=0                       {number of ballistic electrons for imp. ion.
        NSC=0                        {number of scatterings
        IBAL=0                       {flag for ballistic electrons
c
c
c       Zero the energy loss and loss rate variables.
        DO 50 I=1,23
           NE(I)=0                   {number of each type of scattering
           ELOSSRATE(I)=0.0          {eneryy loss rate for each type of scattering (eV/sec)
           ELS(I)=0.0                {energy loss due to each mechanism (eV)
 50     CONTINUE
c
c       Start the electron at an X-minima at real space position (0,0,0).
 2000   DO 60 I=1,3
           KBZ(I)=K0(I)              {start electron at K0
           POS(I)=0.0                {start electron at (0,0,0)
 60     CONTINUE
        INDEX=1                      {start electron in band 1
c
c       Find electron energy corresponding to this k-vector, and dE/dk.
        CALL ENERGY(EN,DEN)
c
c       Start simulation.
c       Determine the electron drift time and find the K state
c       after drifting.  Then, determine the scattering mechanism,
c       the final K state and return to line 1000.  This sequence
```

```
c       is then repeated until the desired accuracy is achieved.
c
c       Keep track of electron energy before a short drift.
 1000   EOLD=EN
c
c       Keep track of electron position before a short drift.
        DO 70 I=1,3
           POLD(I)=POS(I)
 70     CONTINUE

c       Fine adjust time step to some percent of the mean free drift time.
        if(en.le.0.1)then
           dtime=1.0
        else if((en.gt.0.1).and.(en.le.0.6))then
           dtime=0.2
        else if((en.gt.0.6).and.(en.le.1.0))then
           dtime=0.1
        else if((en.gt.1.0).and.(en.lt.1.5))then
           dtime=0.05
        else if(en.gt.1.5)then
           dtime=0.01
        endif

c       The electron will drift for time DTIME, so update total simulation time.
        STIME=STIME+DTIME
c
c       Update electron k-vector. (electrons treated as positively charged)
        DO 80 I=1,3
           KBZ(I)=KBZ(I)+DELK*DTIME*DIR(I)
 80     CONTINUE
c
c       Find the new electron energy, after having been allowed to drift DTIME.
        CALL ENERGY(EN,DEN)
c
c       Zero the square of the total distance of travel during DTIME.
        SCLSQ=0.0
c
c       Zero VEL, the drift velocity of the electron during drift time DTIME,
c       normalized to 1/hbar and DTIME.
        VEL=0.0
c
c       Update electron position, SCLSQ, and VEL.
        DO 90 I=1,3
           POS(I)=POS(I)+DEN(I)*DTIME*TSTEP/HBAR
           SCLSQ=SCLSQ+(POLD(I)-POS(I))**2
           VEL=VEL+DIR(I)*DEN(I)
 90     CONTINUE
c
c       Update the weighted sum of drift velocities (derivative estimator).
        svel=svel+vel*dtime/hbar
c
c       Update the weighted sum of the squared velocity in the field direction.
        svelfsq=svelfsq+(vel/hbar)**2 * dtime
c
c       Update the weighted sum of the squared velocity in the perpendicular direction.
c       Note that this will need to be modified for fields not along the [0,1,0] dir.
        svelpsq=svelpsq+(den(2)/hbar)**2 * dtime
c
c       Update the weighted sum of energy * velocity in field direction.
        sev=sev+vel/hbar*0.5*(EOLD+EN)*DTIME
c
c       Update the total distance the electron travels along its path.
        SCL=SCL+SQRT(SCLSQ)
c
c       Update the time spent in second band, when appropriate.
        IF (INDEX .EQ. 2) THEN
           STIME2=STIME2+DTIME
        ENDIF

c       Update the average energy estimator.
        EAV=EAV+0.5*(EOLD+EN)*DTIME
c
c       Update the sum of energy changes between drifts.
        SDED=SDED+EN-EOLD
c
c       Interpolate to find the scattering rate SRATE.
        MARK=IDINT(EN*100)
        IF (MARK .LT. 1) MARK=1
        IF (MARK .GT. 499) MARK=499
```

```
      SRATE=SLOPE(MARK)*EN+INTCP(MARK)

c     Calculate necessary constants for alpha estimator (Bude formula)
      ETH=1.12
      B=(5E+05-FIELD)/5E+05
      A=FIELD/5E+05
      C(1)=0.2145*B+0.2427*A
      C(2)=-3.6302*B-3.6593*A
      C(3)=21.5126*B+19.2716*A
      C(4)=-12.5451*B-3.9533*A

c     If above impact ionization threshold energy, multiply impact
c     ionization rate by normalized time of drift, to add to SUMRATEII
c     which is equivalent to the number of impact ionization events
c     in time STIME*TSTEP.
      IF (EN .GT. ETH) THEN
         EV=EN+1.12
         Y=C(1)*EV*EV*EV+C(2)*EV*EV+C(3)*EV+C(4)
         SUMRATEII=SUMRATEII+EXP(Y)*DTIME
      ENDIF

c     Continuous energy dissipation estimators for energy loss rates.
c     X-X scatterings
      DO 100 IEVENT=9,14
         ELOSSRATE(IEVENT-8)=ELOSSRATE(IEVENT-8)+DTIME*EPH(IEVENT)
     2   *(RATE(IEVENT,MARK)-RATE(IEVENT-6,MARK))
 100  CONTINUE

c     X-L scatterings

      DO 101 IEVENT=19,22
         ELOSSRATE(IEVENT-12)=ELOSSRATE(IEVENT-12)+DTIME*EPH(IEVENT)
     2   *(RATE(IEVENT,MARK)-RATE(IEVENT-4,MARK))
 101  CONTINUE
c
      CALL RNUM(DSEED,1,RN)
c
c     Check for scattering event.  If no scatter, go to line 1000.
      IF (RN.LE.1-SRATE*DTIME) GOTO 1000
c
c     Electron scattered.  Choose scattering mechanism.
c
 105  CALL RNUM(DSEED,1,RN)
      DO 110 J=1,23
          IF (RATE(J,MARK)-RN) 110,120,120
 110  CONTINUE
      GOTO 130
c
c     Assign the number of the scattering mechanism.
 120  EVENT=J
c
c     If acoustic phonon emission is choosen and the electron doesn't
c     have enough energy to emit a phonon, then try again.
 130  IF (EVENT .EQ. 2 .AND. EN .LT. ES) GOTO 105

      EOLD = EN

c     Update number of scatterings and event counter.
      NE(EVENT)=NE(EVENT)+1
      NSC=NSC+1
c
      IF (EVENT .NE. 23) THEN
c        The scattering mechanism is not impact ionization.
c
c        Limit acoustic phonon wavevector to .3*tpiba
         IF (EVENT.LE.2) THEN
c
c           The next eight lines simply keep track of variables which should not
c           be modified by subroutine FINALK() if the chosen scattering
c           is acoustic intravalley, and the phonon wave-vector is too large.
            TS1=SEAC1
            TS2=SEAC2
            INTEMP=INDEX
            ETEST=EN
            ICOUNT=0
            DO 132 I=1,3
               KTEST(I)=KBZ(I)
 132        CONTINUE
 135        CALL FINALK(EVENT,EN)
```

```
c             Calculate the magnitude squared of the phonon wave-vector
              DK2=(KTEST(1)-KBZ(1))**2+(KTEST(2)-KBZ(2))**2+(KTEST(3)
     &             -KBZ(3))**2
c
c             If the wave-vector is too large, then try again up to 100 times.
              IF (DK2.GT.0.09) THEN
                 DO 137 I=1,3
                    KBZ(I)=KTEST(I)
 137             CONTINUE
c
c                Re-assign all variables which should not have been updated.
                 SEAC1=TS1
                 SEAC2=TS2
                 INDEX=INTEMP
                 EN=ETEST
                 IF (ICOUNT.LT.100) THEN
                    ICOUNT=ICOUNT+1
                    GO TO 135
                 ELSE
                    NSC=NSC-1
                    NE(EVENT)=NE(EVENT)-1
                    GO TO 105
                 ENDIF
              ENDIF
           ELSE
c
c             The scattering mechanism is not acoustic intravalley.
c             Find a proper final state for the electron.
              CALL FINALK(EVENT,EN)
           ENDIF
c
c          Update the energy loss vector.
           ELS(EVENT)=ELS(EVENT)+(EOLD-EN)
           IBAL=1                                     !not ballistic
c
c          Find the electron energy corresponding to its new k-vector, and dE/dk.
           CALL ENERGY(EN,DEN)

        ELSE

c          The scattering mechanism is impact ionization.
c          Update the event counter.
           NIMP=NIMP+1
c
           IF(IBAL .EQ. 0) NBAL=NBAL+1
           IBAL=0
c
c          Update the sum of drift distance in the field direction.
           DFD=0.0
           DO 261 I=1,3
             DFD=DFD+DIR(I)*POS(I)
 261       CONTINUE
           SDFD=SDFD+DFD

c          It is assumed that the electron losses all its energy in
c          impact ionization, and it is returned to an energy minimum.

           ELS(23)=ELS(23)+EN
        ENDIF
c
c       Print out the statistics every NLOOK scattering events.
        IF (INT(NSC/NLOOK)*NLOOK .EQ. NSC) goto 4000

        IF (NSC .GE. NTOT) GOTO 4000

 226    IF(EVENT .NE. 23) THEN
           GOTO 1000
        ELSE
           GOTO 2000
        ENDIF

cccccccccccccccccccccccccccccccccccccccccccccccccccccccccccccccccccccccc


c       Calculate output quantities.

 4000   CONTINUE
```

```
c       OUTPUT VARIABLE LIST

c       ATII  - Average time between impact ionizations
c       ATSC  - Average time between scattering events
c       ASCL  - Mean free path (average scattering length)
c       ASLFD - Mean free path (field direction)
c       ALPHA - Impact ionization coefficient
c       ADSC  - Average distance between scatterings in field direction
c       APACA - Average phonon energy for acoustic absorption
c       APACE - Average phonon energy for acoustic emission
c       APXXA - Average phonon energy for X-X absorption
c       APXXE - Average phonon energy for X-X emission
c       APXLA - Average phonon energy for X-L absorption
c       APXLE - Average phonon energy for X-L emission
c       AEE   - Average electron energy
c       ELXXA - Average energy loss X-X absorption
c       ELXXE - Average energy loss X-X emission
c       ELXLA - Average energy loss X-L absorption
c       ELXLE - Average energy loss X-L emission
c       ELACA - Average energy loss acoustic absorption
c       ELACE - Average energy loss acoustic emission
c       ELII  - Average energy loss impact ionization
c       ERT   - Energy relaxation time
c       VD1   - 1st drift velocity estimator, Derivative
c       VD2   - 2nd drift velocity estimator, Energy Gain
c       VD3   - 3rd drift velocity estimator, Energy Dissipation
c       NACA  - Number of acoustic phonon absorptions
c       NACE  - Number of acoustic phonon emissions
c       NXXA  - Number of X-X intervalley absorptions
c       NXXE  - Number of X-X intervalley emissions
c       NXLA  - Number of X-L intervalley absorptions
c       NXLE  - Number of X-L intervalley emissions
c       NII   - Number of impact ionization
c       PXXA  - Percent X-X absorptions of total scatterings
c       PXXE  - Percent X-X emissions of total scatterings
c       PXLA  - Percent X-L absorptions of total scatterings
c       PXLE  - Percent X-L emissions of total scatterings
c       PACA  - Percent acoustic absorption of total scatterings
c       PACE  - Percent acoustic emissions of total scatterings
c       PTSB  - Percent time spent in second band
c       PII   - Percent impact ionization
c       EDR1(1:4) - 1st energy dissipation estimator; 1:acoustic, 2:X-X, 3:X-L, 4:II
c                 - based on direct scattering losses
c       EDR2(1:4) - 2nd energy dissipation estimator; 1:acoustic, 2:X-X, 3:X-L, 4:II
c                 - based on commulative itegration
c       TEDR  - total energy dissipation rate, sum of EDR1
c       sev  - sum of energy times velocity in the field direction
c       aev  - average of energy times velocity in the field direction

        FNSC=DBLE(NSC)
        FNIMP=DBLE(NIMP)
c
c       Calculate the average time between impact ionizations.
        IF(NIMP .NE. 0) THEN
            ATII=STIME*TSTEP*1.0E12/FNIMP           {psec
        ELSE
            ATII=0.0
        ENDIF
c
c       Calculate the average time between scattering events.
        ATSC=STIME*TSTEP*1.0E12/FNSC                {psec
c
c       Calculate the average scattering length
        ASCL=SCL/FNSC                               {cm
c
c       Calculate the average scattering length in direction of field
        DFD1=0.0
        DO 4100 I=1,3
          DFD1=DFD1+DIR(I)*POS(I)
 4100   CONTINUE
        ASLFD=(SDFD+DFD1)/FNSC                      {cm

c       Derivitive estimator for drift velocity
        VD1=SVEL/STIME

c       Energy gain estimator for drift velocity
        VD2=SDED/(STIME*TSTEP*FIELD)
c
c       Alpha estimator
```

```
      ALPHA=SUMRATEII/(STIME*VD1)
c
c     Average electron energy estimator.
      AEE=EAV/STIME

c     Hydrodynamic estimators
      aev=sev/stime              {average energy * velocity
      avfsq=svelfsq/stime        {average squared velocity in field direction
      avpsq=svelpsq/stime        {average squared velocity in ppd. direction

c     Time spent in second band.
      PTSB=100.0*STIME2/STIME
c
c     Set counters to zero
      NXXA=0
      NXXE=0
      NXLA=0
      NXLE=0
c
c     Set Number of Acoustic Absorptions and Emissions
      NACA=NE(1)
      NACE=NE(2)
c
c     Set Number of Impact Ionizations
      NII=NE(23)
c
c     Calculate the percent of all scatterings due to acoustic
c     emission and absorption
      PACA=100.0*DBLE(NE(1))/FNSC
      PACE=100.0*DBLE(NE(2))/FNSC
c
c     Calculate average phonon energy and average energy loss
c     due to acoustic emission and absorption
      IF(NE(1).NE.0) THEN
         APACA=SEAC1/DBLE(NE(1))
         ELACA=ELS(1)/DBLE(NE(1))
      ENDIF
      IF(NE(2).NE.0) THEN
         APACE=-SEAC2/DBLE(NE(2))
         ELACE=ELS(2)/DBLE(NE(2))
      ENDIF
c
c     Calculate the percent of scatterings which were impact ionizations
      PII=DBLE(NE(23))/FNSC
c
c     Calculate the total energy loss due to impact ionization
      IF(NE(23) .NE. 0) ELII=ELS(23)/DBLE(NE(23))

      APXXA=0.0
      APXXE=0.0
      APXLA=0.0
      APXLE=0.0
      ELXXA=0.0
      ELXXE=0.0
      ELXLA=0.0
      ELXLE=0.0

c     Calculate the number of X-X emissions and absorptions, plus
c     corresponding sum of all phonon energies and energy losses
      DO 4010 I=3,8
         NXXA=NXXA+NE(I)
         NXXE=NXXE+NE(I+6)
         APXXA=APXXA+EPH(I)*DBLE(NE(I))
         APXXE=APXXE-EPH(I+6)*DBLE(NE(I+6))
         ELXXA=ELXXA+ELS(I)
         ELXXE=ELXXE+ELS(I+6)
 4010 CONTINUE
c
c     Calculate the energy dissipation rate for X-X scattering
      DXX=APXXE-APXXA
      EDR1(2)=DXX*1.0E-12/(STIME*TSTEP)
c
c     Calculate the average X-X phonon energy and energy gain (loss)
      IF(NXXA .NE. 0) THEN
         APXXA=APXXA/DBLE(NXXA)
         ELXXA=ELXXA/DBLE(NXXA)
      ENDIF
      IF(NXXE .NE. 0) THEN
         APXXE=APXXE/DBLE(NXXE)
```

```
            ELXXE=ELXXE/DBLE(NXXE)
         ENDIF
C
C        Determine the percentage of X-X emissions and absorptions
         PXXA=100.0*DBLE(NXXA)/FNSC
         PXXE=100.0*DBLE(NXXE)/FNSC
C
C        Calculate the number of X-L scattering events, and the sum
C        of all the X-L phonon energies and energy losses for both
C        emission and absorption
         DO 4020 I=15,18
            NXLA=NXLA+NE(I)
            NXLE=NXLE+NE(I+4)
            APXLA=APXLA+EPH(I)*DBLE(NE(I))
            APXLE=APXLE-EPH(I+4)*DBLE(NE(I+4))
            ELXLA=ELXLA+ELS(I)
            ELXLE=ELXLE+ELS(I+4)
 4020    CONTINUE
C
C        Determine the energy dissipation rate for X-L scattering
         DXL=APXLE-APXLA
         EDR1(3)=DXL*1.0E-12/(STIME*TSTEP)
C
C        Calculate the average phonon energy and energy loss due to
C        X-L scattering, for both emission and absorption
         IF(NXLA .NE. 0) THEN
            APXLA=APXLA/DBLE(NXLA)
            ELXLA=ELXLA/DBLE(NXLA)
         ENDIF
         IF(NXLE .NE. 0) THEN
            APXLE=APXLE/DBLE(NXLE)
            ELXLE=ELXLE/DBLE(NXLE)
         ENDIF
C
C        Determine the percentage of X-L absorptions and emissions
         PXLA=100.0*DBLE(NXLA)/FNSC
         PXLE=100.0*DBLE(NXLE)/FNSC
C
C        Calculate the energy dissipation rate for acoustic intravalley phonons
         EDR1(1)=(-SEAC2-SEAC1)*1.0E-12/(STIME*TSTEP)
C
C        Calculate the energy dissipation rate for impact ionization
         EDR1(4)=ELS(23)*1.0E-12/(STIME*TSTEP)

C
C        The previous energy dissipation rates (EDR1) were calculated based on
C        phonon energies.  Now use actual losses.
         DO 4030 I=1,6
            EDR2(2)=EDR2(2)+ELOSSRATE(I)
 4030    CONTINUE
         DO 4040 I=1,4
            EDR2(3)=EDR2(3)+ELOSSRATE(I)
 4040    CONTINUE

         EDR2(2)=EDR2(2)*1.0E-12/STIME
         EDR2(3)=EDR2(3)*1.0E-12/STIME

C
C        Calculate total energy dissipation rate, using phonon energy estimators
         TEDR=EDR1(1)+EDR1(2)+EDR1(3)+EDR1(4)
C
C        Energy relaxation time calculation.
         ERT=(AEE-1.5*KB*T)/TEDR
C
C        Energy dissipation estimator of drift velocity
C        This estimator comes directly from 2nd moment of Boltzmann eq'n
         VD3=TEDR*1.0E12/FIELD

C        Write out results to file.
         OPEN(unit=20, file='simc.out')
         WRITE(20,4998)
 4998    FORMAT(////'RESULTS OF SI SINGLE PARTICLE MONTE CARLO')
         F1=FIELD/(1000.0)
         TT=STIME*TSTEP*1.0E12

         IDV(1)=IDINT(DV(1))
         IDV(2)=IDINT(DV(2))
         IDV(3)=IDINT(DV(3))
         WRITE(20,5000) F1,IDV(1),IDV(2),IDV(3),TT,T,NTOT,NLMT
```

```
5000   FORMAT(//'ELECTRIC FIELD:',T40,F7.2,2X,'kV/cm'/
    2  'DIRECTION:',T40,'  <',I1,I1,I1,'>'/
    3  'TIME OF SIMULATION:',T40,F7.1,2X,'psec'/
    4  'LATTICE TEMPERATURE:',T40,F7.2,2X,'K'/
    5  'MAXIMUM NUMBER OF SCATTERINGS:',T40,I7/
    6  'MAXIMUM NUMBER OF IMPACT IONIZATIONS:',T40,I7//)

       WRITE(20,5005) AEE
5005   FORMAT('AVERAGE ELECTRON ENERGY:',T40,F7.4,2X,'eV')
       AET=AEE/(1.5*KB)
       WRITE(20,5010) AET
5010   FORMAT('AVERAGE ELECTRON TEMPERATURE:',T40,F7.1,2X,'K'/)
       beta=aev/(aee*vd1)
       write(20,5015) beta
5015   format('<E.v> / <E><v> ; DIMENSIONLESS PARAMETER BETA'
    2         ,t50,1e12.3/)
       write(20,5017) avfsq
5017   format('AVERAGE OF SQUARED VELOCITY, FLD DIR.:',T40,1E12.3/)
       write(20,5018) avpsq
5018   format('AVERAGE OF SQUARED VELOCITY, PPD DIR.:',T40,1E12.3/)
       WRITE(20,5020) VD1,VD2,VD3
5020   FORMAT('DRIFT VELOCITY ESTIMATORS'/
    2  T5,'DERIVATIVE ESTIMATOR:',T40,E10.4,2X,'cm/sec'/
    3  T5,'ENERGY GAIN ESTIMATOR:',T40,E10.4,2X,'cm/sec'/
    4  T5,'ENERGY DISSIPATION ESTIMATOR:',T40,E10.4,2X,'cm/sec'/)
       WRITE(20,5030) ASCL
5030   FORMAT('MEAN FREE PATH:',T40,E10.4,2X,'cm')
       WRITE(20,5033) ASLFD
5033   FORMAT('MEAN FREE PATH/FIELD DIRECTION:',T40,E10.4,2X,'cm')
       WRITE(20,5040) ATSC
5040   FORMAT('MEAN TIME BETWEEN SCATTERINGS:',T40,E10.4,2X,'psec')
       WRITE(20,5050) ATII
5050   FORMAT('MEAN TIME BETWEEN IMPACT IONIZATIONS:',T40,E10.4,2X,
    2  'psec'/)
       WRITE(20,5055) PTSB
5055   FORMAT('PERCENT OF TIME SPENT IN SECOND BAND',T40,F10.4//)

       WRITE(20,5059)
5059   FORMAT('SCATTERING STATISTICS')
       WRITE(20,5060) NSC
5060   FORMAT('TOTAL NUMBER OF SCATTERING EVENTS:',T40,I7//)
       WRITE(20,5065)
5065   FORMAT(T7,'type of scattering',T35,'number',T46,'percent',T58,
    2  'average',T73,'average')
       WRITE(20,5070)
5070   FORMAT(T34,'of event',T56,'energy loss',T70,'phonon energy')
       WRITE(20,5080) NACA,PACA,ELACA,APACA
5080   FORMAT('acoustic',T20,'absorption',T35,I6,T47,F5.2,T59,F7.4,
    2  T74,F7.4)
       WRITE(20,5090) NACE,PACE,ELACE,APACE
5090   FORMAT(T20,'emission',T35,I6,T47,F5.2,T59,F7.4,T74,F7.4)
       WRITE(20,5100) NXXA,PXXA,ELXXA,APXXA
5100   FORMAT('intervalley',T14,'X-X',T20,'absorption',T35,I6,T47,F5.2,
    2  T59,F7.4,T74,F7.4)
       WRITE(20,5110) NXXE,PXXE,ELXXE,APXXE
5110   FORMAT(T20,'emission',T35,I6,T47,F5.2,T59,F7.4,T74,F7.4)
       WRITE(20,5120) NXLA,PXLA,ELXLA,APXLA
5120   FORMAT('intervalley',T14,'X-L',T20,'absorption',T35,I6,T47,F5.2,
    2  T59,F7.4,T74,F7.4)
       WRITE(20,5130) NXLE,PXLE,ELXLE,APXLE
5130   FORMAT(T20,'emission',T35,I6,T47,F5.2,T59,F7.4,T74,F7.4)
       WRITE(20,5140) NII,PII,ELII
5140   FORMAT('impact ionization',T35,I6,T47,F5.2,T59,F7.4/)
       WRITE(20,5142)
5142   FORMAT(T8,'Energy Dissipation Rates')
       WRITE(20,5143) EDR1(1)
5143   FORMAT('Acoustic:',T40,F7.3,2x,'eV/psec')
       WRITE(20,5144) EDR1(2)
5144   FORMAT('Intervalley (X-X):',T40,F7.3,2x,'eV/psec')
       WRITE(20,5145) EDR1(3)
5145   FORMAT('Intervalley (X-L):',T40,F7.3,2x,'eV/psec')
       WRITE(20,5146) EDR1(4)
5146   FORMAT('Impact Ionization:',T40,F7.3,2x,'eV/psec')
       WRITE(20,5147) TEDR
5147   FORMAT('TOTAL:',T40,F7.3,2X,'eV/psec')
       WRITE(20,5148) ERT
5148   FORMAT('ENERGY RELAXATION TIME',T40,F7.3,2X,'psec'//)
       WRITE(20,5150)
5150   FORMAT('IMPACT IONIZATION STATISTICS')
```

```
      WRITE(20,5160) NIMP
 5160 FORMAT('NUMBER OF IMPACT IONIZATIONS:',T40,I7)
      WRITE(20,5165) NBAL
 5165 FORMAT('NUMBER OF BALLISTIC ELECTRONS:',T40,I7)
      WRITE(20,5170) ALPHA
 5170 FORMAT('IMPACT IONIZATION PARAMETER ALPHA:',T40,E10.4,2X,'1/cm')
c
      CLOSE(UNIT=20)
      IF(NSC.EQ.NTOT) GOTO 6000
      GOTO 226
 6000 STOP
      END
c
c ******************************************************************
c
c     SUBROUTINE 1
c          Read in data for run.
c
      SUBROUTINE INDAT
      IMPLICIT DOUBLE PRECISION (A-H)
      IMPLICIT DOUBLE PRECISION (O-Z)

      DOUBLE PRECISION KB, T, MTX, MLX, MTL, MLL, M0, KORD(3,5584)
      DOUBLE PRECISION LATTICE
      DOUBLE PRECISION a(2), b(3,2),k(3),LATT
      DIMENSION TPHXX(3,2), TPHXL(4), XXIVD(3,2), XLIVD(4), Z(3)
     2     ,EXT(2,0:41,0:41,0:41), DEXT(3,2,0:41,0:41,0:41)
     3     ,EORD(5584), EPH(22), DV(3)
      INTEGER BAND(5584)
      PARAMETER (LATT=1.157E8)
      COMMON/BAND/EXT, DEXT
      COMMON/BAND2/EORD, KORD, BAND
      COMMON/CONSTANT1/KB, PI, T, HBAR, RHO, VS, Q, ES
      COMMON/CONSTANT2/LATTICE, MTX, MLX, MTL, MLL, M0, ACD, XXIVD
     2     ,XLIVD, ALPHA, Z, P, ETH, XLDIFF, TPHXX, TPHXL
      COMMON/CONSTANT3/EPH, SRPA
      COMMON/CONSTANT4/TSTEP, PMFDT, FIELD, DV, NTOT, NLMT, NLOOK
c
c     Open input data file 'data' and read in material parameters.
c
      OPEN (UNIT=15, FILE='data')
      REWIND 15
      READ(15,100) T                  {absolute temperature in K.
      READ(15,105) LATTICE            {lattice constant in angstroms
      READ(15,110) MTX                {transverse mass in X
      READ(15,110) MLX                {longitudinal mass in X
      READ(15,110) MTL                {transverse mass in L
      READ(15,110) MLL                {longitudinal mass in L
      READ(15,120) ALPHA              {nonparabolicity factor
      READ(15,120) ACD                {acoustic deformation potential
      READ(15,130) P                  {Keldysh factor
      READ(15,140) ETH                {impact ionization threshold
      LATTICE=LATTICE*1.0E-8          {convert angstroms to cm
c
c     Read in phonon temperatures and deformation potentials for
c     X - X equivilent intervalley scattering.
c
      DO 10 J=1,2
         DO 20 I=1,3
            READ(15,150) TPHXX(I,J), XXIVD(I,J)
 20      CONTINUE
 10   CONTINUE
c
c     Read in phonon temperatures and deformation potentials for
c     X - L intervalley scattering.
c
      DO 30 J=1,4
         READ(15,150) TPHXL(J), XLIVD(J)
 30   CONTINUE
c
c     Create an array EPH(22) which contains the phonon energy
c     corresponding to each scattering event.
c
      DO 40 I=1,3
         DO 50 J=1,2
            EPH(2+3*(J-1)+I)=KB*TPHXX(I,J)
            EPH(8+3*(J-1)+I)=-KB*TPHXX(I,J)
 50      CONTINUE
 40   CONTINUE
```

```
      DO 60 I=1,4
         EPH(14+I)=KB*TPHXL(I)
         EPH(18+I)=-KB*TPHXL(I)
60    CONTINUE
c
c     Read in runtime parameters.
c
      READ(15,200) FIELD          {electric field magnitude
      READ(15,210) DV             {field directional vector
      READ(15,220) TSTEP          {time step in seconds
      READ(15,240) PMFDT          {% of mean free drift time
      READ(15,230) NTOT           {total number of scatterings
      READ(15,230) NLMT           {total number of impact ionizations
      READ(15,230) NLOOK          {interval between printouts
      READ(15,240) SRPA           {energy search-range parameter

      FIELD=FIELD*1000.0          {electric field in V/cm
      PMFDT=PMFDT*0.01
c
c     Read in emprical pseudopotential band structure data.
c     EXT holds the energy of a given point in the irreducible wedge.
c     DEXT holds the gradient at a given point in the irreducible wedge.
c     In this program, the irreducible wedge is defined to be that volume
c     of the first Brillouin zone defined by the following equations:
c     Ky >= Kx >= Kz
c     Ky,Kx,Kz >= 0
c     Ky + Kx + Kz <= 1.5*(2*pi/a)        where a = lattice constant
c
c     Read in ordered E-K relation, lowest E first.

      OPEN(UNIT=69,FILE='Ek_der2.dat')
      do 102 i=1,7413
         read(69,*) (k(j),j=1,3),a,(b(j,1),b(j,2),j=1,3)
         ix=IDINT((k(1))/0.025+0.0001)
         iy=IDINT((k(2))/0.025+0.0001)
         iz=IDINT((k(3))/0.025+0.0001)
         ext(1,ix,iy,iz)=a(1)
         ext(2,ix,iy,iz)=a(2)
         do 202 j=1,3
            dext(j,1,ix,iy,iz)=b(j,1)/LATT
            dext(j,2,ix,iy,iz)=b(j,2)/LATT
202      continue
102   continue
      close(unit=69)
c
c     Read in sorted E-K relation, for final k-vector determination
      OPEN(UNIT=13,FILE='eord.si')
      REWIND 13
      DO 80 J=1,5584
         READ(13,170) (KORD(I,J),I=1,3),EORD(J),BAND(J)
80    CONTINUE
c
c
c     Input format statements.
c
100   FORMAT(///42X,F3.0)
105   FORMAT(42X,F10.6/)
110   FORMAT(42X,F10.6)
120   FORMAT(/42X,F10.8)
130   FORMAT(/42X,F8.4)
140   FORMAT(42X,F10.8/////)
150   FORMAT(19X,F3.0,20X,E10.1)
160   FORMAT(24X,4(1X,E11.5))        {for band structure
170   FORMAT(4(F8.4,X),I2)  {ordered E-K relation
200   FORMAT(////42X,F10.3)
210   FORMAT(/42X,F10.8,1X,F10.8,1X,F10.8/)
220   FORMAT(42X,E8.2)
230   FORMAT(42X,I9)
240   FORMAT(42X,F10.3)
c
c
      RETURN
      END
c
c **********************************************************************
c
c     SUBROUTINE 2
c          Calculate the energy for a given K point.  If the K point is
c          sufficiently close to one of known energy, the energy is looked
```

```
c           up.  Otherwise, the energy and its derivatives are interpolated
c           from the known values.
c
      SUBROUTINE ENERGY (EN,DEN)
      IMPLICIT DOUBLE PRECISION (A-H)
      IMPLICIT DOUBLE PRECISION (O-Z)
      DOUBLE PRECISION K1, K2, KI, EXT(2,0:41,0:41,0:41), KRZ(3)
     2     , EN,ETEMP, DEXT(3,2,0:41,0:41,0:41), KBZ(3),LATTICE
      DIMENSION FIX(3), DEN(3), E(2), DE(2,3), DEDGE(4,3), EEDGE(4)
     2     , DTEMP(3), ET(2), DT(2,3)
      INTEGER IX(3), KPER(3)
      PARAMETER (LATTICE=1.157E8)
      COMMON/BAND/EXT, DEXT
      COMMON/KVECTOR/KBZ, KPER, INDEX
      common/krz/krz
c
c     EN - energy of point KRED
c     DERVK - array of directional derivatives of E at KRED
c     K1 - component along interpolation of first known K point
c     K2 - component along interpolation of second known K point
c     KI - component along interpolation of KRED
c     IX - integer equivalent of known point whose coordinates are all
c          smaller than the corresponding coordinates of KRED
c     INDEX - band index
c
c     Find an equivalent vector to K in the irreducible wedge and place K in the
c     first Brillouin zone.
c
      CALL NEWK
c
c     Since the bandstructure is tabulated on a regular grid in K-space, the
c     electron wave-vector KRZ() will lie inside some cube in K-space for which
c     the energies at the corners are tabulated.  Since an interpolation is
c     needed to find the energy corresponding to the wave-vector KRZ(), the
c     next four lines find the bandstructure array index corresponding to one
c     of these corners for which the energy has been tabulated, for use as a
c     reference.
c
      DO 10 I=1,3
         IX(I)=IDINT(KRZ(I)/0.025)
         FIX(I)=0.025*DBLE(IX(I))
 10   CONTINUE
c
c     If this given K point is close enough to a known K point, then look it up.
c     Otherwise, interpolate.
c
      IF((KRZ(1)-FIX(1)) .LE. 0.0001 .AND. (KRZ(2)-FIX(2)) .LE. 0.0001
     2   .AND. (KRZ(3)-FIX(3)) .LE. 0.0001) GO TO 200
c
      DO 20 I=1,3
         IF((KRZ(I)-FIX(I)) .LE. 0.0001) IX(I)=IX(I)-1
         IF(IX(I) .LT. 0) IX(I)=IX(I)+1
 20   CONTINUE
c
c     IX now holds the closest point of smaller components in integer form.
c     This point will form the origin of a cube around K.  Seven
c     interpolations will be done, 4 edge, 2 face, 1 center to determine
c     E and DE at K.  The interpolations on the energy will be cubic, though
c     the interpolations on dE/dK are linear.
c
      DO 30 J=0,1
c
c        The J loop changes the 2-coordinate.
c
         DO 40 I=0,1
c
c           The I loop interpolates 2 points with fixed 2-coordinate
c           in the 3-coordinate direction.
c
            DO 50 K=1,2
               DO 60 M=1,3
                  DE(K,M)=DEXT(M,INDEX,IX(1)+I,IX(2)+J,IX(3)+K-1)
 60            CONTINUE
               E(K)=EXT(INDEX,IX(1)+I,IX(2)+J,IX(3)+K-1)
 50         CONTINUE
            K1=0.025*DBLE(IX(3))
            K2=K1+0.025
            KI=KRZ(3)
            CALL INTP(K1,K2,E,DE,KI,ETEMP,DTEMP,3)
```

```
c
c             Store the energy and its derivatives for each edge interpolation.
c
              DO 70 K=1,3
                 DEDGE(I+1+J*2,K)=DTEMP(K)
 70           CONTINUE
              EEDGE(I+1+J*2)=ETEMP
 40         CONTINUE
c
c           Interpolate the edges along 2 faces.
c
            K1=0.025*DBLE(IX(1))
            K2=K1+0.025
            KI=KRZ(1)
            E(1)=EEDGE(J*2+1)
            E(2)=EEDGE(J*2+2)
            DO 80 K=1,3
               DE(1,K)=DEDGE(J*2+1,K)
               DE(2,K)=DEDGE(J*2+2,K)
 80         CONTINUE
            CALL INTP(K1,K2,E,DE,KI,ETEMP,DTEMP,1)
c
c           Store the energy and its derivatives for each face interpolation.
c
            DO 90 K=1,3
               DT(J+1,K)=DTEMP(K)
 90         CONTINUE
            ET(J+1)=ETEMP
 30       CONTINUE
c
c         Interpolate between the 2 face points to find the desired K point.
c
          K1=0.025*DBLE(IX(2))
          K2=K1+0.025
          KI=KRZ(2)
          CALL INTP(K1,K2,ET,DT,KI,EN,DTEMP,2)
c
c         Close to an energy minimum, ie. E=0, this interpolation may give
c         a very small negative energy instead of E=0.0.  To avoid this
c         and other round off problems, round the energy to the nearest 0.0001eV.
c         NOTE: after the cubic interpolation was added, such problems have not
c         been encountered.  However, the next line was left in the code for safety.
c
          IF(EN .LT. 0.0) EN=ABS(EN)
c
          DO 110 I=1,3
             DEN(KPER(I))=DTEMP(I)
             IF (KBZ(KPER(I)) .LT. 0.0) DEN(KPER(I))=-1.0*DEN(KPER(I))
 110      CONTINUE
          RETURN
c
c         The K point is close enough to a known value to look up its energy.
c
 200      EN=EXT(INDEX,IX(1),IX(2),IX(3))
          DO 100 I=1,3
             DTEMP(I)=DEXT(I,INDEX,IX(1),IX(2),IX(3))
             DEN(KPER(I))=DTEMP(I)
             IF (KBZ(KPER(I)) .LT. 0.0) DEN(KPER(I))=-1.0*DEN(KPER(I))
 100      CONTINUE
          RETURN
          END
c
c ***********************************************************************
c
c         SUBROUTINE 3
c              Interpolate cubically between 2 points with known energy
c              and energy derivatives.  The energy is returned in EN and its
c              3 directional derivatives are returned in DEN.
c
          SUBROUTINE INTP(XA,XB,E,DE,XI,EN,DEN,ID)
          IMPLICIT DOUBLE PRECISION (A-H)
          IMPLICIT DOUBLE PRECISION (O-Z)
          DOUBLE PRECISION E(2), DE(2,3), EN, DEN(3), LATTICE
          PARAMETER (LATTICE=1.157E8)
c
c         XA, XB - one particular coordinate of two known K points
c         E - array holding the energies at XA and XB
c         DE - matrix holding directional derivatives of E at XA and XB
c         XI - one particular coordinate of K point whose energy will be
```

```
c            interpolated
c        EN - energy of interpolant point
c        DEN - directional derivatives of energy of interpolant point
c        ID - coordinate number along which interpolation will take place
c        LATTICE - 2*PI/a, a is the lattice constant in cm
c        DYA - DE normalized to the dimension of a mesh unit.
c        YA, YB -  E(XA) and E(XB)
c
c
         DYA=DE(1,ID)*0.025*LATTICE
         DYB=DE(2,ID)*0.025*LATTICE
         YA=E(1)
         YB=E(2)
c
c        Find coefficients such that E(XI) = A*X^3 + B*X^2 + C*X + D
c            where X is the fractional distance of XI between XA and XB.
         A=DYB-YB-YB+DYA+YA+YA
         C=DYA
         D=YA
         B=YB-A-C-D
         X=(XI-XA)/(XB-XA)
c
c        Assign energy
         EN=((A*X+B)*X+C)*X+D
c
c        Calculate directional derivatives.  One of the derivatives can
c        be interpolated parabolically from above coefficients, but the
c        other two must be computed with a linear interpolation.
         DO 10 I=1,3
         IF (ID.NE.I) THEN
             DEN(I)=X*(DE(2,I)-DE(1,I))+DE(1,I)
         ELSE
             DEN(I)=((3*A*X+2*B)*X+C)/LATTICE/0.025
         ENDIF
c
 10      CONTINUE
         RETURN
         END

c
c ********************************************************************
c
c        SUBROUTINE 4
c             Displace K-vector KBZ into the first Brillouin zone and
c             return a vector KRZ in the irreducible wedge, equivalent
c             to KBZ through symmetry transformations.  The array KPER
c             is also returned holding the coordinate permutations
c             used to move KBZ into the irreducible wedge.
c
         SUBROUTINE NEWK
         IMPLICIT DOUBLE PRECISION (A-H)
         IMPLICIT DOUBLE PRECISION (O-Z)
         DOUBLE PRECISION KBZ(3), KRZ(3), KRZT(3)
         INTEGER KPER(3), KPERT(3), FLAG
         COMMON/KVECTOR/KBZ, KPER, INDEX
         common/krz/krz
c
c        KBZ - vector moved into the first Brillouin zone
c        KRZ - vector equivalent of KBZ in the irreducible wedge
c        KPER - array holding coordinate permutations from KRZ to KBZ
c        KRZT, KPERT - temporary arrays
c
 10      DO 20 I=1,3
            KRZT(I)=ABS(KBZ(I))
 20      CONTINUE
c
c        Determine whether KRZT is in first Brillouin zone.  FLAG = 1 if not.
         FLAG=0
         IF(KRZT(1).GT.1.0001) FLAG=1
         IF(KRZT(2).GT.1.0001) FLAG=1
         IF(KRZT(3).GT.1.0001) FLAG=1
         IF((KRZT(1)+KRZT(2)+KRZT(3)).GT.1.5001) FLAG=1
c
         IF (FLAG .EQ. 1) THEN
c
c           Add reciprocal lattice vectors until KBZ is in first B.Z.
c
            DO 30 I=1,3
               IF (KBZ(I) .GT. 0.0) THEN
```

```
                  KBZ(I)=KBZ(I)-1.0
              ELSE
                  KBZ(I)=KBZ(I)+1.0
              ENDIF
 30        CONTINUE
        ELSE
c
c          KBZ is in first B.Z., so create KRZ (the equivalent k-vector
c          in the irreducible wedge) by simply sorting the magnitude of components.
c
           CALL SORT(KRZT,KPERT,3)
           KRZ(1)=KRZT(2)
           KRZ(2)=KRZT(3)
           KRZ(3)=KRZT(1)
           KPER(1)=KPERT(2)
           KPER(2)=KPERT(3)
           KPER(3)=KPERT(1)
c
c          KRZ is in order ky > kx > kz to be in the irreducible wedge.
c          KPER(x) holds the number of the coordinate of KBZ that
c          KRZ(x) came from.
c
           RETURN
        ENDIF
        GOTO 10
        END
c *******************************************************************
c
c       SUBROUTINE 5
c            Generate a vector R of random numbers uniformally distributed
c            between 0 and 1 with length N.
c
        SUBROUTINE RNUM(DSEED,N,R)
        IMPLICIT DOUBLE PRECISION (A-H)
        IMPLICIT DOUBLE PRECISION (O-Z)
        DOUBLE PRECISION DSEED, A, B, DMOD
        DIMENSION R(N)
c
c       R - vector of random numbers
c       N - number of random numbers in vector R
c       DSEED - seed number for random number generator
c
        DO 10 I=1,N
           A=DSEED*7.**5
           B=2.**31-1
           DSEED=DMOD(A,B)
           R(I)=1./2.**31*DSEED
 10     CONTINUE
        RETURN
        END
c
c *******************************************************************
c
c       SUBROUTINE 6
c            Sort vector v of length N in increasing order.  Return
c            vector IPER holding index pointing to the original
c            position of the element.
c
        SUBROUTINE SORT(V,IPER,N)
        IMPLICIT DOUBLE PRECISION (A-H)
        IMPLICIT DOUBLE PRECISION (O-Z)
        DIMENSION V(N), IPER(N)
c
c       V - vector to be sorted
c       N - length of vector V
c       IPER - index array holding original position of each element
c
        DO 10 J=1,N
           IPER(J)=J
 10     CONTINUE
        DO 20 J=1,N-1
           DO 30 I=J+1,N
              IF (V(J) .GT. V(I))THEN
                VTEMP=V(J)
                V(J)=V(I)
                V(I)=VTEMP
                ITEMP=IPER(J)
                IPER(J)=IPER(I)
                IPER(I)=ITEMP
```

```
              ENDIF
 30         CONTINUE
 20       CONTINUE
          RETURN
          END
c
c ******************************************************************
c         Jan 29, 1988
c         SUBROUTINE 7
c              Calculate the scattering rates for each scattering mechanism and
c              create a probability table from these scattering rates.
c              All scattering rates (except for impact ionization) are
c              calculated using non-parabolic effective mass bands.
c              Impact ionization is calculated using the Bude formula.
c
c              See :
c              Jeff Bude, "Field Assisted Impact Ionization in Semiconductors,"
c              in Computational Electronics, Kluwar Academic Publishers,
c              pp. 127-131, 1990, edited by K. Hess, J.P. LeBurton, and
c              U. Ravaioli.
c
c
c
          SUBROUTINE TAU(RATE,TRATE)
          IMPLICIT DOUBLE PRECISION (A-H)
          IMPLICIT DOUBLE PRECISION (O-Z)

          DOUBLE PRECISION KB,KT,LATTICE,MTX,MLX,MTL,MLL,M0,MD,NUM
          DOUBLE PRECISION C(4), A,B,EV,Y
          DIMENSION RATE(24,500), TPHXX(3,2), TPHXL(4), XXIVD(3,2)
     2            ,XLIVD(4), Z(3), TRATE(500)
          COMMON/CONSTANT1/KB, PI, T, HBAR, RHO, VS, Q, ES
          COMMON/CONSTANT2/LATTICE, MTX, MLX, MTL, MLL, M0, ACD, XXIVD
     2        ,XLIVD, ALPHA, Z, P, ETH, XLDIFF, TPHXX, TPHXL
          COMMON/II/RTOT
          COMMON/CONSTANT4/TSTEP, PMFDT, FIELD, DV, NTOT, NLMT, NLOOK

c
c         RATE - a scattering probability lookup table.  The first 22 columns
c                hold the partial sums of the different scattering rates.
c                RATE(23,E) holds the complete sum at each energy.  The
c                energy scale goes in units of 0.01 eV from 0.01 to 5.0 eV.
c                All rates up to RATE(22,E) are normalized to the total
c                scattering rate at each energy, RATE(23,E).
c
c         These are the indices for the different scattering mechanisms.
c
c         1,2              Acoustic phonons - absorption, emission
c         3,4,5            X - X intervalley f - absorption
c         6,7,8            X - X intervalley g - absorption
c         9,10,11          X - X intervalley f - emission
c         12,13,14         X - X intervalley g - emission
c         15,16,17,18      X - L intervalley - absorption
c         19,20,21,22      X - L intervalley - emission
c         23               Impact Ionization
c
c
c         Initialize the look-up table.
c
          DO 10 I=1,24
             DO 20 K=1,500
                RATE(I,K)=0.0
 20          CONTINUE
 10       CONTINUE
c
c         Note, all constants have the following units:
c         All energies are in eV.
c         All variables but the electron rest mass M0 are in cgs units.
c         M0 is in kg.
c
c         Calculate acoustic phonon intravalley scattering.
c         RATE(1,E) is for absorbtion and RATE(2,E) is for emission.
c         The expression for the phonon occupation number has been
c         approximated by a polynomial.  This makes integration of
c         the phonon occupation number simple.
c         The variables X1 and X2 hold the limits of integration.
c         See:
c         C. Canali et. al., "Electron drift velocity in silicon,"
c         Phys. Rev. B., vol. 12, pp. 2265-2284
```

```
c
      MD=(MTX*MTX*MLX)**(1./3.)*M0

      KT=KB*T

      ES=5E-5*MD*VS*VS/Q
      AA=((MD/2.0)**(0.5))*((KT/(HBAR*VS))**(3.0))
     2   *((ACD*ACD)/(4.0*HBAR*VS*RHO*PI))*1E5*SQRT(Q)
c
      ESSR=SQRT(ES)
      DO 30 J=1,2
         DO 40 I=1,500
            E=0.01*DBLE(I)
            ESR=SQRT(E)
            IF (E .LE. ES) THEN
                X1=((4*ESSR)/KT)*(ESSR-ESR)
                X2=((4*ESSR)/KT)*(ESSR+ESR)
            ELSE IF ((E .GT. ES) .AND. (J .EQ. 1)) THEN
                X1=0.0
                X2=((4*ESSR)/KT)*(ESR+ESSR)
            ELSE IF ((E .GT. ES) .AND. (J .EQ. 2)) THEN
                X1=0.0
                X2=((4*ESSR)/KT)*(ESR-ESSR)
            ENDIF
            IF (J .EQ. 1) THEN
              F1=(X2*X2 -X1*X1)/2.0-(X2**3-X1**3)/6.0+(X2**4-X1**4)/48.0
     2           -(X2**6-X1**6)/4320.0
              F2=(X2**3-X1**3)/3.0-(X2**4-X1**4)/8.0+(X2**5-X1**5)/60.0
     2           -(X2**7-X1**7)/5040.0+(X2**9-X1**9)/272160.0
            ELSE IF ((J .EQ. 2) .AND. (E .GT. ES)) THEN
              G1=(X2*X2)/2.0+(X2**3)/6.0+(X2**4)/48.0-(X2**6)/4320.0
              G2=(X2**3)/3.0+(X2**4)/8.0+(X2**5)/60.0
     2           -(X2**7)/5040.0+(X2**9)/272160.0
            ENDIF
            IF (J .EQ. 1) THEN
                RATE(J,I)=AA*((E*(1+ALPHA*E))**(-0.5))*((1+2*ALPHA*E)
     2             *F1+2*ALPHA*KT*F2)
            ELSE IF ((J .EQ. 2) .AND. (E .LE. ES)) THEN
                RATE(J,I)=0.0
            ELSE IF ((J .EQ. 2) .AND. (E .GT. ES)) THEN
                RATE(J,I)=AA*((E*(1+ALPHA*E))**(-0.5))*((1+2*ALPHA*E)
     2             *G1-2*ALPHA*KT*G2)
            ENDIF
 40      CONTINUE
 30   CONTINUE
c
c     X - X intervalley scattering.
c
      DO 50 J=1,2
         DO 60 I=1,3
            INDEX=2+I+(J-1)*3
            HW=KB*TPHXX(I,J)
            EX=EXP(HW/KT)
            NUM=XXIVD(I,J)*XXIVD(I,J)*MD*Z(J)
            NUM=NUM*SQRT(MD)
            DEN=SQRT(2.)*PI*RHO*HBAR*HBAR*HW
            QUO=10.0*(Q**(-0.5))*NUM/DEN
            DO 70 K=1,500
               EN=0.01*DBLE(K)
               ENEW=EN+HW
               PROD=SQRT(ENEW*(1.+ALPHA*ENEW))*(1.+2.*ALPHA*ENEW)
               RATE(INDEX,K)=QUO*PROD/(EX-1.)
               IF(EN .GE. HW) THEN
                  ENEW=EN-HW
                  PROD=SQRT(ENEW*(1.+2.*ALPHA*ENEW)) {check '2' factor
     2                 *(1.+2.*ALPHA*ENEW)
                  RATE(INDEX+6,K)=QUO*PROD*EX/(EX-1.)
               ENDIF
 70         CONTINUE
 60      CONTINUE
 50   CONTINUE
c
c     X - L intervalley scattering.
c
      MD=((MTL*MTL*MLL)**(1./3.))*M0
      DO 80 I=1,4
         HW=KB*TPHXL(I)
         EX=EXP(HW/KT)
         NUM=XLIVD(I)*XLIVD(I)*MD*Z(3)
```

```
            NUM=NUM*SQRT(MD)
            DEN=(2.**(0.5))*PI*RHO*HBAR*HBAR*HW
            QUO=10.0*(Q**(-0.5))*NUM/DEN
            DO 90 K=1,500
               EN=0.01*DBLE(K)
               IF (EN .GE. XLDIFF) THEN
                  RATE(14+I,K)=QUO*SQRT(EN+HW-XLDIFF)/(EX-1.)
                  IF (EN-HW .GE. XLDIFF) THEN
                     RATE(18+I,K)=QUO*SQRT(EN-HW-XLDIFF)*EX/(EX-1.)
                  ENDIF
               ENDIF
 90         CONTINUE
 80      CONTINUE
c
c        Create partial sums table.
c
         DO 100 J=1,500
            DO 110 I=2,22
               RATE(I,J)=RATE(I,J)+RATE(I-1,J)
 110        CONTINUE
c
c           Adjust for true density of states above 2 eV.
c
            ED=DBLE(J)*0.01
            IF (J.GE.200.AND.J.LT.300) THEN
               RFAC=-1.1733*ED*ED*ED+10.64*ED*ED-30.5867*ED+29.016
            ELSE IF(J.GE.300) THEN
               RFAC=-0.2397*ED*ED*ED+3.2240*ED*ED-14.6444*ED+22.6603
            ENDIF
            RFAC=RATE(22,200)*RFAC/RATE(22,J)
            IF(J.GE.200) THEN
C               RFAC=(RATE(22,200)*0.0625*(0.01*DBLE(J)-6.)**2)
C    &                  /RATE(22,J)
               DO 120 I=1,22
                  RATE(I,J)=RFAC*RATE(I,J)
 120           CONTINUE
            ENDIF
c
 100     CONTINUE
c
c        Impact ionization rate, Bude formula.
c
c
         ETH=1.12
         B=(5E+05-FIELD)/5E+05
         A=FIELD/5E+05
         C(1)=0.2145*B+0.2427*A
         C(2)=-3.6302*B-3.6593*A
         C(3)=21.5126*B+19.2716*A
         C(4)=-12.5451*B-3.9533*A
         RTOT=110

         DO 130 K=1,500
            EN=0.01*DBLE(K)
            IF (EN .GE. ETH) THEN
                EV=EN+1.12
                Y=C(1)*EV*EV*EV+C(2)*EV*EV+C(3)*EV+C(4)
                RATE(23,K)=EXP(Y)
            ENDIF
            RATE(23,K)=RATE(22,K)+RATE(23,K)
            TRATE(K)=RATE(23,K)
 130     CONTINUE
c
c
c        Normalize all RATE's up to RATE(23,J). TRATE(J) will
c        hold the total rate, normalized to 1/TSTEP.
c
         DO 140 I=1,23
            DO 150 J=1,500
               RATE(I,J)=RATE(I,J)/TRATE(J)
 150        CONTINUE
 140     CONTINUE
         DO 160 J=1,500
            TRATE(J)=TRATE(J)*TSTEP
 160     CONTINUE

         RETURN
         END
c
```

```
c
c ***********************************************************************
c
c       Feb 17, 1988
c       SUBROUTINE 8
c            Find the final k state and final energy for a particlar scattering
c            mechanism.
c
        SUBROUTINE FINALK(EVENT,E)
        IMPLICIT DOUBLE PRECISION (A-H)
        IMPLICIT DOUBLE PRECISION (O-Z)
        DOUBLE PRECISION KORD(3,5584),EORD(5584),KTMP(3)
        DOUBLE PRECISION KB,KT,NORM,KBZ(3)
        DIMENSION EPH(22), RAN(2)
        DOUBLE PRECISION DSEED
        INTEGER EVENT, BAND(5584), KPER(3)
        COMMON/BAND2/EORD, KORD, BAND
        COMMON/CONSTANT1/KB, PI, T, HBAR, RHO, VS, Q, ES
        COMMON/CONSTANT3/EPH, SRPA
        COMMON/DISSIPATION/SEAC1, SEAC2
        COMMON/KVECTOR/KBZ, KPER, INDEX
        COMMON/RANDOM/DSEED
        common/check/nsc
c
c       Find the electron energy AFTER scattering.
c
        KT=KB*T
        IF(EVENT .LE. 2) THEN
c
c       Acoustic phonon absorption and emission (inelastic scattering).
c       Although there is a continuous distribution of final state electron
c       energies, an average final state electron energy is calculated based
c       on the average acoustic phonon energy.
c
c       es=7.461e-5
        if(event.eq.1)then
          x=4.0*sqrt(es)/kt*(sqrt(e)+sqrt(es))
          if(x.gt.3.5)then
            x=3.5
          endif
          rn=1./3.*x**3-1./8.*x**4+1./60.*x**5-1./5040.*x**7
          rd=0.5*x**2-1./6.*x**3+1./48.*x**4-1./4320.*x**6
          eac=kt*rn/rd
          seac1=seac1+eac
          en=eac+e
        else
          x=4.*sqrt(es)/kt*(sqrt(e)-sqrt(es))
          if(x.gt.3.5)then
             x=3.5
          endif
          rn=1./3.*x**3+1./8.*x**4+1./60.*x**5-1./5040.*x**7
          rd=0.5*x**2+1./6.*x**3+1./48.*x**4-1./4320.*x**6
          eac=-1.0*kt*rn/rd
          seac2=seac2+eac
          en=eac+e
        endif


        ELSE IF(EVENT .GT. 2) THEN
c
c       The scattering is intervalley.  Add the energy of the inervalley
c       phonon to the initial electron energy to get the final energy.
c
           EN=E+EPH(EVENT)
        ENDIF
c
c       Find the spread of energy values within SRPA of the
c       final electron energy-introduce a search range tolerance.
c
c       First make sure that if the energy is VERY low that the
c       normal energy selection procedure is bypassed.  Leave energy
c       as it is.
        IF((EN+SRPA-EORD(1)) .LE. 0.0) THEN
           IFINAL=1
           GOTO 85
        ENDIF
c
c       Make a quick scan through the ordered list of energies to
c       find an index which corresponds roughly to the electron energy.
```

```
      DO 30 I=1,110
         IF(EN-SRPA-EORD(50*(I-1)+1)) 40,40,30
 30   CONTINUE
 40   IM=(I-1)*50+1
c
c     Now find the lowest energy which is still within SRPA of the
c     electron energy (plus the phonon energy).
      DO 50 I=IM,1,-1
         IF(EN-SRPA-EORD(I)) 50,50,60
 50   CONTINUE
 60   IMIN=I+1
      IF(IMIN .LE. 0) THEN
          IMIN=1
      ENDIF
c
c     Find the highest energy which is still within SRPA of the
c     electron energy (plus the phonon energy).
      DO 70 J=IMIN, 5584
         IF(EN+SRPA-EORD(J)) 80,70,70
 70   CONTINUE
 80   IMAX=J-1
      IF(IMAX .LE. 0) IMAX=1
c
c     Determine the final state in the irreducible wedge by selecting
c     the final state energy with a Lorentzian probability distribution.
c     ( The integral of a Lorentzian distribution is the arctan function. )
      IF(IMIN .GE. IMAX) THEN
         IFINAL=IMIN
      ELSE
         CALL RNUM(DSEED,1,RN)
         RN=0.25+RN*0.5
         ARG=(RN-0.5)*PI
         RN=0.5+0.5*TAN(ARG)
         IFINAL=IDINT(DBLE(IMIN)+RN*(DBLE(IMAX-IMIN)+0.999))
      ENDIF
c
c     The index IFINAL gives the final state of the electron.
c     Assign the electron's band index.
 85   INDEX=BAND(IFINAL)
c
c     Assign the electron's final state energy.
      ENEW=EORD(IFINAL)
c
c     Assign the electron's final state k-vector (in the reduced wedge).
      DO 90 I=1,3
         KTMP(I)=KORD(I,IFINAL)
 90   CONTINUE
c
c     Determine the actual final state in the first Brillouin zone by
c     randomally choosing a permutation of KTMP.
c
      CALL SHUFF(KTMP,EVENT)
      DO 100 I=1,3
         KBZ(I)=KTMP(I)
 100  CONTINUE
      E=ENEW
      RETURN
      END
c
c ******************************************************************
c
c      Feb 19, 1988
c      SUBROUTINE 9
c           Determine the true final K state from the state in the
c           irreducible wedge which was chosen by
c           energy conservation.  Consideration is made of the
c           particular scattering mechanism chosen.  Also, the input
c           k-vector is in the irreducible wedge  and must be mapped
c           to one of the possible final states.
c
      SUBROUTINE SHUFF(KTMP,EVENT)
      IMPLICIT DOUBLE PRECISION (A-H)
      IMPLICIT DOUBLE PRECISION (O-Z)
      DOUBLE PRECISION KTMP(3), KBZ(3), KNEW(3), KTP, R1(3)
      DIMENSION R(3)
      DOUBLE PRECISION DSEED
      INTEGER EVENT, KPER(3), ITMP(3)
      COMMON/KVECTOR/KBZ, KPER, INDEX
      COMMON/RANDOM/DSEED
```

```
c
      IF(EVENT .GT. 14) THEN
c
c        The scattering is isotropic so employ the full crystal symmetry to
c        map the reduced zone vector KTMP to the final K-vector KNEW.
c
         CALL RNUM(DSEED,3,R1)
         CALL RNUM(DSEED,3,R)
         CALL SORT(R,ITMP,3)
         KNEW(1)=KTMP(ITMP(1))*DSIGN(1.0D+00,(R1(1)-0.5))
         KNEW(2)=KTMP(ITMP(2))*DSIGN(1.0D+00,(R1(2)-0.5))
         KNEW(3)=KTMP(ITMP(3))*DSIGN(1.0D+00,(R1(3)-0.5))

      ELSE IF(EVENT .LE. 2) THEN

c        Acoustic phonon scattering.  Fix the valley but permute
c        the remaining two coordinates.

         KNEW(KPER(2))=KTMP(2)*DSIGN(1.0D+00,KBZ(KPER(2)))
         CALL RNUM(DSEED,3,R)
         IF(R(2) .LE. 0.5)THEN
            KNEW(KPER(1))=KTMP(1)*DSIGN(1.0D+00,(R(1)-0.5))
            KNEW(KPER(3))=KTMP(3)*DSIGN(1.0D+00,(R(3)-0.5))
         ELSE
            KNEW(KPER(3))=KTMP(1)*DSIGN(1.0D+00,(R(1)-0.5))
            KNEW(KPER(1))=KTMP(3)*DSIGN(1.0D+00,(R(3)-0.5))
         ENDIF

      ELSE IF((EVENT .GE. 3 .AND. EVENT .LE. 5) .OR.
     2   (EVENT .GE. 9 .AND. EVENT .LE. 11)) THEN
c
c        X - X, f scattering
c        Permute the reduced zone vector, KTMP so that the largest component is
c        in the direction of the largest component of the initial K-vector, KBZ.
c        Then randomally permute the other two coordinates using symmetry
c        properties.  Then, randomally select one of these two coordinates
c        and exchange its value with the value of the largest component. Now,
c        the new vector will be in one of the X-valleys perpendicular to the
c        initial vector, fulfilling the requirement for f scattering.
c
         CALL RNUM(DSEED,1,RN)
         IF(RN.LE.0.25)THEN
             ITMP(1)=KPER(1)
             ITMP(2)=KPER(3)
             ITMP(3)=KPER(2)
         ELSE IF((RN.GT.0.25) .AND. (RN.LT.0.5))THEN
             ITMP(1)=KPER(3)
             ITMP(2)=KPER(2)
             ITMP(3)=KPER(1)
         ELSE IF((RN.GT.0.5).AND.(RN.LT.0.75))THEN
             ITMP(1)=KPER(3)
             ITMP(2)=KPER(1)
             ITMP(3)=KPER(2)
         ELSE IF(RN.GT.0.75) THEN
             ITMP(1)=KPER(1)
             ITMP(2)=KPER(2)
             ITMP(3)=KPER(3)
         ENDIF

         KNEW(ITMP(1))=KTMP(2)
         KNEW(ITMP(2))=KTMP(1)
         KNEW(ITMP(3))=KTMP(3)
         CALL RNUM(DSEED,3,R)
         KNEW(1)=KNEW(1)*DSIGN(1.0D+00,(R(1)-0.5))
         KNEW(2)=KNEW(2)*DSIGN(1.0D+00,(R(2)-0.5))
         KNEW(3)=KNEW(3)*DSIGN(1.0D+00,(R(3)-0.5))

c
      ELSE IF((EVENT .GE. 6 .AND. EVENT .LE. 8) .OR.
     2   (EVENT .GE. 12 .AND. EVENT .LE. 14)) THEN
c
c     X - X, g scattering
c
c     Permute the reduced zone vector KTMP so that the largest component is in
c     the direction of the largest component of the initial K-vector, KBZ.  Then,
c     cuase it to have the opposite sign of KBZ.  This fulfills the criteria for
c     g scattering -- that the electron is scattered to a different valley on the
c     same axis.  The other two components of KTMP are randomally chosen by
c     symmetry considerations.
```

```
c
           KNEW(KPER(2))=-1.0*KTMP(2)*DSIGN(1.0D+00,KBZ(KPER(2)))
           CALL RNUM(DSEED,3,R)
           IF(R(2) .LE. 0.5)THEN
              KNEW(KPER(1))=KTMP(1)*DSIGN(1.0D+00,(R(1)-0.5))
              KNEW(KPER(3))=KTMP(3)*DSIGN(1.0D+00,(R(3)-0.5))
           ELSE
              KNEW(KPER(3))=KTMP(1)*DSIGN(1.0D+00,(R(1)-0.5))
              KNEW(KPER(1))=KTMP(3)*DSIGN(1.0D+00,(R(3)-0.5))
           ENDIF
c
        ENDIF
c
c       Put the new value of K into KTMP to return to FINALK.
c
        DO 10 I=1,3
           KTMP(I)=KNEW(I)
 10     CONTINUE
        RETURN
        END
c
c ******************************************************************************
```

INDEX